Comments about *Charging Ahead*

"Dr. Berger's book ranges from lucid descriptions of how renewable energy technologies work, to vivid histories of pioneering renewable energy firms and their principals, to cogent energy recommendations. This book is rich in data, case studies, and incisive analysis."
—*Pacific Institute Report*

"Berger. . . stirs the imagination with exciting examples of the progress being made in the fields of solar, wind, bioenergy, and geothermal technology and with electric vehicles and hypercars."
—David Rouse, *Booklist*

"An excellent primer on the technologies and companies involved in the political debates. . . . With thorough case studies, Berger provides technical as well as economic backgrounds of the major renewable energy realms. In addition to an introduction to these technologies, Berger provides a portrait of the economic difficulties that many companies have encountered as a result of the failure of domestic energy policy and proposes his own measures to help move our communities towards a renewable energy future."
—*Public Citizen News*

"An important, very readable book on the history, status, and role that conservation and renewables can [have] to impact the economy by providing clean energy in a fully sustainable mode. Dr. Berger has created a definitive work on what America can and must do to exploit the benefits of renewable energy and conservation."
—Paul Maycock, Editor, *Photovoltaic News*

CHARGING AHEAD

Other Books by John J. Berger

Nuclear Power: *The Unviable Option, A Critical Look at Our Energy Alternatives*

Restoring the Earth: *How Americans Are Working to Renew Our Damaged Environment*

Environmental Restoration: *Science and Strategies for Restoring the Earth*

What You Really Ought to Know About Forests (forthcoming)

JOHN J. BERGER

CHARGING
AHEAD

The Business of Renewable Energy and What It Means for America

UNIVERSITY OF
CALIFORNIA PRESS
Berkeley · Los Angeles · London

University of California Press
Berkeley and Los Angeles, California

University of California Press, Ltd.
London, England

Published by arrangement with
Henry Holt and Company, Inc.

Library of Congress Cataloging-in-Publication Data

Berger, John J.
    Charging ahead : the business of renewable energy and what it
means for America / John J. Berger.
        p.   cm.
    Includes bibliographical references and index.
    ISBN 0-520-21614-8 (alk. paper)
    1. Renewable energy sources—United States.   I. Title.
TJ807.9.U6B47   1998
333.79′4′0973—dc21   98-15166
                                                                  CIP

Designed by Victoria Hartman

Printed in the United States of America

1   2   3   4   5   6   7   8   9

The paper used in this publication is both acid-free and totally
chlorine-free (TCF). It meets the minimum requirements of
American Standard for Information Sciences—Permanence of
Paper for Printed Library Materials, ANSI Z39.48-1984. ⊚

CONTENTS

FOREWORD

$\mathbf{M}$any people today are largely unaware of the tremendous contribution that renewable energy and energy efficiency could make to world energy supplies and to our economy, not to mention to the environment. Yet for economic reasons rooted in energy developments of the past 25 years or so, progress in renewable energy today is threatened. In *Charging Ahead*, John J. Berger documents those threats, calls for a fundamental reorientation of national energy policy, and presents powerful economic arguments for renewable energy and energy efficiency. The energy efficiency investments he describes are cost-effective today, and can save the economy hundreds of billions of dollars. Many investments in renewables are already paying off in the form of clean, reliable, commercial generating technologies.

Why is there a tendency today to underestimate the value of renewables and to misunderstand the capabilities and limitations of unregulated markets to produce sound, long-term energy decisions? Clues to these questions can be found in events of the 1970s when two OPEC oil shocks, long lines at gasoline stations, and skyrocketing energy prices made renewable energy a hot topic. Through subsidies and tax credits, governments poured large sums into synthetic fuels as alternatives to conventional fossil fuels. Much of that money was misspent on costly efforts to extract oil from western shale deposits and on the environmentally costly gasification of coal, as well as on poorly designed wind turbines. At the same time, individuals bought solar collectors to heat homes in cold rainy climates and shifted to geothermal heat pumps.

Over the next two decades, however, several developments occurred that reduced government and industry interest in renewable energy. A sharp decline in energy prices occurred during the 1980s. Today in the late 1990s, energy prices in real terms are less than those in 1974 following the first big OPEC oil price hike. In addition, excess electrical generating capacity has accumulated and vast new supplies of fossil fuels have appeared over time. A lot more natural gas than expected was added to reserves and new technologies have emerged that made finding oil more efficient and dramatically raised ultimate recovery rates from existing fields. Finally, the implosion of the USSR caused vast amounts of oil to become available on world markets. In response to the drop in real energy prices of the 1980s, the government reduced its coal gasification research, put oil shale commercialization on hold, and curtailed solar energy and wind energy tax credits. Most individuals also quit buying solar collectors for anything other than heating hot water in sunny dry climates.

While public interest declined in renewable energy options (solar cells, solar thermal electric power, wind, biomass, and geothermal power), these technologies continued their rapid advance and registered major improvements in efficiency and reliability. *Charging Ahead* tells the story of these little-known developments. In most cases costs are down and exciting prospects are on the horizon for widespread use of these technologies. *Charging Ahead* also places the need for renewable energy in the right context. The issue is not natural resource exhaustion. Markets handle mineral exhaustion problems very well. As mineral resources, such as fossil fuels, become less available (something that occurs only slowly), prices rise. This leads users to become more fuel efficient and to cut back on what they buy, suppliers have greater incentives to look for new fields and to develop known deposits that were previously too expensive to profitably utilize, and researchers get the funds, public and private, to invent and perfect new technologies.

In contrast, markets don't adequately handle sudden crises and shortages caused by political or economic events, such as those that underlay the two OPEC oil shocks in the early and late 1970s. They don't have the time they need to adjust and, since what is man-made can be unmade, the circumstances make it very risky to invest in new higher-cost supplies—they may never be used if the economic or political events that underlay the crises vanish. Such was the case in the 1970s. Fossil fuel prices then came down in the 1980s, and the investments predicated on the high prices of the 1970s proved to be mis-investments. Those who made them lost a lot of money. In today's political and economic environment, a repeat of the OPEC oil shock is highly unlikely. OPEC has lost its monopoly on oil supplies.

As *Charging Ahead* properly argues, the case for accelerating the introduction of renewable, non-polluting energy sources depends heavily on pollution and long-term risks to the environment (although there are several other compelling arguments for renewables). If fossil fuels are sold at prices that cover only the economic costs of delivering them to the ultimate users, they are under-priced. Their real costs should include the costs of cleaning up the pollution they produce. And this is of course the economic case for a carbon tax on fossil fuels (proposed but then dropped in reaction to massive lobbying) or for special subsidies or tax breaks for renewable energy sources. Ideally, the costs of any form of energy should reflect the full costs of using that energy form. If this is not possible politically, the second best option is to give non-polluting energy sources tax breaks or subsidies equivalent to the costs of cleaning up the pollution caused by fossil fuels. For that is exactly what their usage creates.

If the issue is current local pollution, say metropolitan air pollution, the right incentives or disincentives are in principle easy to calculate, even if getting the "right" number to insert into the relevant equations is often difficult. If the issue is a form of pollution, such as carbon dioxide, whose effects on global temperature are highly uncertain, distant in time, and essentially irreversible, the right incentives to encourage a faster shift to nonpolluting renewable energy sources are impossible to calculate and a matter of judgment. John Berger believes that the current incentives are not leading us to make this shift fast enough, and he is probably right.

Capitalism does many things well, such as catering to individual wants, but it is myopic when it comes to long-term investments in its own future. Consider a situation where we know that what we do today will make the environment worse 50 years from now. What will a hard-nosed capitalist do using discounted net present values (the standard capitalistic decision-making algorithm)? The answer is nothing, since today's value of a dollar's worth of damage 50 years from now is zero. Fast forward 50 years. Once again the capitalist faces the issue of what he should do to stop the world from getting environmentally worse another 50 years into the future, and once again the answer will be nothing. Fast forward another fifty years, and no matter how bad things are, it will still not be rational to take any remedial actions. Making a sequence of rational

economic decisions, it is possible for a capitalistic market-driven economy to walk off an environmental cliff and commit collective suicide.

But clearly that is not rational from a human perspective. So keeping
track of where renewable energy technologies are going and thinking seriously about what could and should be done to speed up their introduction
into our economy make sense. After reading *Charging Ahead*, anyone will
be in a better position to make those important judgment calls.

Lester C. Thurow
January 10, 1997

PREFACE TO THE PAPERBACK EDITION

Since 1997 when *Charging Ahead* was first published, the need for carbon-free, clean energy systems and energy efficiency has become even more apparent. The year 1997 was the hottest on record. Political leaders, including President Clinton and Vice President Gore, joined thousands of prominent scientists in calling on the United States to reduce its emissions of carbon dioxide and other global warming gases. The President even took the unprecedented step of singling out climate change as the world's "overriding" environmental problem.

Meanwhile, energy prices remain extremely cheap by historic standards. Oil prices actually plummeted to just over $13 a barrel in early 1998. The nation's economic boom continued into its seventh year. National energy use, vehicle miles traveled, and carbon dioxide emissions rose during 1997; national fuel economy standards for vehicles remained lax. More families were able to afford second and third cars, or brawny sport utility vehicles that produce 20,000 pounds of carbon dioxide a year and now, with minivans, account for 45 percent of all new vehicle sales. Not surprisingly, U.S. oil imports climbed 4.4 percent, increasing our dependence on foreign oil as well as our vulnerability to future oil supply disruptions and foreign military entanglements.

The U.S. endured another costly eyeball-to-eyeball confrontation with Iraq from late 1997 through early 1998—nominally over U.N. weapons inspections, but obviously related to our Middle Eastern oil interests. The still-rising price for the recent show of Persian Gulf military force is already about $700 million *above* the costs of maintaining our "normal" military presence there.

Other new international developments reflect a growing worldwide recognition that unabated fossil fuel combustion poses grave and unacceptable risks to the world's climate and environment. The U.S. signed an international climate stabilization treaty in Kyoto, Japan with 159 nations in December 1997. If ratified by the Senate, it will commit the U.S. for the first time to reduce its global warming gas emissions seven percent below 1990 levels from 2008–2012.

As an initial step toward implementation of the Kyoto protocol, President Clinton in 1998 announced a five-year, $6.3 billion package of tax credits and new domestic spending to promote energy efficiency and renewable energy. The credits provide modest incentives for the purchase of more efficient vehicles and homes and for solar energy equipment. They also support the Administration's 1997 Million Solar Roofs Initiative, a federal program that calls for the installation of a million (total) rooftop solar electric and solar heating systems over a five-year period.

Although the new incentives package is very useful, it amounts to less than $5 per person per year for every American. By contrast, the nation grants billions of dollars a year in subsidies to the fossil fuel industries, so we are spending more to exacerbate climate problems than to solve them. Moreover, investing less than the equivalent of a deli lunch per person per year will not provide the money the nation needs to modernize its energy sector and install large amounts of clean generating capacity.

Nonetheless, a gradually intensifying commitment to renewable energy and energy efficiency is evident in the Department of Energy's expanded FY 1999 budget request and in its goals for renewables and efficiency, as well as in DOE's new draft energy strategy. A few examples illustrate the trend:

✦ DOE's Building Systems Program aims to reduce energy use in new homes by 50 percent and by 20 percent in existing dwellings.

✦ Within the next two years, DOE's Geothermal Energy Program expects to reduce the typical cost of geothermally produced electricity from 5–8 cents per kilowatt-hour (kw-hr) to only 3.5 cents, a price very competitive with power from coal (at 4–6 cents/kw-hr) and with some power produced from natural gas, especially when the social costs of natural gas emissions are included.

✦ For photovoltaic (solar cell) electricity, the DOE's National Renewable Energy Laboratory is striving to bring down the cost of thin film solar cells to $1 per peak watt, roughly equivalent to 4 cents/kw-hr in sunny areas.

The Administration's new draft energy strategy has some positive features but ought to include a call for the creation of a clean energy system for America by a specific date with proposed milestones to mark progress toward that goal. Perhaps one day soon that call will be issued by a far-sighted and courageous American president. Unfortunately, the DOE's draft plan still only proposes tiny sums for renewables in relation to the size and strength of the nation's $8 trillion economy and to the hundreds of billions of dollars we spend annually on fossil fuels.

On the critical question of building new renewable energy generating capacity, the energy strategy proposes that the nation double its non-hydro renewable energy capacity to 25,000 megawatts by 2010. That would amount to increasing renewable generating capacity by the equivalent of one large power plant per year—less than a seventh of one percent of the nation's current generating capacity. Rebuilding the nation's entire generating system at that rate would take more than 700 years, assuming that energy demand stayed constant instead of increasing (as it is doing).

Clean energy technologies are obviously capable of growing at a much faster pace than DOE proposes. Since 1997, renewable energy has continued increasing rapidly in efficiency, reliability, and sales, while declining in price. World photovoltaic cell output grew 30 percent in 1997, and many observers now recognize that the global market for solar energy systems eventually will be huge. World wind energy capacity also grew by more than 20 percent in 1997, making wind and solar the world's fastest growing energy technologies. Wind is on an even lower declining cost curve than geothermal power since its cost is already 4 cents/kw-hr (at the windiest sites), and thus it offers excellent competition for coal-fired electricity, nuclear power, oil, and some natural gas facilities, especially if the 1.6 cent/kw-hr production tax credit for wind power is extended by Congress.

Because wind power is a capital-intensive technology, its cost is very sensitive to the range of financing costs encountered by different owner classes. Municipal utilities, which pay no income or property taxes and can finance wind projects using 100 percent tax-free debt, should be able to build wind projects for the astonishingly low cost of only 2.3 cents per kilowatt-hour as early as 2000 in the windiest areas, according to projections by researchers at DOE's National Renewable Energy Laboratory.

These economic factors coupled with the technological improvements noted plus growing worldwide concerns about global warming have attracted large new capital investments by major oil and gas companies to the solar and wind energy industries over the past year.

Companies that have significantly expanded their stakes in the solar business in 1997 include British Petroleum, Royal Dutch Shell, and Enron. Shell in 1997 announced that it will make renewable energy one of its five basic business groups, and that it will invest $500 million a year in renewable energy. The move puts the company's solar, wind, and biomass activities on a par organizationally with Shell's exploration, oil, gas, and chemicals divisions.

Enron, a $19 billion marketer and producer of gas and electricity, formed Enron Renewable Energy Corp (EREC) of Houston, Texas in 1997 to pursue wind and solar energy business. EREC's subsidiaries include Enron Wind Corp (EWC) and Amoco/Enron, a joint venture that produces solar electric panels under the Solarex trademark (see Chapter 11). Most U.S. manufacturers in the solar cell business doubled or tripled their solar cell production in 1997, and Amoco/Enron has been expanding production, too.

Enron also purchased Zond Corporation last year, which by then had become the leading U.S. wind turbine manufacturer-developer (see Chapter 16), and the bankrupt Tacke Windtechnik GmbH of Germany, the world's fifth largest wind turbine maker. (FloWind Corporation of San Rafael, California, another major wind energy company discussed in *Charging Ahead,* declared bankruptcy in the past year, leaving unresolved the future of its subsidiary, Advanced Wind Turbines, Inc. (see Chapter 16).

British Petroleum's subsidiary, BP Solar, last year purchased the former Advanced Photovoltaic Systems, Inc. solar cell plant discussed in Chapter 10, and is converting it to manufacture state-of-the-art, thin-film solar modules made with cadmium telluride. The ribbon-cutting was attended by U.S. Vice President Al Gore and British Petroleum Chairman and CEO John Browne.

Chairman Browne received worldwide attention last year when he became the first prominent oil industry executive to acknowledge global climate change as a significant international concern requiring action. BP announced plans to expand its annual solar equipment sales tenfold over the next decade to $1 billion a year.

Just as renewable energy and energy efficiency have continued their remarkable advances in the past year, clean transportation technologies have also made enormous progress. Virtually all the major auto makers are now actively marketing or introducing new electric vehicles, and the vehicles themselves are available with more advanced batteries of nickel metal-hydride, lithium, and other compounds that offer longer range and accept an 80 percent charge in 10 minutes. General Motors is now projecting that in less than ten years 25 percent of its sales may be electric vehicles.

Even though government and the business community are beginning to recognize the impressive technological advances made in renewable energy as well as its bright (long-term) financial prospects, the technologies remain vastly underutilized, and their installed capacity is growing far too slowly to substantially reduce global warming. Although this may sound paradoxical given all the evidence of rapid progress just presented, there is no contradiction: While the global wind and solar industries have been increasing their manufacturing capacity and installed capacity rapidly *in percentage terms* in some parts of the world, the increases have been tiny in absolute terms relative to the enormous magnitude of global energy production and consumption. In addition, much of the domestic photovoltaic industry's production is for export; the domestic wind industry has been stagnant for the past few years; and a significant amount of domestic hydro and biomass capacity has shut down during the same period. All this is taking place not only in the context of extremely low prices for fossil fuels that compete with renewable energy, but amidst the turmoil of ongoing nationwide utility industry restructuring, which to date has prompted utilities to reduce their investments in energy efficiency and renewables.

Furthermore, in many developing nations where renewable energy systems and energy efficiency clearly ought to be top energy policy priorities, international investors are lining up instead to provide capital for hundreds of new coal-burning power plants in countries such as China, even as new climate research is revealing the suddenness with which dramatic climate change has occurred historically and could recur.

For all these reasons, a fundamental transformation—a paradigm shift—is needed in U.S. energy policy to make the creation of a clean energy and transportation system a top national priority. Once that commitment is made, the U.S. will be in a far more powerful position to urge other nations to emulate those climate-safe energy policies.

But given the strength of entrenched conventional energy industries, how can the U.S. be mobilized to forgo the immediate gratification of cheap polluting technologies in favor of those that in the longterm offer freedom from pollution, price stability, and supply security? In the past, renewable energy technologies lacked a powerful, stable political and institutional base of support capable of advancing their cause. That is why an intensive and protracted public education campaign should be launched to create a critical mass of American voters who understand the fearsome hidden costs and risks of perpetuating reliance on dirty fossil fuel technologies and who comprehend the capabilities of renewables, their benefits, and their environmental necessity.

This public education campaign would be challenging and expensive, but it would be a lot less expensive for society than pursuing a "business as usual" energy strategy. As a practical and strategic matter, the campaign's chances of success would be increased by better coordination and collaboration between pro-renewable energy constituencies, including the public health community, the broad-based environmental organizations, and the individual safe energy industries—solar, wind, biomass, geothermal, and hydro—which often work in relative isolation from each other. The property insurance industry and other corporations that stand to lose from an increase in extreme weather events also ought to be tightly woven into a broad national coalition. I hope foundations, individuals, and enlightened corporations can be counted on to strongly support a massive public education campaign to bring informed debate on energy policy to the town halls and college campuses of America.

In addition to the policy recommendations outlined in Chapter 25, Americans also now urgently need to press the Federal government to include powerful incentives for clean energy sources in pending utility industry restructuring bills to stimulate renewables' market penetration. Government at all levels also should be urged to buy a much larger share of its energy from renewable sources while shifting from internal combustion vehicles to clean transportation technologies.

Support for renewable energy today often vacillates when the White House changes hands or a new Congress is seated. Yet providing the nation with clean, safe, and secure energy supplies is not a Republican or a Democratic issue, but a transcendent global necessity on which the health of the planet and the welfare of the world depends. Education can make that clear to the public, and an aroused and informed public CAN bring the changes needed!

—JJB,
Point Richmond, CA
March 10, 1998

INTRODUCTION

E nergy is essential to every phase of our lives, and we pay a princely sum for it—$505 billion a year in the United States.[1] American families spend an average of $2,000 a year for their fuel and electricity. Yet we often take energy for granted—that is, until our energy supply is interrupted. Only then do we recognize how directly our economic security depends on energy: During the past quarter-century, recessions have followed each major oil supply interruption and price shock (in 1974, 1979, and 1990).

Despite the fact that energy deeply affects not only our economy but our environment and national security, some people argue that the rude hand of government should stay out of the energy business: Let "free markets" rule and make all our energy decisions. But "free markets" in energy are a myth; real energy markets do not work perfectly and are fraught with barriers. Government involvement in energy is profound and unavoidable. Our only choice is whether government involvement is benign and constructive or damaging.

The nation's Strategic Oil Reserve contains only a 37-day oil supply. We cannot afford to be passive or complacent about energy. We need energy policies that are both proactive and farsighted. Precisely because we are not currently in an energy crisis, now is the time to reexamine energy issues and set a new course.

The thrust of this book is that we should set that course by the sun. We must steer decisively toward a renewable energy economy, away from the dark whirlpools of fossil fuels, the ominous clouds of acid rain, and the gathering signs of global warming.

Charging Ahead provides a look at the status, struggles, and promise of today's renewable energy industry. Industry pioneers currently are creating tomorrow's renewable energy economy. We will be going inside their laboratories, workshops, factories, and boardrooms, perhaps sharing a bit of their hopes and frustrations. We will pause from time to time on these journeys for necessary scientific and technical background, or for an industry overview.

While I am a technological optimist about renewable energy technologies today, I recognize that some people are skeptical about them and see renewables as failed technology. They assume that renewables are marginal and inherently costly, capable of doing little to augment our energy supplies. They forget how heavily we have subsidized fossil fuels and overlook the market power of today's dominant fuels, which gives fossil fuel companies so much political clout. The chapters that follow show just how wrong and simplistic the skeptics' conclusions are.

The renewables featured in this book are some of the world's most advanced energy technologies, developed with the help of microelectronics, new materials, fast computers, laser beams, and robotic production lines.[2] Wind turbines, solar electric (photovoltaic) cells, solar thermal electric power plants, and geothermal heat pumps and power plants are technological success stories. This new generation of energy technology can make large and economically valuable contributions to our energy supplies, while freeing us from pollution and other environmental risks. I hope *Charging Ahead* increases awareness of these exciting possibilities and lays the groundwork for new energy policies that foster renewable energy.

Why have I chosen to include energy efficiency technology and electric vehicles in a book on renewable energy? Electric vehicles are included because electric propulsion is a "bridge" technology that extends renewable energy into the transportation sector. While we can't plug today's gasoline-powered cars and trucks into the electric power flowing from a wind turbine generator or from a solar electric panel (or solar power plant), electric vehicles *can* be connected to renewable electricity easily. Similarly, alcohol and diesel fuels derived from the solar energy stored in plants can operate common internal combustion engines. Energy efficiency is included because it reduces both energy demand and the environmental impacts of energy use. Energy expert Amory Lovins and others have demonstrated that the U.S. economy has saved more than $2 trillion in energy costs over the past 20 years (while growing 65 percent) through energy efficiency, fuel substitution, and other means.

Reducing on-site energy demand—and therefore the size of the energy system needed locally—lowers system costs and thereby increases the market for renewable energy systems. In the words of Senior Scientist Dr. Donald Aitkin at the Union of Concerned Scientists, "efficiency makes renewables affordable." By contrast, without energy efficiency, ever-growing energy demand will drive energy costs up while undermining environmental progress toward cleaner air.

Despite the doubts of some and the opposition of vested interests, an unstoppable revolution in energy technology has indeed begun. The impending changes are as profound as the world's earlier shifts from wood to coal, from coal to oil, and, more recently, from oil to natural gas and nuclear power.

The next shift is from fossil and nuclear energy to solar-based renewables. Fossil fuel use may peak by the year 2030. By 2050, or even earlier if we wish, a very substantial portion of the world's energy will be solar-derived—beyond the already significant contributions of biomass and hydropower. According to a recent study by Shell International Petroleum Corporation, renewables by 2050 may be providing almost as much energy as all fossil and nuclear energy combined. Not long thereafter in historical terms—certainly by A.D. 2100, and probably far sooner—the transition to a pollution-free renewable energy economy will be complete. By then, today's fossil fuel and nuclear energy technologies will be museum pieces.

On the way to that future, trillions of dollars will be earned worldwide through sales of renewable energy and energy efficiency technologies—and from the energy they will produce. Many of the renewable technologies likely to be most responsible for those sales are described in the pages that follow. Since these technologies will evolve greatly over time, the best we can do is identify the technological "lineages" that seem most likely to produce victorious descendants in the twenty-first century. *Charging Ahead*, however, may provide useful perspectives with which to play the extremely challenging game of picking technological winners from among the more numerous losers.

The book's concluding chapter outlines policies that will increase the chances that the United States will benefit as much as possible from the billions of dollars in new business and hundreds of thousands of new jobs that renewables and efficiency technology are just beginning to create. My discussion of what we could do to advance renewable energy points toward the limitations of current energy policies—shortcomings to which many of the stories in the book attest.

One of my chief goals in *Charging Ahead* was to chronicle the encouraging first steps we are taking toward practical renewable energy systems and to highlight the obstacles faced. I naturally intend the chapters to be useful for understanding the new technologies and future energy trends. More importantly, however, I hope *Charging Ahead* will spotlight the capacity of renewable energy, energy efficiency, and electric vehicles—when coordinated together—to transform our world through clean energy systems, clean industries, a clean transportation sector, and a clean environment. I also hope *Charging Ahead* adds to your appreciation of renewables for their beauty and elegance as well as for their astounding potential to improve our lives.

PART I

A GREAT
TRANSFORMATION

1

THE UPHILL STRUGGLE

As noted, within about 50 years—and possibly sooner, if we choose—the world will greatly reduce its massive reliance on fossil fuel and nuclear energy by shifting to clean and sustainable renewable energy sources.

These abundant eternal sources are manifestations of the sun's radiant energy: sunlight itself, wind power, water power, near-surface heat within the Earth, various oceanic energy sources, and the energy stored in plant tissue. Energy systems drawing on these natural sources can operate forever on their energy income, without ever depleting the Earth's finite coal, oil, uranium, or natural gas resources.

THE EXCITEMENT ABOUT RENEWABLES

While most renewables have been used for millennia, modern science and engineering technology have of late made them much more efficient, convenient, and economical. We could now for the first time provide for modern civilization's energy needs virtually without pollution. That is an exciting and inspiring prospect.

A renewable energy economy based on solar energy would improve the nation's economy, environment, and public health for several reasons:

+ Because renewables do not use fossil fuels (most are entirely fuel-free) they are largely immune to the threat of future oil or gas shortages and fossil fuel price hikes.
+ For the same reason, because most renewable technologies require no combustion, they are far kinder to the environment than coal, oil,

and natural gas. Smog and acid rain could be eliminated with renewables. The collective lungs of America could breathe a sigh of relief.
✦ Finally, global climate threats could be dispelled.

Renewable technologies also offer other advantages and possibilities. Because renewables are domestically available in vast quantities, they are secure from geopolitical instability and oil embargoes.

Moreover, as a renewable energy economy engenders new industries, they in turn will create new jobs in construction, research, manufacturing, and administration. The United States could eventually retain hundreds of billions of dollars that we now export overseas each decade to pay for foreign energy supplies.

Simultaneously, a vigorous national program of energy efficiency could save the nation additional hundreds of billions of dollars in scarce capital that is now spent on inefficient energy use in homes, offices, industry, and transportation.

With foresight, sufficient investment, and a well-designed national energy plan, the United States, as the world's foremost economic power, could become the world's undisputed leader both in renewable energy and in energy efficiency technology.

The sooner we make a high priority of creating a renewable energy economy, the sooner we can begin enjoying all its economic advantages in trade, energy security, employment, and domestic investment.

From a historical perspective, the world today is already on the verge of a monumental energy transformation. A significant number of the wind and solar technologies that will be crucial to the twenty-first century's power supplies are already operating commercially.

While many of the technologies are proven and ready for use, our renewable industries are still either in their infancy, just fledging, struggling, or stagnant. The solar age has dawned, but the sun's rays are still just peaking over the horizon. We have not yet made the required national commitment to nurture these industries to maturity so they can deliver their full benefits to us.

DESPITE OVERWHELMING NEW EVIDENCE
ABOUT RENEWABLES, SKEPTICISM PERSISTS

While survey after survey has shown strong public support for renewables, there is a stubborn countercurrent to that enthusiasm. Some opinion leaders and members of the public believe that renewables don't work, can't meet our energy needs on the scale required, and cost too much. Many people still cling to outdated images of broken wind turbines, boondoggle wind farm tax shelters, and leaky solar hot-water heaters.

Some may have dealt personally with a fly-by-night solar contractor who oversold a poorly designed solar heating system. Others may have been beguiled in the 1970s by fanciful scenarios of how fossil fuels were about to run out, and solar energy was about to displace them. The golden age of solar energy supposedly had arrived in the 1970s. In hindsight, this proved not to be the case, and the over-optimism tarnished solar energy's reputation.

Today, skeptics often portray solar technology as an engineering curiosity, an heirloom of the 1970s, a pet technology for environmentalists, or a pricey ornament for luxury homes. But as suggested, solar energy has been maturing technologically and making tremendous economic strides since the 1970s.

Solar cell panels that cost $1,000 a peak watt in the 1960s and $30 in the early 1970s were only $4 by the mid-1990s, and costs are still decreasing rapidly. Wind power that cost on the order of 30 cents a kilowatt-hour in the late 1970s cost only 5–7 cents in 1995, with new capacity available at just 3.5–4 cents a kilowatt-hour in the windiest areas.

As the most cost-competitive of the renewables, wind power is now cheaper than nuclear power or electrical power from petroleum, and its costs are comparable to those of new coal plants. The costs of other renewables are being brought down as well. Electricity from biomass—specifically, from burning whole-tree energy crops—is expected to cost only 4.6 cents per kilowatt-hour by the year 2000. Ethanol, a liquid biomass fuel, today costs only $1.20 a gallon to produce—an impressive 70 percent decline since 1980.

While renewable technologies have made terrific progress over the past two decades, renewables, except for hydropower and wood combustion, still make but a tiny contribution to U.S. energy supplies.[1] This may seem odd, since some renewables are now cost-competitive with certain fossil fuels, and there is no solar energy shortage; renewable energy resources far exceed our energy needs. Why then is their contribution to our energy supplies so small?

ENERGY MARKETS AND
THEIR MISLEADING PRICE SIGNALS

Renewables have been held back by enormous economic and political obstacles. Taxpayer subsidies have helped make fossil fuels inexpensive and difficult competition for renewables. The fossil fuel and nuclear industries today command more than 90 percent of the U.S. energy market, and their dominance has enabled them to grow rich and powerful. Their extraordinary political influence has helped maintain the nation's generally pro-fossil and pro-nuclear energy policy. Though technical obstacles, such as the intermittency of wind and sunshine, also play a role in hampering the adoption of some renewables, they are less serious.

The biggest challenge for developers of renewables is to reduce their cost relative to competing fossil fuel technologies and to get future fuel price risks incorporated into fossil fuel prices. As noted, those prices have remained very low, allowing coal and gas companies to undersell new renewable energy technologies by large margins for baseload (around-the-clock) utility power.

Renewable technologies are more expensive in some instances than older established fossil fuel technologies, particularly when compared on price alone, especially short-term price. Yet the full environmental costs of the conventional technologies are generally not reflected in those short-term prices. So fossil fuel power appears artificially cheap. And the very significant environmental benefits produced by renewables are not properly valued by the marketplace, because these benefits are largely public goods, such as cleaner air and water. We do not often, as private individuals, volunteer to pay for environmental amenities to which all have free access.

Whereas seven state public utility commissions in recent years have accorded renewables some credit for their environmental benefits, the credits ordinarily are still inadequate. In addition, because of the newness of many renewable technologies and the relatively small scale at which they are being manufactured and installed, renewable energy power plants and equipment cost more than they would if mass produced in large quantities.

FINANCIAL OBSTACLES TO RENEWABLES

Although most renewables require no fuels, capital costs of renewables, such as plant and equipment, are substantial, and these costs generally must be financed at the start of a project. Yet the financial benefits of renewables are realized only over the life of the plant or equipment, usually in the form of zero fuel costs and, of course, freedom from future fuel price shocks, as well as in low operation and maintenance costs. By contrast, capital investment is not such a large proportion of the total costs of natural gas plants, and the costs of fossil fuels—a very significant cost component of total fossil fuel power costs—ordinarily can be passed through directly to utility consumers.

In addition to the large capital requirements for many renewables, investors who are asked to supply the capital to independent (non-utility) power producers (IPPs) for renewable projects demand a risk premium in the form of higher rates of returns, because of the relative newness of the technologies, the uncertain regulatory climate in which renewable energy technologies operate, and often because of the fragility of the IPPs themselves.

The investment picture is complicated by the current restructuring and deregulation of the electric utility industry, including the entry of new independent power producers to the marketplace. (Utility merger and acquisition activity also is on the upsurge.) In addition, utilities are struggling to hold down costs, accelerate the depreciation of older plants, and diversify into unregulated business areas. All these developments destabilize the electric power market, create financial uncertainty, and make it riskier to develop new renewable energy technologies and thus more expensive to finance them.

Fossil fuels, meanwhile, have enjoyed decades of lavish federal subsidies. When combined with their market domination, long operating histories, and price advantages, it is easy to see why most renewable energy sources still make small contributions to our energy supply. While no grand conspiracy is responsible for this result, lobbyists for the established oil, coal, gas, and nuclear power industries and their political allies in Congress understandably do pursue their own interests vigorously. They are quite willing to block or weaken legislation that would spur renewables, such as a carbon tax, or to absorb critical federal dollars needed to bring down the costs of renewables. Over the years, those subsidies have amounted to hundreds of billions in R&D funding and tax breaks. But that is only part of the story.

Competition between traditional energy sources is intense. The nation is shifting from the more polluting and expensive fossil fuels, like coal and oil, to cleaner, cheaper natural gas, muting opposition to fossil fuels. And by using cleanup technology, such as power-plant scrubbers and fly-ash precipitators, utilities have been making coal power cleaner and its pollution less visible, undercutting arguments for renewables.

In addition, because investments in energy efficiency technologies tend to be more cost-effective than investments in new power generation of any kind, overall energy demand growth has slowed in the United

States and in other advanced industrial nations. This has contributed to overcapacity in the energy sector, which has reduced demand for new energy supplies, making it still harder for new renewables to "break in."

UNCLE SAM'S WAVERING COMMITMENT

Although dependable government support for renewables could serve as a steady counterweight to the forces retarding them, federal aid instead has vacillated. R&D funding levels oscillate from one year to the next, and from one presidential administration to another, creating additional economic uncertainty. National energy policy—especially as reflected in the tax code—changes its emphasis too quickly to provide the long-term stable planning horizon that major new renewable energy investments require.

Foreign governments, meanwhile, give their renewable energy industries more generous and longer-term support than does the United States, therefore providing a more predictable operating environment, creating stiff competition for American renewable energy companies. European governments also work together cooperatively on renewable energy R&D, again leading to tough competition.

For all these reasons, the public's long-term interests in swiftly bringing a renewable energy economy into being are neglected. This has meant lost opportunities for the environment, the economy, and the people who have been valiantly trying to create a pollution-free, sustainable energy economy.

While formidable, the barriers to the more rapid spread of renewables are not insurmountable. All could be overcome by enlightened national energy policies that counter special interests, make necessary institutional changes, stimulate the market for renewables and efficiency technology, provide adequate R&D support, and share the risk of small- and medium-sized innovative companies.

To institute these measures, a broad, active, and potent political constituency is needed. Until it emerges, the renewables and efficiency industries will continue delivering only a fraction of the benefits of which they are capable, regardless of the technological progress they are making today.

This story of unfulfilled potential could be written differently in the future, if the renewable and efficiency industries were to receive strong political support. But the impressive achievements of the renewable energy and efficiency pioneers in the chapters ahead are all the more laudable, and their setbacks all the more understandable or poignant, given the barriers they still face.

2

TODAY'S ENERGY FOLLIES

Our energy policy is to take the cheapest oil in the world, regardless of source, and take our chances that it will continue to be
cheap and available.
—William K. Tell, Jr.,
Senior Vice President, Texaco

In the mid-1970s, the United States experienced an "energy crisis."
People nervously waited for hours in gasoline lines hoping the pumps
would not go dry. Today, energy seems cheap. Supplies seem secure. We
no longer seem to have energy problems, much less a crisis. But the idea
that we have solved our energy problems is an illusion. Our energy policies quietly exact tremendous costs from our economy and inflict great
damage on the environment.

HOW FOSSIL FOOLISHNESS
SHORTCHANGES AMERICA

Since 1970, we've spent more than a trillion dollars just buying foreign
oil.[1] Our transportation sector gets more than 97 percent of its energy
from petroleum today.[2] More than 50 percent of the oil we use is imported. Those ever-growing imports take precious dollars beyond our
borders. They deprive us of money we could otherwise use to rebuild our
industries and transportation systems, and to fund our medical care and
educational systems.

Meanwhile, Persian Gulf states spend large portions of their oil revenues on weapons of war. We can easily get drawn into Middle Eastern

hostilities, as demonstrated by our 1990 war with Iraq, following Iraq's invasion of Kuwait. As long as we remain so heavily dependent on foreign oil, we will be vulnerable to future oil price shocks and to foreign conflicts that could disrupt oil pipelines and international tanker traffic.

Our overdependence on oil is shortsighted. The nation's domestic petroleum reserves are declining sharply. If current energy trends continue, we will be dependent on imports—at unknown prices—for virtually 100 percent of our oil in just 15 years.[3] We could certainly argue about exactly how many years cheap oil will last. Maybe it will even be 20 years, or 30. But the fact is that the world's cheapest oil tends to be used first, and since world supplies are limited and irreplaceable, we will one day have used all the inexpensive, easily recovered supplies.

Plainly, since we cannot depend on oil forever, because of the resource and economic constraints mentioned, a petroleum economy is ultimately not sustainable. And a global fossil-fuel economy is also not sustainable for environmental reasons, especially in a world likely to contain ten billion people by the year 2070.

What about the current costs of our dependency on oil and other fossil fuels? We spend at least $25 billion a year for the military defense of our oil interests in the Middle East.[4] Tax breaks and other government subsidies to the fossil fuel industries cost another $20 billion or so a year.[5] We spend another $56 billion a year on our oil imports.[6] Then we experience about $150 billion in annual damages from fossil-fuel–induced air pollution.[7] The total comes to more than $250 billion a year.

By using renewable-energy and energy-efficiency technology to cut our oil use sharply, particularly in the transportation sector, not only will we reduce that liability, but a major decrease in oil use by the United States would significantly reduce world oil prices, saving oil-importing nations tens of billions of additional dollars. Our success in reducing demand would also encourage others to do likewise.

Moreover, in addition to the money spent on fossil fuel, its subsidization, and its pollution, we also waste hundreds of billions of dollars worth of energy in our energy-inefficient buildings, appliances, and transportation systems.[8] At least half of all electricity we generate could actually be saved through the use of readily available energy-efficiency technology—such as more efficient motors and improved lights, as studies by energy expert Amory Lovins and others show (see Chapter 21).[9] That's far more than the energy we could produce from the oil and gas reserves in our Outer Continental Shelf and the Alaskan Wildlife Refuge combined. If this seems like a lot of money and energy, it is, but it's still only part of the cost.

THE TRAGIC CONSEQUENCES
OF FOSSIL FUEL DEPENDENCY

Burning fossil fuels creates 30 billion tons of pollution worldwide in the form of soot, ash, carbon dioxide, and other gases.[10] Studies show that just the neglected environmental costs of this combustion—effects known as "externalities"—may be equal to, or greater than, the nominal prices we pay for coal or oil as fuel.[11]

To put the issue less abstractly, studies show that air pollution makes it harder for people with respiratory problems to breathe, aggravates heart conditions, and shortens average life expectancies for all people living in polluted urban areas.

Coal burning, which we rely on for 57 percent of U.S. electricity, assails our eyes, noses, and wallets in the form of air pollution, industrial accidents, tax subsidies, and wasted natural resources. Underground coal miners are subject to black lung disease and emphysema, accidental injury, and death.

Strip miners rip up the coal-bearing lands of Appalachia and the West. Thousands of miles of rivers and streams are polluted by acid drainage from coal mining. High-sulfur coal smoke tarnishes pristine skies and creates acid rain, killing forests, lakes, and wildlife.

Oil has not exactly been kind to the environment either. While a few big oil spills make headlines—like the *Exxon Valdez* disaster that fouled Alaskan bays and shores—the vast oceanic food chain, too, has become contaminated by thousands of routine oil spills. Oil-soaked birds, seals, and otters are but the more visible evidence of the ecological carnage that devastates the tiny plankton on which fish and all higher marine life ultimately depend.

Despite the costs and impacts, our fossil fuel dependency is quite understandable. Fossil fuels are convenient, concentrated, and easily transported. They have fueled the last 200 years of industrial development. Most importantly, they look inexpensive at the gas pump or the electricity meter. But we have not actually looked at their true costs nor faced their real consequences.

We say oil is inexpensive now, because the price is $18 to $20 a barrel. But once we use a barrel of oil, or a ton of coal or uranium, a peculiar thing happens. We discover the large hidden costs mentioned, such as those related to pollution, subsidies, and energy insecurity (for oil). For nuclear energy, some of the unwelcome costs hitchhiking along are the risks and costs of nuclear wastes, decommissioning of old power plants, nuclear accidents, nuclear terrorism, and nuclear proliferation.

Instead of looking mainly at current per-kilowatt-hour energy prices, we need to base energy decisions more on holistic comparisons of energy options. We ought to consider all phases of each energy production process, including probable future costs and their social and environmental impacts over the life spans of the technologies. This fuller, long-range accounting suggests that most renewables are already inexpensive by comparison to coal, oil, and uranium. That result will be even more apparent in the future as the costs of historically declining-cost renewables fall further once renewable energy systems are deployed on a large scale, driving their manufacturing costs down.

As I do the sums, the hundreds of billions of dollars in costs we incur annually through energy waste and damage caused by fossil fuels mount up to trillions of dollars a decade. Additional hundreds of billions have been squandered on the nuclear power industry, as discussed in Chapter 3.

The whole colossal multitrillion-dollar legacy of energy waste, unwise investments, and environmental damage then still needs to be adjusted upward by a huge, though probably incalculable, amount for the risk of climate destabilization due to excess carbon dioxide and other "greenhouse" gases, largely from fossil fuel burning and deforestation. We will look at these issues more closely in Chapter 3, and in Chapters 21 and 22 we will discuss further why energy markets have not worked better to prevent so many wasteful energy investments.

The enormous challenge we now face in energy policy is to eliminate energy waste and the hemorrhagic spending on obsolete conventional energy technologies, so we can redirect our financial resources to create a modern, energy-efficient, renewable energy economy.

3

THE ENERGY-CLIMATE
CONNECTION

If you start destabilizing large-scale natural systems, you are
actually tinkering with the very foundations of life support.
—Tony McMichael, epidemiologist,
London School of Hygiene and Tropical Medicine

LIFE IN A GLOBAL GREENHOUSE

This chapter explores some of the tremendous risks to which we may be
needlessly exposing the world and our descendants by our current energy
policies. Scientists for years have been warning about the buildup of
climate-destabilizing "greenhouse gases" in the atmosphere—carbon
dioxide, methane, nitrous oxide, and other compounds. We are currently
adding these gases to the lower atmosphere much faster than nature can
remove them. Once aloft, they remain in circulation for decades to centuries, and the effect is global warming. No quick, practical way to get rid
of them exists. The term "greenhouse gas" is actually a poor choice of
words, since "greenhouse" has a warm and pleasant connotation. *Heat-
trapping gases* is a more appropriate description for molecules that raise
the earth's temperature.

The consequences of planetary overheating are well recognized by the
scientific community, and are well described elsewhere. Suffice it to say
that sea level would rise from the melting of polar ice and from thermal
expansion of the oceans. Changes would occur in the circulation of upper
atmospheric winds as well as in ocean currents, leading to violent storms,
prolonged droughts in some regions, and catastrophic floods in coastal
areas. Offshore, deep seas could stagnate, zooplankton could decline, and

massive die-offs of higher sea life could take place. Populations of edible fish could plummet, and coral reefs could die.

On land, irrigation-dependent regions would be particularly hard-hit. Aridity might claim productive grain-growing regions of the United States. To the far north, thawing permafrost could release large quantities of methane, which would accelerate global warming. The rapidity of all these ecological changes on land and sea, and in the air and fresh water, would outpace the adaptive abilities of many organisms. Climate-sensitive ecosystems and many endangered species would perish. Unfortunately, once atmospheric circulation and ocean currents were destabilized, the effects could take decades or longer to subside, even if atmospheric gases were returned to preindustrial levels and additional perturbations ceased.

Given all the risks to human beings, ecosystems, and species, it would be fatuous to attempt to ascribe a dollar value to them. A single severe hurricane can cost hundreds of lives and cause billions of dollars in property damage alone. But a climate-induced increase in extreme weather conditions could mean *many* excess hurricanes and other related aberrations over time. Shouldn't we invest adequate resources in sound energy programs and policies that prevent climate disasters rather than spend hundreds of billions later for disaster relief?

Let's briefly take a closer look at the link between energy use and climate. Carbon dioxide is the heat-trapping gas of greatest concern as a global warming agent. Formed during the combustion of fossil fuels and by other means, more than 5.5 billion tons of carbon dioxide are discharged into the air throughout the world every year[1]—that's about a ton for every person alive. Smaller emissions of even more potent methane, nitrogen oxides, hydrocarbons and chlorofluorocarbons, as well as ozone also contribute to global warming.

Thanks to the size of the U.S. economy, coupled with its high-energy consumption and energy inefficiency, we are currently responsible for producing about a quarter of the world's carbon dioxide, plus a disproportionate share of the other dangerous gases. Each year, we generate 1.6 billion metric tons of all heat-trapping gases, and the total is increasing today at perhaps 1 percent per year.[2] Our per person emissions are currently about 500 times those of developing nations like China and India.

Because the capacity of the atmosphere to absorb carbon dioxide without overheating is limited, one nation's indulgence limits the options of others for disposing of waste gases produced by combustion. The United

States has yet to formulate energy policies that acknowledge and move to significantly redress this inequity.

GAMBLING WITH THE WORLD'S CLIMATE

No doubt exists in the scientific community that energy absorbed by naturally occurring atmospheric gases warms the Earth. We would freeze without this process. Scientists disagree, however, about the exact magnitude of the consequences of adding additional heat-absorbing gases to the air and about the precise nature of the regional impacts. But as computerized models of global climate have become increasingly sophisticated and reliable, scientific consensus has been growing on the extent of human responsibility for an already observed average global temperature increase of about one degree Fahrenheit over the past 100 years.[3]

Thus most scientists would agree that a doubling of atmospheric carbon dioxide concentrations would eventually raise the Earth's average surface temperatures by 2–6 degrees Fahrenheit. The National Academy of Sciences estimated the response would be 2–9°F. (Unfortunately, the estimates are not more precise than that.) Although a 2–9°F increase may not sound like much, the consequences would be dramatic.

The world has not experienced a warming of more than 3.6°F in more than 10,000 years, although that increase is near the lower end of the range of expected warming in response to a doubling of carbon dioxide compared to preindustrial levels. A 9°F average temperature rise would be a greater increase than anything experienced on Earth in the past million years. All these increases in global average temperature would be magnified two- or threefold in polar latitudes. (By contrast, at an average global temperature of only 5–9°F *less* than present, much of the Earth's surface was freezing in the last Ice Age.) Yet even maintaining current levels of carbon dioxide emissions will bring about a doubling of atmospheric carbon dioxide within a hundred years. And if the current exponential rate of carbon dioxide emissions increase continued for an additional century, climate models predict eventual global temperature increases of as much as 18°F,[4] a situation that would cause environmental damage of grotesque proportions.

Only by significantly reducing carbon dioxide emissions to below 1990 levels can we begin to bring down atmospheric concentrations. The prestigious Intergovernmental Panel on Climate Change (IPCC), a 2,500-member scientific body, states that no matter what we do now, the average world temperature is inexorably going to get 1–3.6°F hotter because of carbon dioxide already in the atmosphere from the burning of hydrocarbons.[5] The IPCC has also concluded that carbon dioxide emissions must be cut by 60 percent just to stabilize carbon dioxide concentrations at present levels. Obviously, it is imperative that the nations of the world immediately begin serious efforts to make major emissions reductions.

Those reductions may be a long time coming. Eight of the world's advanced industrial nations that in 1992 signed the UN Framework Convention on Climate Change, an international climate stabilization treaty, now appear unlikely to make even the modest reductions in carbon dioxide emissions called for in that agreement. The treaty's 157 signatories had agreed in principle to hold carbon dioxide emissions to 1990 levels by the year 2000. But most of them are still far from observing a "no regrets" energy policy that could insure against climate disruption by lowering emissions. In fact, current projections of world energy use show that ominous increase in carbon dioxide emissions are on the horizon.

THE SPECTER HAUNTING
CLIMATE-PROTECTION EFFORTS

If rapidly developing populous nations like China and India are to grow economically by vast expansions of coal and other fossil fuel combustion, as now planned, they will undermine efforts elsewhere to decrease carbon dioxide emissions. China is already the world's largest coal producer, mining 1.2 billion tons a year. Within 25 years, that amount may nearly triple. By then China might well surpass the United States as the world's greatest carbon dioxide polluter.[6] If China and India are to join in reducing carbon dioxide emissions without crippling their development, they will need to turn to renewable energy and energy efficiency.

The nations of the world, primarily the advanced industrialized ones, have already managed to significantly increase atmospheric carbon dioxide since the industrial revolution. Thus the risks that we and our progeny face are obviously not science fiction.[7] And with Earth's population increasing very rapidly, especially in the developing world, while people everywhere are striving for better living standards, the demand for energy services will unquestionably continue to grow.

The way to meet these rapidly swelling demands without overtaxing environmental systems or inflicting economic pain is through a major expansion in our reliance on energy efficiency technology and renewable energy. Those partners are the most practical way to cut emissions as drastically as they must be cut. Odd as it may sound, codevelopment of efficiency and renewables is a planetary insurance policy to protect against possible global climate catastrophe.

Consider the alternatives: If the proponents of more coal, oil, and natural gas use are wrong, the dire consequences those fuels may cause cannot readily be reversed. However, if proponents of efficiency and renewables are wrong, the world—by erring on the side of prudence—will nonetheless enjoy a cleaner and more efficient energy economy than we have today. Moreover, the longer we feed the world's fossil-fuel addiction, the harder it will be to overcome. Why indulge in irreversible global climate experiments? Far better to switch as soon as practicable to carbon dioxide–free renewables than to learn too late that we have irrevocably altered weather patterns and destabilized our only climate.

Anyone who finds it difficult to take the idea seriously now that we could disrupt such enormous energy flows as those of global climate should recall that the partial destruction of the world's stratospheric ozone layer, which protects us from skin cancers and cataracts, also sounded highly implausible when first predicted in the 1970s. Skepticism also greeted early concerns about acid rain, which sterilized thousands of lakes and about DDT, which wiped out millions of songbirds. If global warming is indeed the menace that the vast majority of scientists believe, then efficiency and renewables are not just our best hope; they are the world's *only* hope.

NUCLEAR FIASCOS AND FUSION ILLUSIONS

Despite earlier dreams and claims by their proponents, nuclear power and nuclear fusion are clearly neither the answer to our energy problems nor to climate threats. Nuclear fusion does promise large quantities of energy using hydrogen nuclei fuel obtainable from water. But, although the fuel may be cheap, fusion reactors will be extremely expensive. And unlike solar, wind, or geothermal energy, fusion will produce very significant amounts of radiation. The neutron-irradiated reactor itself, for example, will eventually need to be disposed of as nuclear waste. Those who have sought to portray fusion as a clean technology have been fostering an illusion.

In addition, fusion is still in an early stage of scientific and technical development and has not yet even produced more energy than the fusion reaction itself requires. Even fusion's greatest proponents do not expect commercial fusion reactors to be available until at least 2030. Meanwhile, fusion will probably require billions more in public funds just to try and reach the market readiness by 2030 that various renewable technologies have already attained by 1996.

Conventional nuclear reactors, the kind that split instead of fuse atoms, also present a host of nasty problems. Their cost overruns, the thousands of nuclear "incidents," the catastrophic accident risks, and the unresolved radioactive waste issues all weigh heavily against nuclear fission. Far from being a solution to our fossil fuel dependency, nuclear power is such a bad deal economically that no new nuclear power plants have been ordered in the United States since the 1970s,[8] and old U.S. reactors are going out of service faster than new ones are being commissioned.[9] U.S. Energy Secretary Hazel O'Leary, a former nuclear utility executive, observed recently, "My sense for the near term is that there's not one board of directors of any investor-owned utility in the U.S. who would spend a nickel to build a nuclear reactor."[10]

Despite this grim and now widely shared appraisal, the nuclear option lamentably has taken such a large share of U.S. and European energy research-and-development dollars that it has starved renewable energy development.[11] In 1990, the U.S. Department of Energy spent over five times more on civilian nuclear power than on all the renewable technologies combined.[12] Civilian nuclear power has cost U.S. taxpayers, consumers, and investors $492 billion (1990 dollars) from 1950 to 1990, just in readily quantifiable costs.[13] That tab in 1992 was still increasing at about $50 billion a year for a technology that produces electricity at an average 1990 cost of 10 cents a kilowatt-hour—twice the cost of fossil fuels and far more than wind power.[14] Without doubt, the stupendous avalanche of money poured into nuclear power could have been spent far more productively on renewables and energy efficiency.

While we have derived a great deal of economic benefit from nonrenewable energy systems, their many poisonous by-products clearly have also cost us dearly. Nonetheless, we will need to use energy from the less toxic and more benign of the nonrenewables, especially natural gas, to build the hardware for the renewable energy economy. Paradoxically, right now—when fossil energy is affordable—is the best time to start building the renewable economy, though the need for renewables naturally appears least acute now. All else being equal, the lower that world energy prices are, the less costly the construction of our new renewable energy systems will be. Yet the very conditions that make the enterprise economical for the long term make it seem less attractive in the marketplace where short-term thinking rules.

Precisely in situations like this, where great public benefits can accrue from long-term strategic action, government market stimuli to advance renewables and efficiency are most helpful and justified. Petroleum, coal, aircraft, and electronics are only a few of the major U.S. industries that have benefited enormously from hundreds of billions of taxpayer dollars.

CHANGING OUR "ENERGY MIX"

Some people have worried that shifting to renewables would be a huge disruption for the nation. Fortunately, great changes can be made in our energy systems without consumers even being aware of them. After all, you can warm your tomato soup and toast your bread just as easily with renewable power as with energy from a coal plant or nuclear reactor.

To those who are skeptical that we can really make a major shift, the case of California should be enlightening. As recently as 1978, California depended on fossil fuels for 81 percent of its electricity. Today, the state uses fossil fuels for only 35 percent of its electric generation (in years with normal rainfall) and gets 20 percent of its electricity from renewables.[15] Maine gets even more—35 percent. Moreover, prior to California's electric utility restructuring in 1996, the California Energy Commission projected that three-quarters of the state's new electrical demand in the next decade would be met by energy efficiency.[16] Although most of California's renewable energy is currently hydropower, California could easily tap other renewable resources.[17]

California's conversion caused no wrenching economic hardship. The state now has more than 700 companies in renewable-energy and energy-efficiency businesses. In 1992 they employed 35,000 people and contributed $1.2 billion to the state in taxes. If California can reduce its fossil fuel dependency by 46 percent within 15 years, other states, too, can make major reductions.[18]

The United States urgently needs a clean, secure energy system that can provide cost-competitive power forever. Such an energy system is no pipe dream. Nor does it require alchemy or radical breakthroughs in fundamental science. You can find the key to tomorrow's energy just by glancing up in the sky on a clear day. That brilliant star that warms our planet and provides us solar, wind, and hydropower can help us make a seemingly magical transformation that will revolutionize our energy future. Wind, sunlight, and other energy sources, even solar energy in green plants, are ready to provide the nation with clean, ample, renewable power. The stories of energy pioneers in Parts II through VI of *Charging Ahead* show how we can get it, and, equally important, what stands in the way.

While these next chapters are about new energy systems and the significant new industries based on them, the chapters are also about people with strength of character and determination who defied engineering and financial odds to advance the frontiers of new technology. I hope you will enjoy their remarkable stories.

PART II

SHINING POWER

4

※

ANSWERS FROM THE DESERT

The question isn't how you get the renewable energy industry to be
the heavyweight champion of the world, but how you get certain
industries out of the intensive care unit and others out of diapers.
—Mike Lotker, former vice president
of Business Development,
LUZ Development
and Finance Corporation

Visionaries have long wondered how the gift of sunlight bathing our
solar system could be captured, converted to electricity, and put to work.
An answer came in the 1980s, from a small, adventuresome start-up company. Today its nine gleaming solar thermal power plants in California's
Mojave Desert generate more than 350 megawatts of electricity—enough
to meet the residential electric needs of a small city. No one else anywhere
in the world has matched this accomplishment.

The plants were designed by Luz International Ltd. (LUZ) and were
built through five subsidiaries and the partnerships they set up. All told,
more than a billion dollars was invested in LUZ and its ventures, much of
it by utility and insurance companies. LUZ's first plant, Solar Electric
Generating System I (SEGS I), was completed in 1984. Other LUZ plants
soon followed.

The LUZ plants make electricity just as any conventional power plant
does—except that most of their heat is supplied not by fuel, but by pure
sunlight. However, LUZ cleverly designed the plants as solar-fossil hybrids, so they would be able to operate even when the sun is not shining,
by burning natural gas.

23

Not only did LUZ develop the new solar technology, but the firm scaled up its technology rapidly from ingenious small-scale prototypes to more efficient installations covering square miles of inhospitable desert. And LUZ brought the costs of solar thermal power down rapidly. As we shall see, though, that was not enough to guarantee LUZ's financial success.

So carefully were LUZ's financing arrangements devised, however, and so reliably did its plants operate, that—despite a minefield of major risks—investors poured large sums into LUZ projects in the 1980s and made respectable returns. People knowledgeable about cutting-edge energy technology were awed by LUZ's accomplishments. Besides developing and commercializing their technology, LUZ landed power-purchase contracts with Southern California Edison Company (SCE) and San Diego Gas & Electric Company (SDG&E). LUZ then built its unique, complex, precision power plants on unbelievably short schedules.

One or two new LUZ plants a year appeared under a merciless Mojave Desert sun from 1984 to 1991. LUZ meanwhile led a charmed life, always a step ahead of its enemies: collapsing fossil fuel prices, declining tax benefits, and unpredictable government energy policies. But in the summer of 1991, a series of events occurred that tested the company to its utmost and revealed much about the struggles of renewable technologies.

LUZ: THE VISION, THE BOOK,
THE CITY, AND THE COMPANY

The idea of LUZ (which rhymes with "blues") originated with a brilliant but eccentric inventor and philosopher named Arnold Goldman, who founded LUZ in 1979. Along with Goldman, the driving forces behind the company included Patrick François, a globe-trotting marketing entrepreneur, and Newton Becker, the wealthy head of an international chain of accounting schools, reputedly the world's largest individual investor in Israeli technology.

Arnold Goldman is a shy and visionary American from Providence, Rhode Island, who now lives in Jerusalem. After earning a master's degree in electrical engineering from the University of Southern California, Goldman went to work for Litton Industries in 1967, but left to found a company called Lexitron. Goldman had recognized that word processing, which at the time existed only on mainframe computers, could be adapted to the personal computer and readily introduced into offices.

Today, some people still refer to Goldman as the father of the modern word processor.

In 1977, Goldman left Lexitron and became financially secure by selling his Lexitron stock. Goldman then relocated to Israel, where he began working in earnest on a book of philosophy and social theory called *Working Paper on Project Luz*, which he had started 18 years earlier.

Working Paper on Project Luz was an enormously ambitious effort to produce a unified theory of knowledge, ranging widely over cosmology, mathematics, physics, ethics, education, economics, and religion. It culminates in a comprehensive vision of a utopian "eternal city" called Luz. The city was to have a high standard of living so its members would have enough time to develop into "whole, multi-dimensional human beings."

Goldman envisioned LUZ as 12 component communities surrounded by an aesthetically appealing yet functional wall composed "of a long series of parabolas which focus light on an energy-absorption tube," so as to gather much of the energy the community would need. Two major industries would contribute to the city's economy. The first was a Solar Modular Housing Corporation that was to produce kits enabling families to design and build their own environmentally sound homes. The second was to be an International Solar Equipment Corporation that would produce solar energy–generation equipment.

For Goldman, part of systematic philosophical thought was to test the validity of his ideas in the real world. When he finished his book in 1979, he decided that a business environment would be the most challenging place to explore the "integrative understandings" he had developed. So in 1979, he started a company called Independent Household Products (IHP).

Goldman initially saw IHP as a city-building company and began searching for investors. Financiers, however, were unpersuaded when he talked to them about the new firm. The project was too large and difficult for them to grasp, and would have required massive funding.

In response to investors' rejections, Goldman took the solar energy portion of his vision and modified his book so it could be a guide to the creation of a solar energy company, which in turn, could be the economic foundation of Luz, the city. The word "luz" symbolized both light and consciousness for Goldman. Luz in the book of Genesis is the place where Jacob goes to sleep and dreams of angels ascending a ladder to heaven. Jacob awoke there and renamed the place Bethel, which means "House of God." The company LUZ was to have been a kind of connection between Heaven and Earth, or at least between Earth and the sun.

Goldman hoped the firm would also make money for investors, produce a useful product for customers, and provide a supportive workplace for employees. Understandably, after two decades of thought and preparation, he was eager to bring LUZ into being.

OIL EMBARGO FEARS

When Goldman found it difficult to enlist others in the IHP solar cities cause, he changed the company's name to Luz International Ltd. At the time (1979), the Yom Kippur War of 1973 and the ensuing Arab oil embargo were still fresh in many people's minds, especially those of American Jewry. Oil prices had jumped from $3 to $12 a barrel in 1973, causing inflation and gasoline shortages. Then the Iranian Revolution, which drove the Shah from power in 1979, brought another disruption in world oil supplies. The ensuing panic caused oil prices to skyrocket from $13 to $34 a barrel.[1] The Jewish community feared that—through the threat of oil supply cutoffs—America might be blackmailed by Arab nations to reduce U.S. support for Israel. Some in the Jewish community therefore believed that alternatives to oil had to be developed. And what better way than to develop indigenous sources of energy, like solar?

Goldman also had another agenda. He was looking for a new technology that would help Israel by creating jobs there. He surveyed a number of renewable energy technologies—wind, photovoltaics, and solar thermal power—and concluded that solar thermal power was closest to being economically competitive. While oil companies drove pipes into the ground for energy, he would use rays of light to tap into celestial power. Goldman at first intended LUZ to build solar power plants to generate process steam for industry, a market he believed was worth billions. He had no idea at the time that solar plants could even be used to generate electricity.

Living in Israel, Goldman knew the Israeli government had programs to subsidize the development of new technology and would provide subsidized export financing. To help him take advantage of these programs, Goldman in 1979 retained Patrick François, a Frenchman living in Israel who had experience starting companies. "We really had a blend of personalities," Goldman said. "He started picking up the marketing end. Then we built a business plan, traveled together to the States, and met a few of the venture capital groups that had been involved with Lexitron."

Goldman and his principal investors decided to base LUZ's R&D and manufacturing in Israel. They saw it as a sound business decision, because skilled labor was cheaper there, and the Israeli government was willing to share in LUZ's R&D costs, believing that LUZ would indeed create domestic jobs and economic activity. They proved right; Luz Industries Israel, a wholly owned LUZ subsidiary, became one of Israel's top ten industrial exporters from 1986 to 1990.

FINANCING THE RISKY DREAM

Arnold Goldman assumed responsibility for advancing LUZ's technology and headed the Israeli R&D and manufacturing. Patrick François essentially took charge of the business side of LUZ.

Goldman at first tried to raise money from American venture capitalists, but couldn't get anywhere. "People didn't understand solar technology and solar energy," Goldman found. "No establishment company would have anything to do with you," he said. "They considered solar to be off the wall." Moreover, they weren't interested in dealing with a start-up offshore company.

In disappointment, Goldman went back to Israel empty-handed and restructured his business plan for presentation to risk-taking individual American investors, specifically to past Lexitron sponsors and members of the Jewish community.

Newton Becker, a certified public accountant (CPA), is the founder and owner of a chain of 130 schools that teach accountants how to pass the CPA certification exam. He is a short, outspoken man with graying hair, grayish-blue eyes, and a great enthusiasm for new energy technology. Goldman strode into Becker's luxurious Bel Air home in December 1979 carrying a two-and-a-half-foot-long model of the parabolic glass trough that would serve both as the energy collector and concentrator for his proposed power plant. "When Arnold came to me," Becker said, "I felt that here I could be in on the beginning of the solar age."

Becker believed that the world's energy supply would inevitably have to be solar, unless some unanticipated new energy technologies were developed. He was also impressed by Goldman's past accomplishments in the word-processing industry and by his cost-consciousness. "I never thought of failure at the beginning," Becker said. "Later on, I knew it was risky."

In raising the R&D capital through December and January, Goldman promised his investors that he would have a working model to show them by the summer of 1980. They could then come and see it and decide whether they wanted to help with Phase Two funding. During the first eight months or so of R&D, François and Goldman assembled a technology development team. They focused on creating a working solar energy collector and concentrator with a computerized control system to keep it aimed at the sun. The device was basically a sheet of mirrored glass bent into a parabolic arc. A frame held it in position, and a hollow tube was positioned at its focal line. A computerized control and single-axis tracking system kept the unit aimed precisely at the sun. When the sun's rays fell on the mirror, concentrated sunlight would be focused on the tube. Fluid flowing in the tube would be heated. The heat could then be used to produce steam and electricity.

A strong esprit de corps developed around the engineering effort. The engineers produced a prototype and mounted the apparatus on the roof of a building. The model worked and was finished on schedule. "By the standards of those days, it was a good-sized model," said Goldman. "Typically, companies take two or three years to put something like that together." Though still in swaddling clothes, LUZ had produced its prototype in only months.

Investors were impressed. Becker, for one, provided more money, becoming the largest investor in the next and larger round of financing. This money helped pay for organizational development and the initial effort to propel LUZ into the solar process heat business.

THE SOLAR HEAT MARKET EVAPORATES

LUZ now tested its solar collector by using it in building a small, prototype industrial process-steam generator for a kibbutz in Israel. It was technologically successful, and the LUZ team learned much about solar thermal system operation. Buoyed by this experience, LUZ soon afterward succeeded in selling and building the initial stages of three pilot industrial steam systems for textile companies in the southeastern United States.

Then the company faltered. Try as they might, LUZ could not get the required business volume to make the southeastern industrial process heat a commercial success. Worse, the sales soon led to an even more profoundly shocking discovery: LUZ had no market. Period.

The founders had imagined that cattle feedlots, meat-processing plants, and other industries would purchase solar steam. Their solar collector, however, required direct sun without any cloud cover whatso-

Close-up of parabolic tracking solar collector for one of LUZ's utility-scale solar thermal electric power plants. *Courtesy of LUZ International Ltd.*

ever—diffuse light cannot be focused. But cloudless skies were all too rare in the Southeast, where they had hoped to market additional units. And in the Los Angeles vicinity, where there was adequate sun and plenty of potential industrial customers, land for the plants' solar collectors was simply too expensive. The ideal physical conditions for solar thermal energy generation were in the deserts of California, as well as in the Southwest, and in west Texas. But solar process steam customers were not sitting out in the desert. And it was not economically feasible to transport the steam long distances from the desert to users.

At first, Becker did not want to accept the conclusion that there was no future for LUZ in industrial process steam. He and Goldman tried selling systems to hospitals and other possible customers. "We squandered two million dollars," Becker recalls. As the process heat initiative fizzled, the company ran out of money. "We had to lay off that whole sales force," said Becker. "We cut back the company dramatically." Becker gave LUZ two offices in his Los Angeles firm from which to conduct sales.

Then, in the mysterious ways that the world works, their foundering solar energy venture was rescued by an oil company. Accurex, a subsidiary of Phillips Petroleum, was also in the process steam market and had the bright idea of using the steam from solar collector modules to make electricity. They obtained a contract about 1983 to sell the power to Southern California Edison Company, but then proved unable to raise the funds to build the plant. Becker and Goldman grabbed hold of the idea. After talking to Southern California Edison (SCE), they excitedly concluded that LUZ *did* have a future—in making solar electricity for utilities.

REFINING THE TECHNOLOGY:
ENGINEERING BREAKTHROUGHS

While LUZ was searching for financing and seeking markets, Goldman was busy developing LUZ's technology. A major difficulty with the early plant designs was radiative losses of solar heat, before it could be used. Despite this, even the early LUZ plants achieved 80 percent of their projected electric output and 90 percent of their projected revenues. No one had come anywhere near these levels of performance in a solar facility prior to LUZ, said Goldman.

In LUZ's first plant, steam temperatures were limited by the performance of the plant's synthetic heat-transfer oil, a compound known as Caloria. That was replaced in later plants with a fluid called Therminol, which could be used at higher temperatures. Plant operations were also limited by the conventional black chrome coating used on the heat collector pipes. Goldman and his team eventually developed a unique ceramic-metallic coating and vacuum insulation for the heat pipes, and a new technology for depositing the coating efficiently. These breakthroughs allowed later plants to operate at higher temperatures. That made it possible for them to achieve higher thermodynamic efficiency, which meant more power for every photon of sunlight received. Newer LUZ plants therefore could meet their power-production goals with smaller, less expensive solar-collector fields. Successive improvements in the computer-controlled collector positioning system also translated into lower costs.

Besides the early heat losses, another significant problem was the massive bulk of the solar collector. "It was like a tank," Newton Becker said. "You needed a crane to move a three-foot section." This led to huge fabrication and shipping costs.

Becker was not the kind of investor willing to sit back and let his money dissipate, however. Becker the financier put on an engineer's hat and drew a more elegant solar-collector design. The new module would use molded glass that could support its own weight and therefore would require a much less massive frame and could be shipped economically in stacks. The new collector could be built for a fraction of its predecessor's cost.

Once the new prototype solar collector was built, most of the other technology needed for a power plant could be had "off the shelf." The big problem was to find a customer to provide the money so they could commercialize the product. After some negotiation with SCE, LUZ executives confirmed that if they could but finance the plant construction, SCE would buy the power as they were obligated to do under provisions of the Public Utility Regulatory Policy Act of 1978 (PURPA). That was a major piece of the puzzle. Not that SCE had to be hauled kicking and screaming to the deal. The company was very supportive of LUZ and actually provided the land for the first LUZ plant.

Becker and Goldman now studied the economics of the Accurex proposal intensively with the help of an accounting firm and decided that a 10-megawatt plant, such as Accurex had proposed, was not economical, but that a slightly larger plant might be. At this point, they came across a 13.8-megawatt turbine generator at a discount and bought it.

The plant initially was supposed to cost $40 million, all to be raised by the investment banking firm of Goldman Sachs. "When we realized we couldn't build it economically at $40 million," Becker said, "we had to go to a larger plant, they bowed out on us—almost at the last minute. It scared the hell out of all of us." All of a sudden, LUZ was faced with trying to find major investors and 60 million dollars on its own.

TAX CREDITS DRAW INTREPID INVESTORS

Goldman, Becker, and François huddled about the financing problem with Werner Wolfen, an attorney and partner at a prominent Los Angeles law firm. They decided they could finance plants with a mixture of debt and equity designed to minimize the investors' risk. Eventually the typical mix would be 50–60 percent debt, 40–50 percent equity.

The equity investors in LUZ plants would also get the benefit of then-extant 15 percent federal renewable energy tax credits, 10 percent ordinary federal investment tax credits, 25 percent California energy tax credits, and five-year accelerated depreciation for their investments. If everything worked, these and other very generous tax law provisions would enable them to more than recover their out-of-pocket costs in the first year of their investment through tax benefits alone. If oil prices fell, which no one at the time really expected would happen, the equity holders' risk was reduced by the fact that they could still take the tax credits and accelerated depreciation, or sell their investment outright the very next tax year for next to nothing, so as to generate an additional tax credit almost equal to the original investment.

The debt holders' notes, in turn, were secured by the equity and SCE's power purchase contract, and debt holders could take tax write-offs for interest expenses. LUZ also put up letters of credit and capital reserves. Major individual investors were by now so confident in LUZ's technology that they took the bold step of personally guaranteeing $13 million of the first plant's costs. In addition, the plant's construction company, Wismer & Becker (not a relation to Newton Becker), was short of work due to a dearth of nuclear power plant construction orders and therefore was willing to defer payment on construction of LUZ's first plant.

Two huge risks nonetheless remained for investors. The biggest was the technological risk of the plant failing to work or underperforming significantly. "If it did not work technologically," Newton Becker explained, "the [investors would] be out everything," because in order to get the solar energy tax credits, the plant had to produce enough solar power to earn revenue sufficient to make payments on the debt. No one really knew whether that could be done.

The other daunting and related risk was that the plant would fail to meet its deadline by not going into production before December 31, 1984, the day on which the year's tax credits expired. That would make the equity investors ineligible for the credits and would blow-dry investors' profits. Needless to say, that would have throttled the infant solar thermal power industry. However, the flip side, explained Becker, was that "if the system worked, there was almost no downside risk . . . that's the only reason we got people to put in $60 million."

Most of the investors in the early plants made modest returns. And their responsibilities virtually ended once they signed the necessary papers and parted with their money. Not so for the LUZ team. Once the money for the plant was committed, LUZ faced big new challenges in getting SEGS I into operation on time at their Daggett, California, site.

Construction on the plant did not even begin until October 1983. The start-up of a new power plant is normally a complex and difficult process.

Yet so tight was LUZ's schedule and so intense the pressure that, despite the December 31 deadline, start-up procedures were not scheduled to begin until December 20.

A FLIRTATION WITH DISASTER

The engineering team from Mitsubishi, which built the turbine, was on-site, as were LUZ personnel and engineering representatives from architect-engineer Wismer and Becker and its subcontractor, CH2M Hill, the companies responsible for engineering procurement and construction.

"It was touch and go to check out the [heat] storage tanks so they were ready to operate," said engineer David Kearney, a LUZ vice president. "It was touch and go to finish the installation of the solar field and to get the control system operating." Unfortunately, the weather was intermittently cloudy and rainy around December 20.

In LUZ's solar plant design, desert sunlight impinging on curved mirrors is reflected by a silver coating on the lower surface of the mirror glass to focus (and thereby to concentrate) the light on a glass vacuum-sealed tube. Each tube surrounds a coated stainless steel pipe filled with heat-transfer oil. Glass housing and stainless steel pipe together constitute the plant's heat receiver, LUZ's proprietary technology. It works a little like a thermos bottle: The large concentric glass vacuum tube surrounding the steel pipe prevents heat losses by convection and conduction. The selective coating on the pipe maximizes heat absorption of high-frequency solar rays and at the same time minimizes heat loss through radiation.

Inside the vacuum enclosure of the heat receiver is another LUZ innovation, an unobtrusive little stainless steel mesh that protects the vacuum by scavenging any stray air molecules that penetrate it. Finally, a little proprietary box about eight inches by eight inches containing solid-state circuitry controls the motors that track the sun and aim the solar collectors toward it to maximize energy capture.

During plant operation, pumps circulate the synthetic oil through the plant's mirror field. Like a liquid sponge, the oil soaks up solar energy as it flows in pipes within the heat receivers from mirror to mirror. The farther it goes, the hotter the oil gets. By the time it reaches the mirror field outlet, it can attain 735 degrees Fahrenheit in the most advanced plants. The oil then flows through a steam generator—a vessel containing a bundle of curved, thin-walled tubes that operate like a radiator in reverse. As the hot oil travels through these narrow tubes, feedwater flowing over the tubes' exterior is flashed to steam by the intense heat. The steam is then piped away and superheated before spinning a turbine-generator to produce electricity.

After leaving the turbine, where the steam gives up some of its energy, the steam in the more advanced SEGS designs is piped through a reheat cycle before being sent back to the turbine to maximize the capture of useful energy. The steam exhausted from the turbine then flows to a condensor—another closed-loop heat exchanger—in which it is recondensed to water. Finally, it is piped back in a closed loop to the solar field to be reheated and superheated, starting the whole process over again.

In the first SEGS plant, natural-gas combustion is used only to raise input steam temperatures to the plant's turbine and accounts for less than a fifth of the plant's electrical output. In later, more advanced plants, which have a gas boiler and two-stage turbines, gas burning can substitute for solar heat entirely, for limited periods.

When the sun shone on the first plant, it would heat the Caloria oil as it flowed through the mirror field. Cold oil was stored in one 850,000-gallon tank; hot Caloria accumulated in the other. From there it could be withdrawn as needed. Operators could use that heat to help start up the plant, to boost output, or to keep the plant generating power during brief periods of cloudiness. Because natural gas alone could not be used to fire a boiler independently and make steam in the first plant, everything hinged on the availability of enough late-December sunlight. The inclement weather, however, was making it so difficult to heat the hot-oil storage tank that the whole project flirted with disaster. The start-up crew was forced to wait nervously for the sun to shine in order to warm the oil. Not until December 27, sometime around 10 or 11 at night, did start-up of SEGS I even seem to be within reach.

THE BLOODY THING WORKS

Getting the plant into service meant generating enough power so that its turbine output could be matched in voltage and frequency with southern California's power grid. Only then could the plant be connected. Mitsubishi had provided the plant's turbine-generator. The head of the Mitsubishi team was giving orders to his engineers in Japanese. They would bring the turbine up toward their target speed, and suddenly, in the last minute, the turbine would automatically shut off. Or a feedwater pump would fail; an instrument would malfunction.

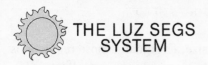
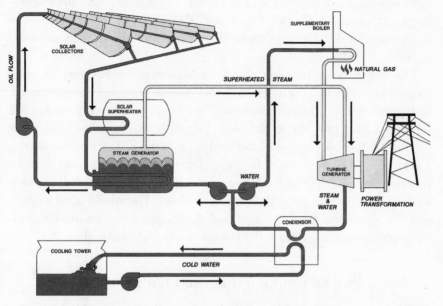

Schematic showing operation of a LUZ Solar Energy Generating System power plant. *Courtesy of LUZ International Ltd.*

"They'd get to the point where it was clear that they're almost ready to synchronize," said Kearney, "and something would go wrong." The air was filled with tension. At last, the plant was generating enough power. A synchronization gauge clicked into place. A tremendous roar of triumph—and relief—went up from everyone in the room. The world's first large commercial solar energy power plant was on-line.

"When we got the first contract with Edison," Goldman said, "I don't think they ever expected we would get the second contract. [But] the bloody thing worked," said Goldman. "We were actually achieving what the government is talking about achieving in four to five years [with its solar technology]." Actually, the plant did not achieve its design performance but operated well enough for LUZ to win a contract for a second plant twice the size of the first.

That plant joined SEGS I at Daggett and performed better, but also did not fully meet expectations. Then came five more 30-megawatt plants at Kramer Junction, California, also in the Mojave. Although larger plants would have been much less expensive, because of economies of scale, the

PURPA regulations limited the plants' size. Eventually, those regulations were modified, allowing LUZ to build two much more cost-effective 80-megawatt plants at Harper Dry Lake, about 12 miles east of Kramer Junction. All told, a total of more than a million solar mirrors were installed in the Mojave.

The five Kramer Junction plants are at a single contiguous site surrounded by a harsh desert studded with sparse gray-green sage, yuccas, and greasewood. Pinkish mountains are visible from Interstate 395 through a hazy gray sky to the west. Jagged brown hills like monstrous prehistoric teeth jut up out of the landscape in the distance to the north.

The long straight rows of brightly glistening mirrors around Kramer Junction are an impressive sight against a yellow-beige ground accentuated by the concentrators' stark black shadows. The whole plant with its metallic shine and bands of light and dark suggests an enormous computer chip stretching toward the horizon on the flat desert floor.

POWER UNDER DARK AND CLOUDY SKIES

SEGS I uses about 42,000 parabolic glass mirrors assembled into 50-meter-long troughlike solar-concentrator assemblies. One can look at a LUZ power plant, which is operated from a multicolored computerized control panel, and see it purely as a conglomeration of pipes and valves and boilers all chugging along autonomously according to the immutable laws of physics. But plant operators must be highly skilled to respond correctly to large daily variations of incoming solar heat and to operating requirements influenced by the periodic use of natural gas.

On an annual average, the units operate only about 25 percent of the time, since they are "peaking plants," which are designed to respond mainly to a utility's highest or peak demand. By contrast, baseload plants, typically having low operation-and-maintenance costs, are normally operated as continuously as possible. About 75 percent of the plant's yearly energy output comes from solar and 25 percent from natural gas. When needed, a natural gas–fired heater in the more advanced plants can supplement or substitute entirely for the solar-field heating.

Many features of the LUZ plants' design make the plants ideal for utility power generation in SCE's service area. For example, about 80 percent of their power is conveniently produced when SCE needs it most—on hot summer afternoons—to serve the high air-conditioning loads of southern California. But because the solar system is backed up by gas, the plant can also produce electricity even when the sun is down, or when the sky is overcast (a relatively rare event in the southwestern deserts). The SEGS plants are therefore reliable power producers, but—because most of their power is generated from solar heat—their operation produces very little pollution: The greenhouse gas emissions from a solar-gas hybrid plant are a mere fraction of those of the most efficient natural gas plants.

Yet not all utilities' peak power demand coincides so well as SCE's with solar plants' output. If solar plants are ever to compete neck and neck with conventional fossil plants, solar plants must be made capable of producing power even at night and during cloudy weather. By adding a small amount of low-cost heat storage to stabilize power output on a 24-hour basis, and by including natural-gas (or eventually biogas or hydrogen fuel) backup capability for the cloudy periods, solar thermal plants could operate economically in a continuous "load-following" mode.[2] (That means that as the utility system required varying power output levels to meet customer demands, the plant could readily adjust its output.) "Rather than viewing storage and backup fuel as competing alternatives," write David Mills and Bill Keepin in *Energy Policy,* "the optimal system would utilize both in such a way as to minimize the contribution from each." This design configuration has many advantages.

Based on their theoretical analysis, Mills and Keepin also conclude that a hybrid solar and gas plant with energy storage "could produce load-following solar electricity around the clock for a levelized real cost of 5 to 7 cents per kilowatt-hour over large regions of the USA."[3] For regions with high average solar radiation, they project costs would be close to 5 cents a kilowatt-hour—about the same costs projected for LUZ's next generation plants and quite competitive with new (1993) fossil plants. Mills and Keepin expect that, in the relatively near future, solar thermal electricity could be produced more cheaply than conventional electricity, especially if environmental externalities are counted. The 5.5 cent per kilowatt-hour estimated cost is even more impressive relative to LUZ's original cost of 24 cents per kilowatt-hour (in 1988 dollars)[4] because it is in (inflated) 1993 dollars, whereas the earlier costs were in the older, more valuable dollars.

AN ABUNDANCE OF ELECTRICITY

How much new solar thermal electric energy capacity could the United States get? Unfortunately, given low fossil fuel prices, the answer is, "Not another kilowatt more"—until state or federal governments again provide some new incentives in the form of more favorable tax treatment or other mechanisms. Yet if we are willing to accept the cost of these incentives for a limited time and also the plants' local environmental impacts, the solar thermal resource base is very large. The land requirements are only about 7.5 acres per megawatt (for an 80-megawatt plant)—not a serious obstacle.

If state-of-the-art solar thermal plants were deployed in sufficient numbers, and if energy storage, backup fuel capability, and transmission facilities were provided, the new plants could deliver as much electricity as we need. Plentiful desert sites exist in Arizona, southern California, Nevada, Utah, Colorado, Texas, and in Mexico's Gran Desierto, not far from the U.S. border. The sun in all these areas is strong year-round, the skies are clear, and land costs are low. Long-distance power transmission is not a major obstacle: Electricity from hydro plants in British Columbia is currently sent more than 2,000 miles to customers in Los Angeles and the Mid-Atlantic States. Solar plants in central Texas could deliver current to much of the Midwest. As former LUZ chairman Newton Becker points out, "All the electrical capacity of the United States could be replaced by our last-generation solar collectors put onto two places: Edwards Air Force Base, near Los Angeles in the Mojave Desert, which is roughly 20 miles by 20 miles, and the White Sands [New Mexico] proving grounds."

LIVING ON THE EDGE—
LUZ'S CHRONIC CRISIS

Technologically, LUZ's plants achieved unprecedented success in smoothly blending solar and natural-gas technologies to produce power on demand. Because LUZ and its subsidiaries had to meet administrative overhead costs, and usually sold only one project a year, LUZ had a powerful imperative: Newton Becker put it bluntly: "If we didn't finish a sale in any one year, we would be dead." A major difficulty for the sales force, however, was that people typically weren't interested in tax-advantaged investments until the year's end.

Thus, said Becker, life at LUZ "was fraught with danger. Every system was a challenge. We never had a year in which the systems were sold early. . . . All the systems were sold later on, and we had no way of financing the early stages of production. To conserve cash, in Israel they would not order goods until the very last minute. . . . We would not get much cash until the sale was closed . . . which always took place in December." Thus, as one senior executive put it, management had to "bet the company almost every day."

Mike Lotker, a LUZ vice president, was also involved in trying to bring in new business for LUZ. He soon discovered that "senior management and the board were always involved in the crisis of the minute." That made it hard for the company to do long-range planning. "If you're continually bleeding, or in shock, or being scared out of your life, you don't set up a long-term diet and exercise plan," said Lotker. "You just try and survive the next day and the next week. That was very much a problem around LUZ, at the senior management level."

In an effort to make the solar energy tax credits permanent—so LUZ wouldn't be driven to rush its plants into service on impossible deadlines, and so it could introduce some predictability into its financing efforts—Newton Becker assembled a lobbying team that worked on the problem throughout the late 1980s. But the nation's memory of the 1970s energy crises had faded. Congress was becoming acutely conscious of the burgeoning national debt. And the Reagan and Bush administrations were unfriendly to renewable energy. Lawmakers became accustomed to begrudgingly extending the solar energy tax credit only on a year-by-year basis. So reluctant was Congress in its support for renewables that it actually allowed the credits to lapse entirely for a portion of the year, producing nightmares for LUZ. Only in 1991 were Becker's efforts to make the credits permanent rewarded. But by then, as we shall see, it was a Pyrrhic victory. Had the tax credits been made permanent in the early 1980s, "it would have made an incredible difference," founder Arnold Goldman observed.

Nonrecourse debt as a benefit for investors was an important feature of the LUZ financing strategy, as for any major capital equipment project. Nonrecourse debt means that the asset for which the money is loaned serves as the sole security for the debt. The borrower's other assets are not put at risk. Since investing in LUZ plants offered tax advantages that potentially could be abused, the U.S. Treasury moved, through the 1981 federal tax bill, to eliminate entirely tax credits for nonrecourse debt. To forestall that, Newton Becker's team lobbied Congress and U.S. Treasury officials. He proposed an original modification of the law that reduced the opportunity for abuse: The tax credits would become null and void unless the project they financed generated sufficient power sales revenue to service the nonrecourse debt. This compromise proved acceptable both to the Treasury and Congress and so preserved a valuable mechanism for encouraging renewable energy investments.

THE INTRICACIES OF POWER PLANT SALES
AND THE POWER PLANT FINANCE TIGHTROPE

Just getting a LUZ plant financed every year before the tax credits expired continually tested management's fortitude and resourcefulness. Plant construction took twelve months, and equipment had to be ordered six months beforehand. But LUZ was never able to get construction financing in advance. Instead, LUZ had to borrow money from its general contractor and equipment manufacturers or have them carry the cost of their equipment, while LUZ was raising the investment capital.

"We'd start constructing these plants in January–February," said a former employee. "Often we wouldn't close the financing for the project and actually get the cash until November or December. So we had to figure out how to survive from the time we started really incurring expenses, which was January–February, until December." Meanwhile contracts with investors had to be negotiated from a position of weakness. All LUZ offered was an investment vehicle, and investors had many other options.

"LUZ's cost of doing business, both in terms of dollars and just human talent and stomach linings, was enormous," recalls LUZ Vice President Mike Lotker. "We were living on the edge all the time. Your activities on a given day could be the difference between failure and success of the company."

Sometimes this was a euphoriant. "You have no idea what a rush it is to close a deal that you've been working on for months! That *rush* is one of the greatest feelings in the world," said a former executive.

Whereas the financing of SEGS VIII—known to company insiders as "the deal from hell"—was difficult for LUZ, completion of SEGS IX was in some ways even more agonizing, because of a dramatic fall in oil and gas prices to which LUZ's future energy sales revenues were pegged.

Gas reached a peak in the early 1980s of about $6.60 per million BTUs (British Thermal Units). By 1991, the price had fallen to $1.45 in 1980 dollars. By 1993, it was probably closer to a dollar—a mere 15 percent of what it was in the early 1980s.[5] The original long-term power purchase contract, known as Interim Standard Offer 4 (SO-4), that LUZ signed with SCE for SEGS III–VII had been negotiated under regulations established by the Public Utilities Regulatory Policy Act of 1978 (PURPA) at a time when oil was selling in the $30–$40-a-barrel range. Power purchase agreements for renewables then provided that the per kilowatt-hour solar energy prices paid by SCE to LUZ's partnerships were based on SCE's "avoided cost." In utility parlance, "avoided cost" is what a utility would have had to pay to produce an equivalent amount of power, by building a new power plant. The higher the prices of fossil fuel, the greater the utility's average cost and the better the LUZ power purchase contract would be.

For SCE, that alternative was usually gas. At the time of the first power purchase agreements for the SEGS plants, virtually everyone expected fossil fuel prices to remain steady or soar. Therefore, it seemed reasonable to allow renewable energy producers like LUZ to have a thirty-year power purchase agreement with a ten-year stable price schedule, based on then-current (high) avoided costs. The SO-4 contracts thus gave investors an opportunity to earn a predictable (and substantial) revenue stream for the first ten years of the plant's operation with no energy price risk. Obviously these provisions were critical in attracting investors. SO-4 contracts were used for SEGS III–VII, and LUZ partnerships were therefore given qualified guarantees of fixed and escalating payments for the power they produced.

Unfortunately, by the time power purchase contracts were negotiated for SEGS VIII and IX, fossil fuel prices had declined precipitously, and utilities were no longer required or willing to enter into long-term, fixed-price power contracts for renewables—fossil fuel costs might continue falling, and utilities did not want to be stuck paying unnecessarily high energy prices. SEGS VIII and IX were therefore forced to brave the chilling climate of declining fossil fuel prices without SO-4 guarantees. This exposed investors to energy price risk and made them far more wary.

Even the early SEGS plants that did have SO-4 contracts will not remain forever insulated from low fossil fuel prices. The ten-year fixed-price contract provisions will expire year by year in an environment of low oil and gas prices during which utilities' avoided costs are also falling. Thus, solar plant partnerships with expired SO-4 contracts in their eleventh year will receive less than half of their earlier payments. Some may find themselves forced to reorganize financially. The average energy price likely to be received by post-SO-4 plant owners will probably average only about 3 cents per kilowatt-hour. An additional payment known as a capacity credit will roughly double that revenue, but the total will be only half of what would have been earned under an extension of the fixed-price contract.

"When the PUC [Public Utilities Commission] did away with Standard Offer 4 [about 1985], they turned investors in the solar plant into speculators in energy futures," said Newton Becker. "They turned them into commodity traders." Becker had realized very early that if natural gas prices, on which LUZ's revenues depended, fell, "LUZ could not compete and we'd be dead." One way to have protected LUZ and its investors from that contingency would have been the passage of a solar tax credit indexed to vary with oil and gas prices. When fossil fuel prices were low, the credit would have risen to have kept solar enterprises profitable. Becker proposed this idea to Congress, but they rejected it. Neither did they adopt Becker's proposal to make interest from renewable energy loans tax-exempt. That, too, would have made more capital available for renewables.

In addition to solar and investment tax credits, LUZ also depended on property tax exemptions to survive. When California Governor George Deukmajian refused to sign an exemption needed for SEGS X, Becker said, "That's the end of LUZ; we're dead." LUZ was then on the verge of putting $300 million worth of equipment into its latest plant, now scaled up to 80 megawatts in size, and, over the system's 30-year life, property tax on that equipment would have increased the project's cost by $90 million and made it uneconomical.

A TAXING BURDEN:
THE TILT TOWARD FOSSIL FUEL

Becker felt it was unfair to pay tax on solar collectors that "really represented a big tank of oil" when the oil industry had no comparable obligation to pay a property tax on a 30-year supply of oil, and the coal industry paid no property tax on piles of coal waiting to be burned. "Solar had to compete with [what amounted to] one-year depreciation in the oil industry," stated LUZ founder Arnold Goldman. LUZ Vice President Mike Lotker, in an analysis of the barriers LUZ faced, calculated that without a local property tax exemption, an 80-megawatt LUZ plant would have to pay about four times as much state and local tax as an 80-megawatt natural gas plant. This was clearly unfair to the solar facility owners, and was but one of several tax advantages enjoyed by fossil fuels.

Becker argues that the "playing field" was unfairly tilted against LUZ from the start in the energy tax arena. For example, the oil industry can immediately write off "intangible drilling costs," that is, the costs of drilling a well, rather than having to deduct its costs over a well's productive life. And, through the oil depletion allowance, the investment in oil production can be written off using a percentage depletion formula that is based on oil sales and can therefore result in deductions ten times, or more, the actual drilling costs. The oil industry, Becker says, "gets what amounts to the equivalent of a 35 percent tax credit. The solar thermal industry got an on-again, off-again credit worth much less. "Not a level playing field," he concluded.

Both the solar and the oil industries are using solar energy. "We are in effect drilling a hole into the sky, direct to the sun, getting fresh solar energy from up above. They're using fossilized solar energy from below," said Becker. Why should the two industries be treated so differently at tax time?

BETTING THE COMPANY ONE MORE TIME

In 1986 oil prices collapsed and federal renewable energy tax credits and investment tax credits began shrinking. The odds against LUZ were becoming overwhelming. As LUZ attempted to finance and build its ninth plant, it once again had to face congressional failure to extend even the scaled-back solar tax credit in a predictable manner to December 30.[6] Instead, in 1989, Congress extended the credit only to September 30. That deadline obliged LUZ to try and somehow get SEGS IX into service by then to qualify for the tax credits on which its financing depended. LUZ was forced to build a $300 million power plant system in just seven and a half months.

"You can't build an extension to your home in seven and a half months!" exclaimed Newton Becker. "A three hundred million dollar system! You realize the logistics involved? We were the largest construction project in the state of California at that time." As commonly happens whenever a major project must be done in haste, the rush caused LUZ to incur huge cost overruns. Thirty million dollars—two-thirds of LUZ's hard-earned capital—went down the drain. Despite LUZ's series of eight prior successful projects, the struggle to complete SEGS IX thus left LUZ in a financially precarious condition. The strain on all involved may have affected senior executives' judgment when they plunged on into building the solar field for SEGS X in May 1990, even though they hadn't sold any debt or equity. "We were doing it on vendors' money and also the equity in the company," Becker said.

The company's net worth was down to $15 million, the tax credits were again scheduled to expire at year's end, oil prices were continuing to fall, the company had incurred very substantial operating losses in 1990, and there was little cash reserve to draw on for the struggle to build and sell SEGS X. Nonetheless, against the judgment of some senior managers who believed financing could not be accomplished, LUZ decided to bet the company one more time.

LUZ had begun SEGS X in anticipation of a property tax deduction for taxes on LUZ's solar equipment. "But someone got to the governor at the last second," Becker explained, and he failed to sign the legislation over concern about the state's potential tax revenue loss. Becker's team finally proved to the governor's satisfaction that by approving the exemption, the state would realize a $70 million revenue gain over the plant's life. Finally, the governor signed the bill.

But because of the delay in getting the property tax exemption, LUZ lost the participation of one of the two Swiss banks that was underwriting the $220 million debt on the tenth SEGS plant. LUZ ran out of money on July 1, 1991, and was forced to halt construction on SEGS X. In that final tragedy, "The investors lost a fortune, and our vendors lost a fortune," said Becker. "Because we were halfway through [SEGS X] at the end, [the] total loss to our vendors was probably close to $200 million.

"They killed the goose that laid the golden egg," Becker said. "I blame the [state] Finance Department, because if they had brought out their concerns early, we could have addressed their concerns during the committee hearings within both houses. By sneaking this in at the last second, they killed the solar [thermal electric power] industry. . . . If they'd been upfront, we could have addressed their concerns, but they stabbed us in the back instead, through the governor."

When LUZ failed, some at the company felt everything was lost, but that is far too harsh a verdict. LUZ's nine plants are still functioning beautifully and are irrefutable proof that solar thermal power plants work on a commercial scale. No one can take that achievement away from Arnold Goldman, Patrick François, Newton Becker, and their associates.

✦ ✦ ✦

The many ways in which LUZ succeeded and its technology shines made its downfall all the more painful to many former LUZ employees. By the time of SEGS VIII and IX, LUZ's costs had been brought down to 8 to 9 cents a kilowatt-hour, and the company was making significant technological advances. It had designed a combined-cycle gas-solar hybrid plant that united the best of gas technology for baseload generation with solar technology for peaking power. The plant might well have been suitable for handling the peak loads of many western U.S. utilities.

"Our next generation, which would have been direct steam [instead of using synthetic oil to transfer heat], would have brought our costs down from roughly eight cents a kilowatt-hour to six and a half," said Newton Becker. "And if we would have had nontaxable financing, like a municipal bond, we could have brought that cost [for] a pure solar system down to around five, 5.5 cents a kilowatt-hour, which is very respectable. . . . We never had a chance to go into our next generation," he lamented.

Although the company held contracts for four additional solar power plants totaling 300 megawatts of power, LUZ declared bankruptcy in late 1991. The blow to LUZ reverberated across the energy field and drove out anyone else who was thinking about entering the solar thermal industry. LUZ's collapse also sent a signal to prospective renewable energy investors that solar power plants were a treacherous investment. The bankruptcy rewarded bold, creative people not with acclaim and financial security, but with the heartache and humiliation of financial disaster. For not only did investors in Luz International Ltd. lose all their money, but workers' jobs vanished and vendors were badly hurt. The government lost hundreds of millions in foregone tax revenues that future SEGS plants would have produced.

Becker is still bitter: "It was criminal. The government sucked us into an industry, and the minute they didn't need us, they took away the tax credits, and left us out to hang."

BEYOND LUZ: A VICTORY IN DEFEAT
AND THE POWER TOWER PHENOMENON

After the company ran out of money, Becker then took over the task of seeing the company through bankruptcy and trying to revive it, but he was unable to do so. Belgo Industries, a Belgian firm, bought the rights to LUZ's technology from the Israeli government to which they had reverted following LUZ's collapse. Subsequently, a tiny new Israeli subsidiary of Belgo called Solel Industries, with half a dozen employees, has resurfaced at LUZ's old headquarters in Jerusalem. The solar thermal electric power plants LUZ built continue successfully producing electricity under independent operating companies.

Newton Becker was so angry after the bankruptcy that he went to Washington one more time, not specifically on LUZ's behalf, but to lobby for a permanent across-the-board extension of the solar tax credit. Thanks to his efforts, Congress renewed the credits permanently. Thus the LUZ experience yielded some additional benefits for solar and other renewable energy industries. However, although Becker sought a 20 percent solar tax credit, which he felt would have helped level the energy playing field for solar and fossil fuels, Congress granted the industry only a paltry 10 percent.

✦   ✦   ✦

Critics have charged that the company could have chosen not to have embarked on construction of SEGS X when it did and instead should have laid off employees and temporarily mothballed equipment. Yet in Becker's view that only would have finished off the company more quickly. LUZ's assets, he believes, were too illiquid and insufficient to have even kept a management team in place capable of ever restarting the firm. Be that as it may, looking back on LUZ's demise, it seems foolish that, rather than nurturing the company and its technological brilliance, vacillating federal and state government policies battered LUZ into oblivion.[7] Had government policy sustained LUZ a few more years, the company would have gladly given the world a new generation of even more efficient and cost-competitive solar plants.

Despite LUZ's disappearance, solar thermal electric technology is still very much alive. The U.S. Department of Energy (DOE), Sandia National Laboratories, the Solar Energy Industries Association, and the Kramer Junction Company (operator of SEGS III–VII) met in 1994 to discuss the possibilities of reviving LUZ solar trough technology. The DOE and various substantial partners are also actively pursuing development of the so-called Power Tower, another breed of solar thermal technology.

Good as LUZ's trough technology is, what excites solar energy supporters about power towers is that they can produce higher temperatures and thus hold the as-yet-unconfirmed prospect of more efficient solar-to-electric conversion. Ultimately they therefore may be able to produce solar thermal power at costs even lower than those of parabolic trough technology. Currently, power tower costs appear greater than LUZ technology, and they require up to twice the land per megawatt. But progress in power tower designs and test facilities is cause for hope that perhaps

The 10-megawatt Solar One Central Station solar thermal power test facility built at Daggett, California. The plant used more than 1,800 sun-tracking mirrors known as heliostats to heat water in a receiver atop a 300-foot tower. Now being rebuilt as Solar Two, the refurbished plant is being outfitted with molten-salt energy storage to enable it to produce power day or night. *Courtesy of Solar Energy Industries Association.*

within 10 or 15 years, tower power may be quite competitive with fossil fuels. While these new commercial solar thermal plants still seem just over the horizon, another clever approach to producing solar electricity is rapidly moving to center stage, as I'll describe in the next several chapters.

Meanwhile, at the Kramer Junction SEGS plants, Dave Kearney, former LUZ vice president, is carrying out a million-dollar-a-year Sandia National Laboratories study on reducing costs of operating and maintaining trough power plants. "The spirit of LUZ isn't lost," Kearney says.

5

CRYSTAL POWER

Photovoltaic generation of electricity can be a multibillion dollar
industry employing hundreds of thousands of persons and
serv[ing] the energy needs of hundreds of millions of persons.
—Paul D. Maycock, *Photovoltaic Technology,*
Performance Cost and Market Forecast

Solar cells are tiny electric engines that seem miraculously to produce
electricity out of nothing. You don't need to fill them with gas, load them
with nuclear fuel, or rub them between your hands and make a wish.
They have no moving parts to wear out, and they don't run down like
batteries.

Mounted in glass panels, they are strong enough to withstand a hail-
storm but can be made thin and flexible enough to roll on a spool like
newsprint. Through a phenomenon called the "photovoltaic effect," they
can provide power on a milliwatt scale for the delicate movement of a
watch, or for a million-watt power plant that helps meet a city's electrical
needs.

FROM BECQUEREL'S BEAKER
TO OUTER SPACE

"Photovoltaic energy," or "light energy," was first observed by the
French physicist Alexandre-Edmond Becquerel in 1839. He noticed
that when a beaker containing a dilute acidic liquid and two electrodes
was exposed to sunlight, electrical current between the electrodes increased. Becquerel must have been perplexed by this observation, since

the quantum mechanical theory necessary to explain it had not yet been conceived.

Later in the nineteenth century, Willoughby Smith observed that the electrical conductivity of selenium was affected by light, and William G. Adams subsequently proved that it produced electricity when illuminated.[1] Charles Fritts used selenium to create the first solar cell in 1883, but it was so inefficient that it could transform only about 1 percent of incident light to electricity. For the next 70 years after Fritts, no practical way was found to produce efficient solar cells.

Scientific research on transistors and rectifiers and the ensuing revolution in solid-state electronics also brought a breakthrough in solar cell development in the late 1940s and early 1950s. Bell Laboratories scientists experimenting with germanium and silicon found they could create far more efficient solar cells from silicon than selenium.

By 1954, they succeeded in producing a 6 percent–efficient silicon solar cell—roughly a 600 percent improvement over selenium cells. Solar cells suddenly became valuable for making electric power, and within four years, solar cells were rocketing into space aboard America's first satellite, Vanguard I. Soon, because of their lightness, reliability, and fuel-free operation, solar cells became standard power-generating equipment on virtually all military and civilian satellites. An entire industry evolved to produce space solar cells, and that industry served as the incubator from which today's terrestrial solar industry emerged.

Meanwhile, a revolution in computer hardware was also occurring utilizing silicon chips to process data. The intensive study and investment devoted to semiconductors for the computer and space industries laid a firm technological foundation for the advances occurring in solar cell technology. By the mid-1990s, silicon solar cells had attained an impressive 25 percent efficiency. Once restricted to the laboratory, the light meter, and orbiting satellites, solar cells have become common in consumer electronic products, such as calculators and battery chargers, and have infiltrated the utility industry, as well as many "off-grid" niche markets.

EASILY BUILT MODULAR
POWER SYSTEMS

Although large coal and nuclear power plants may take 10 or 12 years to site and construct, a clean solar-cell power plant can be built in less than a year. Thereafter, you can breathe fresh air nearby or sleep soundly next door, with no fear of a meltdown. And although the future costs of fuels like natural gas are likely to increase, solar power just becomes more economical as solar technology advances. The photovoltaic industry's bright future appears to have begun: U.S. solar module production soared 36 percent in 1995 over the previous year's output to 35 million watts, while world production increased nearly 15 percent.[2]

Solar cells are interconnected and laminated into durable sealed units to form solar panels or modules that deliver power in proportion to the number of component cells. Panels themselves may then be wired together into arrays to produce still more power.

Solar cells today are made from single-crystal silicon, semicrystalline silicon (also called polycrystalline—having many tiny crystals), amorphous (noncrystalline or glassy) silicon, and from other nonsilicon semiconductor materials, including gallium arsenide, germanium, cadmium telluride, and copper indium diselenide. Solar cells are still relatively expensive today, not because these materials are rare, but because the cells require extensive processing to remove impurities that would interfere with cell operation.

Solar cell, or photovoltaic (PV), power plants using identical, prefabricated modules are easily and speedily installed on simple mountings. These generally consist of a support structure and, sometimes, a tracking device to keep the modules aimed at the sun as it appears to move across the sky. Electrical wiring and power-conditioning equipment are also necessary for utility interconnection. Energy storage or backup power generation capability, or both, may be used for standalone (utility-independent) systems.

Because of their modularity, entire PV systems can be easily sized to provide any amount of power required, if enough PV manufacturing capacity exists. With PV power systems, a corporation or utility need not make risky guesses about future power demand and commit to buying large power stations to meet those power needs. Instead, modular PV capacity can be built quickly in a factory, as needed, for installation on a home, an apartment house, an office building, or an industrial site.

Once installed, a conventional solar cell power plant operates virtually automatically with the aid of standard electronic control systems, resulting in minimal operation and maintenance costs (less than a penny per kilowatt-hour). It needs no control-room attendant, unlike a fossil-fuel or nuclear power plant.

Large contiguous tracts of dedicated land for central-station power production are also not required, although they are desirable for obtaining the most economical power from some PV technologies, such as large tracking concentrators. Solar cells can generate useful power on arid sunny marginal land and even on wastelands or urban surfaces now lying idle. Land availability is not a significant issue in PV power production. A small fraction of the world's agricultural land would be more than enough to produce all the world's electrical and thermal energy. Neither is solar power resource-limited. The Earth gets more energy from the sun every minute than the population of the world uses in an entire year.[3]

One day, modern cities seen from far above may appear iridescent as if spangled with glittering mica. Where idle asphalt and cement once ruled, surfaces may sparkle with PV jewels. Using nothing more than refined versions of current technology, cities and suburbs equipped with these electrically active solar skins could produce thousands of megawatts of electricity. Not only roofs and walls, but also partially transparent solar windows and skylights could become power plants. So could some covered parking structures, canopies, sidewalks, railroad rights-of-way, and utility transmission corridors. Just as there is competition today for high-quality biomass resources once treated as waste but useful as fuel, so companies in the future may vie for solar cell rooftop, wall, or fence franchise rights.

THE TRANSFORMATION OF LIGHT

Solar cells produce electricity while remaining intact and indefinitely reusable, save for physical degradation unrelated to the photovoltaic process. Rather than creating energy internally, solar cells obtain energy from without by capturing and transforming light from the sun. That light comes streaming toward us in discrete units of energy known as photons. Curiously, these photons exhibit properties of waves as well as particles. If you drew a clothesline with waves oscillating along it from one end to the other and then replaced the solid line with dots, you would have a simple graphic analogy of photons in motion.

Although uniform in speed and seemingly uniform in color, white light is actually a mixture of "colors." A prism will separate the light according to its wavelengths. Within the visible spectrum, our eyes and brain perceive these wavelengths as distinct colors. Solar cells transform this celestial palette of light into electricity by using fundamental atomic properties of matter.

Each of the component colors in white light corresponds to a different energy level, which in turn is associated with a unique wavelength of the solar spectrum. The shorter the photon's wavelength, the higher the photon's energy, and conversely, the longer the wavelength, the lower the energy level.

The energy level of a photon can be derived from its wavelength by means of a simple formula. And because solar cells are semiconductors with well-known physical properties, once the photon's energy level is known, we can calculate just how much of its energy a particular solar cell can transform to electricity. The percentage of incoming energy converted to electricity is the efficiency for that cell type at a given illumination.

Solar cells seem deceptively simple in their operation. They just lie basking happily in the sun while effortlessly producing power. When light strikes dissimilar layers of the solar cell "sandwich," one with a surplus of available electrons and one with a shortage, a voltage is created and a current can be withdrawn from the cell. As electrons then flow from one layer of the photovoltaic sandwich to the other, they are collected at an electrode and provided with an external path or circuit back to the other layer. In the process, useful work—such as lighting a lamp or turning a motor—can be performed. The electron flow (current) can continue indefinitely, because the silicon atoms in the crystalline semiconductor stay put as they emit and receive electrons; only the electrons travel.[4]

DESIGNING FOR EFFICIENCY

Although the simplest solar cell has just two electronically dissimilar layers and one junction between them, more complex cell structures have been created to increase cell efficiency. Advanced cells have multiple junctions and regions, each region sensitive to a different part of the solar spectrum, so that more of the sun's energy can be captured, resulting in higher overall cell efficiency.

Cells also come in a wide variety of designs making use of various coatings, electrical contacts, dopants, and semiconductor materials. The structure, or "architecture," of the cell may be altered by the position of contacts and by texturing the cell surface to produce additional performance variations. Cell output can be boosted by equipping cells with plastic lenses that concentrate sunlight on them. This makes the cell more complicated, but also more efficient.

Although modern crystalline or polycrystalline silicon cells are generally 200–300 micrometers thick (a micrometer is a millionth of a meter), thin films are only a few micrometers thick, and individual layers may be one micrometer or less. Amorphous silicon cells in their entirety can be less than a micrometer thick, because of amorphous silicon's extraordinary ability to absorb sunlight.[5]

Layered "tandem" cells consist of two or more multiple-junction systems atop one another with intentionally introduced impurities called dopants that optimize light absorption. "Stacked" cells consist of two or more separate cells with different light sensitivities joined together by a transparent adhesive. The purpose of both layering and stacking is to capture and use a broader range of the solar spectrum.

Multiple-junction cells may eventually attain efficiencies as high as 45 percent and perhaps as high as 55–60 percent.[6] Typical high-efficiency commercial crystalline silicon cells today have an efficiency of only about 16 percent. One outstandingly efficient laboratory-size cell made of a gallium compound by Boeing has reached a 34 percent efficiency.[7] Efficiency is a crucial determinant of PV costs, which are largely proportional to the cell area required for a given power output. Increasing efficiency reduces so-called "balance of system" (BOS) costs for an equivalent power output. BOS costs include land, support structures, and everything except the PV modules themselves.

Today, most single-junction and multiple-junction cells, whether thin film or solid, are still flat devices mounted in panels at an optimal fixed angle to the sun to use natural unconcentrated sunlight as their energy source. Within ten years, concentrator solar cells with efficiencies of more than 40 percent are very likely.[8] This may render them far less expensive than ordinary cells.

Strategically dispersed throughout a well-chosen utility-company service area to produce power where needed, PV systems can save power companies lots of money on transmission and distribution facility upgrades otherwise needed to meet growing electrical demand.

The Swiss, Germans, and other Europeans are already far ahead of us in integrating solar panels into building facades, roofs, and "curtain walls" (a nonbearing wall or building facade). This integral-design approach allows the cost of a solar electric system to be defrayed to the extent that solar cell modules displace structural building components, such as roofs, walls, and skylights.

THE MANY BENEFITS OF PV POWER

Solar cells are truly a dream power source. They are silent, highly reliable, portable, convenient, risk-free, pollution-free, fuel-free, recyclable, technologically elegant, suitable for utility or customer use, mass producible, long-lived, and thoroughly compatible with a sustainable economy. In addition, solar cell peak output, which occurs when the sun shines brightest, often matches utility summer peak-power-demand patterns for air-conditioning, thereby increasing the value of their energy.

Their main operational flaw is that solar cells often show a slight decrease in power output over time (on the order of 1 percent per year), possibly from a browning of their transparent plastic encapsulant. Amorphous silicon cells experience more significant light-induced degradation to the semiconductor itself; however, progress is being made in stabilizing this particular type of cell. Fortunately, too, most of the power loss in amorphous cells occurs promptly upon exposure to field conditions, and cell performance is quite stable thereafter, so cells can simply be rated lower than their initial peak power to allow for the predicted drop in output.

In addition to this modest shortcoming, a few types of cells incorporate toxic materials, such as cadmium. These elements in general are not acutely toxic and can be recycled. The amounts of cadmium are quite small relative to the amounts routinely disposed of in unrecycled nickel-cadmium batteries for consumer products.

With so few drawbacks and multiple advantages, it's obvious why survey after survey shows strong and sustained public support for solar cells as an energy source. Not only is solar energy favored by public-interest leaders, but a significant number of utility customers have shown a willingness to pay more for it solely in order to enjoy and promote photovoltaic technology.[9] This willingness can hasten the early introduction of PV in utility systems. In the long term, surcharges should not be required, because the costs of renewables will decline significantly, and most fossil and nuclear energy systems are likely to become increasingly expensive. (That trend has already become strikingly apparent for nuclear, oil, and large coal plants.) Moreover, because PV is ideal for decentralized use by utility customers or the utility, users can enjoy significant savings of transmission and distribution costs.

BRINGING COSTS DOWN
AHEAD OF SCHEDULE

Until now, the sole significant drawback to solar cell power has been its relatively high cost per kilowatt-hour: PV has had a reputation for being the most expensive of the renewable electric-generating technologies.

Even as recently as 1994, some conventional utility accounting estimates suggested that solar cells still were three to four times as expensive as wind and fossil fuels for bulk electrical supply on a cash-flow basis.[10] This comparison was based on recent market prices faced by utilities actually acquiring these technologies. Even if we accept these estimates for argument's sake, consider these new solar costs trends.

Enron Corporation of Houston, Texas, the nation's largest purchaser and marketer of long-term natural-gas supplies, announced in late 1994 that it will build a 100-megawatt thin-film solar cell power plant in Nevada in a joint-venture partnership with Amoco Corporation. The goal of the joint venture is to produce electricity for initial sale to the federal government at only 5.5 cents a kilowatt-hour—less than the 5.8-cent average cost the government now pays for conventionally generated electricity.[11] To put that cost range in perspective, the average retail price of electricity in the United States in 1994 was 8 cents per kilowatt-hour and will rise. Some areas of the country are already paying far more; in New York State, for example, the cost is 15 cents, and peak power costs are 20 cents a kilowatt-hour or more in some parts of the country. At costs of just 10 cents per kilowatt-hour, PV is already an appropriate technology for many utility-scale markets.[12]

Power production at the new Enron plant would gradually increase as the venture's solar-panel manufacturing capability expands at a new factory now under construction in James County, Virginia. The partners expect to produce more than ten megawatts per year of a new thin-film module based on large-area, multijunction amorphous silicon technology. The joint venture will assume control of the businesses previously run by Solarex Corporation of Frederick, Maryland, Amoco's solar subsidiary. If the whole project proceeds as planned, it would signal a historic transition to large utility-scale PV plants and to large PV manufacturing facilities that offer the economies-of-scale required for competitive utility power.

If the Enron-Amoco project attains its target cell-production costs, this would reduce the cost of PV electricity for utilities by about two-thirds, making PV broadly competitive with a wide range of conventional fossil and nuclear technologies. Even if Enron-Amoco merely came close to their target, the commercial outlook for PV electricity would be enhanced dramatically. Enron's cost projections agree with those also announced in late 1994 by AMONIX, Inc., of Torrance, California, based on an entirely different technology. The company is developing an ingenious concentrator solar cell that intensifies the solar energy reaching the PV-active cell surface by hundreds of times. (See "Smart Solar Cells" [Chapter 9]).

The imminent prospect of commercially competitive "bulk" solar cell power comes as a great surprise even to ardent solar power proponents. Until the recent series of cost-shattering projections, most solar advocates, including professional experts, expected that—at current rates of progress and production—PV would require another ten to twenty years before it would be competitive with bulk fossil-fuel power. Once the U.S. utility industry rids itself of surplus capacity, power at 6 cents a kilowatt-hour will be highly competitive throughout the United States for numerous utility and private applications. (Excess capacity today causes power to be available for sale on the spot market at 2 cents a kilowatt-hour.) AMONIX's and Enron's plans now bring within reach costs that were expected only beyond the year 2010. The PV industry is clearly beginning to flex its muscles.

Even excluding the AMONIX and Enron-Amoco plans, PV is already the least expensive and most reliable power source for many remote power uses beyond the reach of existing utility lines. That's one reason why about 100,000 American homes already use PV power, most in rural areas beyond utility grids. No knowledgeable person disputes that PV, or in some instances wind, is the least-cost technology for these uses today. Extending an overhead line today in the United States costs $40,000 to $60,000 a mile and as much as $130,000 if the line has to be put underground.[13] Examples of cost-effective remote PV applications include lighting, pumping, signaling, cathodic protection of pipelines, and telecommunications. In addition to comparing favorably to the costs of utility-line extensions, PV is also less expensive than relying on diesel generators and batteries or new central station conventional power plants when their associated new transmission and distribution costs are included.

Whatever the ultimate impacts of the latest cost breakthroughs, PV production has increased tenfold in the United States since 1980, and as costs fall, demand for modules continues to grow. Today solar module manufacturing is a $300 million a year business in the United States and earns $1 billion worldwide. The United States leads the world in the value of modules manufactured. Each decrease in module cost renders solar cells more competitive and in turn opens up ever-larger markets. In response to the increased demands, manufacturers increase production, and their costs fall still further as the full advantages of high-volume production and automation are realized. The Utility Photovoltaic Group (UPVG), an 82-member consortium of electric utilities and their trade associations, estimates that, industrywide, each time solar cell sales double, costs fall by 20–25 percent.[14] At that rate, even using recent recorded market prices of $6–$8 a watt for installed systems, solar module prices will soon reach broadly competitive levels. During the technology's short history—despite vacillating government support—PV module costs have already plummeted from $1,000 per peak watt in the 1950s to about $4 per peak watt in 1996. Costs of $3 per peak watt are expected in 1997. One company, Energy Conversion Devices, was even projecting in 1996 that by mid-1999, its manufacturing process would make solar panels for a mere dollar a watt.

THE IMMENSE MARKETS AWAITING

The DOE-sanctioned UPVG regards $3 per peak watt as a probable threshold for PV to enter vast new markets. According to their study, *Photovoltaics: On the Verge of Commercialization*, at $3 per peak watt, the potential exists for nearly 9,000 megawatts in near-term domestic PV sales in some ten specific markets. That represents potential revenues of $27 billion. For comparison, worldwide PV sales in 1995 were only on the order of 80 megawatts.

In comparison to this tiny base, the UPVG projects exciting international sales potential, which in turn could serve to lower domestic prices. After studying market potentials in 84 developing countries, UPVG estimated that these countries represent an aggregate market of 300–320 megawatts, just over the next five years—a total market value of $900–$960 million at $3 a watt.

Although developing countries generally lack capital, the developing world over time will increasingly turn to renewable energy, with the help of international development banks and national development programs. Electric rates are very high in some developing countries; power costs 35–65 cents a kilowatt-hour in parts of Egypt today, for example.[15] The biggest markets for PV today are in Asia and Latin America, according to James E. Rannels, director of the DOE's National Photovoltaic Program. China, India, Indonesia, Vietnam, Mexico, and Brazil have large market potential, as do Kenya and South Africa.

Turning back to the domestic market potential, 35 U.S. cities already have midday peak power rates above 20 cents per kilowatt-hour.[16] Current PV-system costs already compare favorably with these high rates. The PV Partners Program of the Sacramento Municipal Utilities District (SMUD) currently provides customers with PV power from rooftop panels at only 17 cents per kilowatt-hour.

The UPVG recently launched a promising industry-government collaboration known as Project TEAM-UP in order to facilitate a predictable, sustained pattern of large-volume solar cell orders. The idea is to provide $500 million over a six-year period to purchase 50 megawatts of solar generating capacity from suppliers. TEAM-UP makes cost-shared grant awards to utilities to encourage the early adoption of PV generating systems. The participants hope that the subsidized volume purchases will catalyze additional commercial interest and hasten the start of the PV industry's long-awaited explosive growth stage. That stage is sometimes referred to in the utility industry as "self-sustaining commercialization."

6

A NUCLEAR GRAVEYARD

SMUD will be [using] 54 percent renewables by decade's end.
—Jan Schori, General Manager,
Sacramento Municipal Utility District

The Sacramento Municipal Utility District (SMUD), the nation's fifth largest public utility, is among the utility industry's foremost champions of photovoltaics (PV) and other forms of renewable energy. During district director Ed Smeloff's ongoing tenure, which began in 1986, and especially during the time when S. David Freeman managed the company (1990–1994), SMUD proved that it is possible for a utility to shut down a large nuclear plant—with a net gain in revenue, energy efficiency, and energy supply security. The company has since installed PV energy systems on hundreds of its customers' homes.

A NUCLEAR ALBATROSS

SMUD was once known mainly as the operator of a troubled nuclear plant (a twin of the Three Mile Island nuclear reactor that experienced a core-melt accident in 1979). Today, SMUD is earning a new reputation as a national leader in PV. The district owns the world's largest operating solar cell power plant, the first public solar electric vehicle charging station, and even operates solar-powered streetlights.

Several years ago, however, when the company depended on the Rancho Seco nuclear power station for electricity, the plant's operation and maintenance costs, plus the frequent bills for replacement power when the plant malfunctioned, actually exceeded the company's revenue from

5 9

power sales. Unplanned repairs and unnerving outages were common oc-
currences. "We were running a deficit and balancing our books through
creative bookkeeping," said Smeloff. In a sense, SMUD was deferring
the cost of current power purchases into the future. But when SMUD
tried to leave the nuclear utilities' ranks, a firestorm of resistance
erupted, fanned by millions of nuclear industry dollars, pro-nuclear advertising, and consultants.

SMUD overcame the resistance, however, by marshaling economic
and scientific evidence on the advantages of shutting the plant and rely-
ing instead on an 800-megawatt "conservation power plant," plus 400
megawatts of renewable energy projects. When Rancho Seco finally
closed, its operating and maintenance costs had climbed to $180 million a
year. Once SMUD shut Rancho Seco and began investing in energy efficiency and renewables, however, the utility became profitable again, its
rates stabilized, and its bond ratings improved. SMUD has even been
able to set aside additional revenue in a "balancing account" for use in
offsetting future rate increases.

THE REINVENTION OF SMUD

SMUD's transformation was neither easy nor accidental. Smeloff, along
with General Manager Freeman,[1] fought for and won the right to remake
SMUD into a leader in renewable energy and energy efficiency. Its board
and staff now formulate and articulate renewable energy policies that are
guiding other large American power companies. For example, SMUD is
an influential member of the Utility Photovoltaic Group (UPVG), an 82-
member national utility consortium.

Through UPVG's Project TEAM-UP, participating utilities have committed themselves to spend $500 million over six years on solar cells
starting in fiscal year 1995, thus helping provide the solar industry a reliable, long-term market. The resulting sustained solar cell demand is vital
to the vigorous expansion of the PV industry, which in turn can then
make further significant cost reductions. Project TEAM-UP is predicated
on cost-sharing by the DOE.

SMUD currently owns the world's largest continuously operating PV
power plant and has a total of 2.7 megawatts of PV capacity. The utility
also assists the development of other forms of renewable energy. For
example, SMUD contracted to have 50 megawatts of wind power on-
line by 1996, and SMUD's current general manager expects that by the

year 2000, SMUD will get more than half of its energy from renewable sources. SMUD also is a force behind a national cooperative research project, PVUSA, to test utility-scale PV systems and show how utilities can use solar electricity. PVUSA is supported by the DOE and other entities.

GREEN PRICING AND PSYCHIC SATISFACTION

One of the company's most exciting programs, PV Pioneers, places utility-owned PV systems on customers' rooftops. More than 100 systems, most of them 4 kilowatts in size, have been successfully installed under a "green pricing" program. The program cost $6.3 million in 1994, with the DOE picking up a quarter of SMUD's tab. The term *green pricing* is used for programs in which utility customers voluntarily agree to pay a small additional surcharge for renewably generated electricity.

Under SMUD's version of green pricing, customers agree to provide part of their rooftop area for PV systems, and they pay SMUD a temporary $6 per month surcharge to participate in the program. Despite minimal publicity, more than 2,000 customers have volunteered, apparently for the satisfaction of knowing they are generating clean energy and pioneering PV development. Concurrently, the utility gains experience in installation, operation, bulk solar-cell purchasing, and maintenance. The environment also benefits: Each 4-kilowatt PV Pioneer system on a private home will produce about 263 megawatt-hours of electricity during its 30-year life, according to SMUD. This replaces the combustion of 253,000 pounds of coal, or 27,970,000 cubic feet of natural gas. Based on the national power generation resource mix, SMUD found that each of the 4-kilowatt systems avoided the emission of 433,000 pounds of carbon dioxide, 2,990 pounds of oxides of sulfur, and 1,660 pounds of oxides of nitrogen. The company plans to add about 100 such residential systems every year for the next five years.

LOW-RISK PV INVESTMENTS

SMUD also has built a 200-kilowatt PV power plant at its 472-megawatt Hedge Transmission and Distribution Substation in a rapidly developing area of south Sacramento. The solar plant uses flat-plate PV panels to supply electricity to the grid. The panels track the sun from simple, inexpensive ground mounts built of sawed-off utility poles.

Sacramento Municipal Utility District's 2-megawatt photovoltaic power plant, the nation's largest utility-owned solar cell plant, sits in the shadows of the closed Rancho Seco Nuclear Power Plant. *Courtesy of the Sacramento Municipal Utility District.*

Because the adjacent Hedge substation is nearing capacity, particularly in the hot summer months when air-conditioning loads are high, the solar plant provides much-needed "grid support" to the utility, generating maximum power precisely when summer demand peaks. This coincidence of solar supply and utility need allows SMUD to defer or avoid upgrading the substation. Through programs like UPVG and PV Pioneers, board member Ed Smeloff believes that between 1995 and 2000, prices for large installed PV systems can be driven down to $2–$2.50 per watt, at which point he feels that a self-sustaining market for utility-scale PV power can be created.

Smeloff notes that utilities are in a unique position to stimulate early adoption of technologies like PV, because each utility can commit to relatively small-scale, early PV purchases with imperceptible impacts on its rates. Yet with many utilities making similar purchases, the aggregate effect enables the PV industry to expand its capacity and lower prices, which stimulates additional demand.

Donald E. Osborn, director of SMUD's Solar Program and its Energy Efficiency Department, believes that paying higher costs for early applications of solar energy can be a good investment if they help lower long-term costs and improve PV performance. "When solar investments are selected carefully and in collaboration with other stakeholders in renewable energy development," said Osborn, "they can be among the wisest and, ultimately, the lowest risk investment that can be made, despite their higher initial capital costs."[2]

SMUD's board has made accelerating the commercialization of advanced and renewable energy technology an explicit policy and has adopted a multiyear Advanced and Renewable Technology Development Program funded at about 1.5 percent of the utility's annual budget. In addition to PV, the program includes solar thermal, biomass, geothermal, and fuel cell technologies. SMUD works collaboratively with other utilities, government agencies, and manufacturers to promote early applications of all these technologies.

SMUD spent about 7 percent of its revenue in 1994 on conservation programs—more than any other U.S. utility. The conservation programs aggressively promote the use of solar hot-water heaters and the construction of passive solar homes. Thanks to its overall conservation efforts, SMUD has avoided the need for nearly 300 megawatts of electrical demand, and it expects to displace 640 megawatts of demand by the year 2000.

SMUD's transformation has brought about some dramatic changes in the landscape at the site of the Rancho Seco nuclear power plant. While the nuclear unit sits idle, the surrounding property now gleams with SMUD's 2-megawatt solar power plant, and plenty of room remains for expansion. Meanwhile, installed costs of PV systems are falling rapidly, in part because of early purchases by SMUD and others.

SMUD's public-spirited decision to shoulder some of the higher costs of early PV adoption contrasts with the attitude of many industries that want to defer investing in solar until the next generation of less costly technology miraculously and spontaneously arrives. If everyone adopted that strategy, costs would stay high. Utilities and other firms who defer buying new renewable energy technology until costs fall are "free riders," seeking to benefit from the investments of early adopters.

Photovoltaic panels providing utility company "grid support" by generating peak power near users, thereby reducing strain on transmission and distribution systems. *Courtesy Siemens Solar Industries, Inc.*

Could other utilities today follow SMUD's lead and shift from heavy dependence on nuclear power to greater reliance on energy efficiency and solar? Smeloff believes so. Given the robust market for independent power projects, he feels that "a three- to four-year window [lead time] is enough time to say, 'We will close this plant and replace it with new plants'—and actually get those plants built and operating."

SMUD's commitment to using renewable resources has already "gone a long way toward moving these advanced and renewable technologies to commercial availability," says a 1993 SMUD report.[3] SMUD's energy policies and plans are made for the long-term. SMUD has taken the stand that its next set of power plants must be suitable to serve the company in the twenty-first century and handle an eventual scarcity of natural gas. "The solar, wind, geothermal, and biomass resources that the [Advanced and Renewable Technologies] Plan includes will be able to do it."

7

☀

A PASSION
FOR SILICON

I am a fundamentalist PV man. Nothing will stop me. I'm totally
consumed [by] and committed to this.
  —Ishaq Shahryar, former president,
    Solec International, Inc.

Solec International of Hawthorne, California, has only about 2 percent
of the U.S. photovoltaic market,[1] yet until 1994, it was the only terrestrial-
cell manufacturer in the nation to make a profit, even though private
industry, including some of the world's biggest multinational energy companies, had invested well over $2 billion in solar cell development. Solec
outdid its competition, in part by ignoring the conventional wisdom that
"plain vanilla" silicon cells could not hold their own for long in the PV
market against the newer and much-vaunted thin-film cells. Solec's story
is interesting both for the perspective it offers on the past challenges
faced by the PV industry and on the industry's future. The company today
has increased its market share by close to 300 percent and more than
doubled its workforce since 1993.

A TOUGH, SOPHISTICATED COMPETITOR

Until shortly before this book went to press, Solec International was
headed by founder Ishaq Shahryar, an Afghanistan-born chemist and
semiconductor expert. Through Solec, Shahryar simultaneously advanced single-crystal silicon manufacturing technology and wedged open

new markets for solar cell products. Shahryar is the first solar scientist-entrepreneur to build a profitable small solar-cell company specializing in cells for terrestrial uses. Although he started with virtually no money, this lone, foreign-born entrepreneur succeeded in founding and guiding a pioneering solar manufacturing enterprise unharmed through jagged reefs of solar industry competition that proved too much even for some of the world's largest corporations.

Some competitors lost $100 million or more. Most of them gave up or sold out their solar cell operations over the years while Shahryar persevered. Solec's customers today include telephone giant GTE Government Information Services, the U.S. Department of Defense, and Southern California Edison, and Shahryar licensed his proprietary cell manufacturing technology to joint ventures in France, India, Italy, and Japan. In late 1994, hoping to increase Solec's market share, Shahryar sold controlling interests in Solec to Sanyo and Sumitomo Corporations, while remaining president.

Solec's operation still consists of one small manufacturing plant. The president's suite was at the front of the plant, adjoining the reception area. It is a plain, unpretentious office with a formica conference table, a "white board" resting on an easel, and two grayish cylindrical silicon ingots, each standing several feet high and weighing 60 kilograms. Silicon, element number 14 in the Periodic Table, is the stuff from which Solec's cells are made, and Shahryar's thoughts never stray far from it.

Although small, Solec is technologically sophisticated and vertically integrated: It grows its own ingots of pure silicon, slices its own silicon wafers, manufactures solar cells, and assembles modules ranging from 6.5 watts to 100 watts. Solec also produces custom photovoltaic systems and makes solar-powered products, such as outdoor lighting suitable for parks, parking lots, and streets, and solar panels to power many of the nation's freeway emergency call boxes.

Today Shahryar holds a patent on a new manufacturing technique that he expects will enable Solec to produce and market single-crystal cells with an efficiency of 20 percent, which is about 50 percent higher than the cells Solec sold in 1995. Those cells are already 14 percent efficient—relatively high for commercial modules by contemporary industry standards. Two other important Solec patents are pending.

Most single-crystal silicon ingots are produced today by the Czochralski (CZ) process in which a single crystal of silicon is dipped into a melt of pure silicon and slowly pulled from the vat under controlled temperature and pressure to produce a massive single-crystal ingot, like the ones in

Chemist and semiconductor expert Ishaq M. Shahryar, founder and former president of Solec International, Inc., seated in an electric car powered by Solec's solar cells. Shahryar resigned from Solec in 1996 to found a new firm, Solar Utility Company, of which he is chairman and chief executive officer. *Courtesy of Ishaq M. Shahryar.*

Shahryar's old office. The ingot is then cooled and sliced with an inside-diameter diamond saw into disks a millimeter thick.[2] These are then doped, textured, interconnected, mounted on panels, and finally tested and encapsulated to make modules.

Shahryar's modified CZ process produces polycrystalline rather than single-crystal silicon, and the ingot can be withdrawn from the melt 60 percent faster, resulting in a substantial savings in labor, time, and energy. Because the polycrystalline material is also stronger than the single crystal, it can be sliced to one-third the thickness of single-crystal material, thereby doubling Solec's silicon-wafer output, further reducing material, labor, and energy costs.

Through the use of fast-pulled, thin-sliced, high-efficiency "poly," Solec by 1998 expects to meet the U.S. Department of Energy's (DOE's) ambitious goal of bringing the cost of PV power to $2 per watt for modules. That would render solar cells broadly competitive in many situations with most conventional power-generating technologies without even granting any special credit to PV for its unique environmental attributes.

Shahryar, a millionaire today, came to the United States in 1956 from Afghanistan on a merit scholarship from the Afghanistani government. He enrolled in the University of California at Santa Barbara and earned a bachelor's degree in chemistry. On graduation, he began his professional career working on diodes and transistors at Teledyne Semiconductors, where—after only five or six years, and while still in his twenties—he was promoted to production superintendent.

After several employment stints in aerospace, Shahryar joined Textron Corporation's Spectrolab subsidiary, which was making space solar cells, but was then planning on developing low-cost terrestrial cells. While at Spectrolab, he helped introduce manufacturing innovations, such as screen printing electrical contacts onto cells, that brought cell cost down from $500 to $30 a watt.

Then, in the early 1970s, Shahryar told Spectrolab president Bill Yerkes that although the United States had ample supplies of electricity, the Middle East and Southeast Asia might be a better market for photovoltaics.

The power grid had been unreliable in Afghanistan, even in the capital of Kabul; power failures of three to five hours a night were not uncommon. Shahryar, as a student, sometimes had to study by candlelight or kerosene lamp. He had vowed then to provide reliable power to villagers in remote areas and to others suffering from inadequate electrical service. "My goal was rural electrification—a light for students to read by at night."

Although Shahryar was instrumental in identifying Third World electrification as a major impending market for the PV industry, his long-term goal was to develop a utility market for PV in developed countries, such as the United States, Japan, and European nations.

When Textron sold Spectrolab to Hughes Aircraft, Spectrolab essentially abandoned its terrestrial solar cell business, and Shahryar decided to venture into business on his own. Although this was a risky career decision, Shahryar says he never felt fear or had second thoughts. "It was like someone was telling me, 'Go for it. Nothing is going to go wrong.' "

Money, says Shahryar, was not his primary goal. "My motivation was what service I could do for other people. I think with that in mind, I didn't care what happened to me." The year was 1972, he was unmarried, and his family was still in Afghanistan. Shahryar sold his car and was able to raise $200,000 in investment capital and negotiate a $500,000 loan, secured by his house, from the Small Business Administration.

THE SINGLE-CRYSTAL STRATEGY

His strategy was to take single-crystal silicon technology, the solar cell industry workhorse, and make it as efficient and cost-effective as possible. He deliberately chose not to develop other less well-known semiconductor materials that held exciting promises of major future cost reductions, but that still required expensive research and development.

Shahryar says, "I never thought that thin-film technology would be developed to be compatible in efficiency and cost with silicon technology. I have not had any reason to change that view since 1982."

At first, times were lean at Solec, and life was challenging. Shahryar was the president, salesman, engineer, and marketer. Sometimes he had to ration himself to one meal a day. At times, he did not know where his next payroll was coming from. He flew "standby" when traveling to find customers or license his technology. "Back then, twenty hours a day was a regular schedule; even today my whole life is consumed with PV," says Shahryar.

Gradually, however, the business grew and was able to supply modules to the Jet Propulsion Laboratory and to many projects funded under the Carter administration's photovoltaic market stimulation program. The U.S. Army, Air Force, and Navy all purchased Solec modules. He also sold to utilities. Things were going so well, in fact, that Shahryar in 1981 had offers for the company from General Electric and Pilkington Glass Company of England.

"I sold 80 percent of the company [to Pilkington] and held 20 percent," says Shahryar, who remained as president and ran the company for Pilkington over the next five years. Solec was not profitable during those years, however, and the heavy investments that competing oil companies were making in solar cells further dampened Pilkington's enthusiasm for the solar industry. Not wanting to compete with the big oil companies, Pilkington sold Solec back to Shahryar.

After repurchasing Solec, Shahryar found the late 1980s the roughest time of the company's life. The grueling schedule, sleepless nights, and fear of not being able to make his payroll began all over again. "The market was very, very slow. ARCO [a competitor] was very aggressive. . . . Many companies went bankrupt," Shahryar says.

"Competition was tremendous. Everyone would go after the same customer, even if it was only for $5,000 in sales. Those days," says Shahryar, "I was like a father who kept all his feelings to himself." He would walk through his little plant with his sales manager and act as if nothing were wrong. "Let's build this module," he would say, "and if we don't sell it today, we'll sell it tomorrow." If sales were flat, he would say, "The wonderful thing is, this gives us time to do research and development!"

"God was with us," Shahryar says. "The next day or so, we'd receive an order, or we'd find a transfer of technology out of the blue. Many nights I lost sleep. I lost my hair. So it's not been an easy route. . . . But I became much stronger out of it. I knew there was light at the end of the tunnel—which is here today. . . . I never considered quitting, not ever."

Good luck and strength of character helped Shahryar through this crisis. In 1987 Solec provided the PV cells for a solar electric race car being built for entrepreneur and environmentalist John Paul DeJoria, a self-made multimillionaire. At the time, Shahryar badly needed a financial partner to help support the development of new solar cell production technology. In their second meeting—without ever visiting the Solec plant or examining the company's books—DeJoria wrote Shahryar a check for $1 million. When Shahryar later asked DeJoria why he had chosen to invest in the company with no questions asked, DeJoria answered, "Because I invested in you." In time, DeJoria met all of the company's other capital needs. Nonetheless, DeJoria's partnership was not the end of Shahryar's difficulties.

A TRIUMPH OF PERSISTENCE

Knowing he couldn't compete head-on with the oil companies, Shahryar went after "niche markets" in the navigation and communications fields and developed integrated solar panel systems for lighting bus shelters, streets, and call boxes, and for water pumping. He also licensed his technology profitably to other companies through personal sales efforts.

Solec thereby survived the 1980s, and by the end of the decade, with the help of a new major investor, the company became profitable in an industry so competitive and unpromising in the short-term that even giants like Exxon, SOHIO, Polaroid, ARCO Solar, and Mobil bowed out.

When Shahryar was interviewed in early 1994, Solec was the only profitable PV company, apart from the firms specializing in space solar cells.

Solec had so many orders it was moving to a larger plant and was seeking a third, and larger, partner, which it found in the subsequent alliance with Sanyo and Sumitomo. In 1994 the firm had just won a highly publicized competitive bid to provide solar modules to the Sacramento Municipal Utility District (SMUD). The deal brought $3.4 million to the company, and the $6.60 per watt price Solec charged SMUD became an industry benchmark that other companies then strained to attain.

Solar cells will play an important role in future utility applications, Shahryar predicted. And those applications, he claimed, are going to be "so large you can't believe it."

In Europe, "conventional electricity is already 20 or 25 cents per kilowatt-hour. I think within the next five years, photovoltaic power will be very competitive [with it]. We'll probably be around 15 cents a kilowatt-hour by 1998."

Solec never enjoyed exotic government contracts for leading-edge non-silicon products because it was concentrating on improving the solar industry "workhorse"—single-crystal silicon. While other companies failed to focus their industrial resources narrowly on single-crystal technology, Solec relentlessly drove its single-crystal cell costs down by increasing manufacturing productivity and cell efficiency. It invested some of its revenues in R&D that focused narrowly on the next generation of single-crystal cells. Meanwhile, some of its erstwhile competitors, especially the large oil companies, underestimated the difficulty of developing new thin films and poured hundreds of millions of dollars into thin-film R&D. Shahryar knew from his semiconductor research experience that these newer cells would be far more time-consuming and difficult to produce than generally expected.

In staying with single-crystal silicon technology, Solec also found niche markets that would pay higher prices, and slowly increased its sales volume. Although at first glance Solec's faith in single-crystal silicon might seem shortsighted and risky given the megabucks riding on thin films, the next generation of concentrator solar cells will be the silicon cell's life extender and may well give single-crystal silicon a competitive edge over thin films to the year 2010. An 18 or 20 percent efficient solar cell equipped with a ten- or twentyfold concentrator would reduce costs to 6 or 7 cents per kilowatt-hour, Shahryar estimates. "But I envision that by 2020 the solar industry may be able to concentrate solar cells to 500 times; so you are talking about cheap, cheap electricity."

Before departing from Solec to found another photovoltaic enterprise (Solar Utility Company of Los Angeles, California), Shahryar left Solec in good shape technologically: He had reduced the cost of the silicon ingots Solec manufactures by 50 percent and had identified advanced silicon-slicing machinery that would enable Solec to double its output of silicon cells from each silicon ingot.

Will new evolving cell technology using thin films of copper indium diselenide (CIS), or cadmium telluride, or other nonsilicon materials eventually threaten the prosperity that Shahryar and Solec had finally attained with silicon technology? Shahryar is not worried and believes improvements in thick silicon cells will stay ahead of the competition. "Let's assume in five or ten years," he says, "that thin-film technology, such as CIS, amorphous [silicon], double, triple junctions, comes out. I think the market will be so large there will be room for everybody and every technology, as long as you are cost-effective. But I believe silicon technology, until 2010, will be the dominant technology. . . . I would say 55 to 60 percent of the industry [will still be using] silicon technology by 2010."

Experts believe the solar cell industry will have no problem meeting ambitious production cost goals by building concentrating cells. The overall cost of a concentrator cell is relatively insensitive to the price of silicon. Thus, to a significant extent, the concentrator neutralizes one of the major production advantages of thin-film cells—their sparse consumption of silicon or other semiconductor material. The semiconductor of a concentrator cell occupies only a small proportion of the total cell area. Therefore, the cost of the silicon raw material becomes a minuscule proportion of the total cell cost. Even $50-a-kilogram silicon will be affordable because the amount of silicon wafer required to generate 200 watts with concentrators will be a small fraction of the amount needed today to generate similar power using unconcentrated cells.

If in ten or fifteen years a five-hundred-sun concentration becomes possible, the same amount of silicon that today produces one watt would produce 500 watts with little additional cost. Today the cost of a solar panel per peak watt (measured under standard outdoor operating conditions) is $3.60. At a five-hundred-sun concentration, silicon costs per watt would be on the order of one-five hundredth of today's. Advanced silicon cells are now achieving 30 percent efficiency in the laboratory, though at great cost per area of silicon. Yet even a costly 30 percent efficient cell concentrated just a hundred times would become very economical, because the silicon costs per watt would fall by two orders of magnitude.

At Spectrolab Shahryar had helped reduce the cost of terrestrial solar cells by more than sixteenfold; at Solec he brought it down another 500 percent to $6.60 per peak watt for installed systems and $4 for panels. Therefore, when Shahryar today says he expects solid silicon solar cell prices to fall to $2 per peak watt for panels within four years, people listen. The $2 goal with installed system costs of $3 would lead to bulk electricity costs of only about 12 cents per kilowatt-hour, depending on climate and other factors. Those installed costs for residential or commercial rooftop PV in California and the Sunbelt would be competitive with grid power for many applications. As noted, the average 1995 price for power in the United States was 8 cents per kilowatt-hour, with prices of 15 cents per kilowatt-hour in many areas of the Sunbelt and elsewhere.

AN UNWAVERING COMMITMENT

Conversing with Shahryar leaves you feeling that the future of PV cells is brighter than you could ever have imagined. Everyone who "pooh-poohed" single-crystal silicon as likely to be overtaken by new thin-film technology may rue the day they abandoned the single-crystal market to Ishaq Shahryar and others.

Silicon technology has another huge advantage over all the other semiconductors now entering or about to enter the power module market, such as CIS, cadmium telluride, and amorphous silicon. Because silicon semiconductors are used by the computer industry and for space applications, billions of dollars have been invested in basic silicon R&D, making silicon the best understood of the cell materials. Other technologies have a long way to go to catch up. Thin-film modules are largely unavailable commercially for power generation today, and their current production costs do not yet reflect the future cost savings that excite industry analysts.

PV, meanwhile, has become a religion and a fulfilling way of life for Shahryar. "I am a fundamentalist PV man," he says, summing up his life. "Nothing will stop me. I'm totally committed to this." At home, he sees the benefits around his house in the form of Solec moonlights, pathway lights, and even nearby streetlights. "Everytime I go through the freeways and see my call-box panels there, I blossom. When I go to Europe, or even Indonesia, I blossom. When I got my 20 percent efficient solar cell patent, I blossomed. That was more than if someone had given me $20 million. . . . I'm driven by commitment. Whatever you do, if you are committed to it, you can do it." After 30 years in the PV business, Shahryar declares he is as happy and excited about it now as he was on day one.

Today the integration of Solec with Sanyo and Sumitomo is progressing. These two giants, already among the world's leading solar-cell makers, can provide Solec with high-powered technical assistance and could give it the commercial and financial support to market the company's products worldwide, taking advantage of burgeoning demand for solar power systems. They have the kind of "deep pockets" that most solar entrepreneurs can only dream of. Time will tell, however, whether the two multinationals will use Shahryar's single-crystal technology wisely, slash its costs, and help trigger the explosive global demand that Shahryar envisions.

8

※

SOLAR CELLS AND TACO BELLS

> One of the things I liked about the silicon solar cells in the begin-
> ning was—you could eat off them. I mean, they didn't have any-
> thing on them. Some glass and a little silver and stuff—it'd be
> great for dinner.
> —Bill Yerkes, founder, Solar Technology
> International, and past president, ARCO Solar

While Ishaq Shahryar created a small solar cell manufacturing enterprise that attracted a friendly multinational takeover, the efforts of his
Spectrolab colleague, Bill Yerkes, burgeoned into the world's largest solar
cell–making operation. Curiously, Yerkes's feat was achieved with a little
help from Taco Bell, the restaurant chain.

What's the connection between high-tech solar cells and humble,
fast-food tacos? For one thing, a typical tortilla (from which a taco is
made) and Bill Yerkes's first solar cell were similar in size. More importantly, producing a great solar cell and a great taco require standardized
preparation methods and careful quality control. Though the resem-
blance ends there, Yerkes's firm, Solar Technology International (STI),
found it expedient to recruit its initial workforce from Taco Bell's
kitchens.

The story of STI is partly the tale of a resourceful individual of modest
means who overcame a sudden career reversal and—by turning it to his
advantage—created a successful solar cell manufacturing company. But
the subsequent saga of STI's transformation into ARCO Solar and eventually into Siemens Solar Industries, also provides a revealing inside look
at how a powerful multinational oil company that began dabbling in solar

energy soon found itself in over its head and made some of the solar cell industry's worst technological blunders. In tracing these events, we will also be illuminating the pros and cons of thin-film solar cells versus conventional crystalline cells.

THE SELLING OF SPECTROLAB: A CAPTAIN OF INDUSTRY IS JILTED; A TOUGH COMPETITOR IS BORN

STI founder Bill Yerkes is an energetic, white-haired mechanical engineer who, when interviewed, was managing the Electronics-Prototype Laboratory for Boeing Corporation in Seattle, turning new technology into commercial products. Yerkes today holds a top job with Bill Gates's Teledisk Corporation, where he is responsible for providing 7–8 megawatts of photovoltaic power for the company's planned network of 800 telecomputer communications satellites. After working at Boeing in engineering and management during the 1960s, Yerkes eventually became president of Textron Corporation's Spectrolab subsidiary, which makes space solar cells in Sylmar, north of Los Angeles. Although Spectrolab had become very profitable under his leadership, Textron sold the company to Hughes Corporation in 1975 after senior management decided to shed the company's smaller divisions. Once Hughes installed its own management, Yerkes got a year's salary and an hour to clean out his office. From being a well-respected "captain of industry," he was tossed unceremoniously on the corporate scrap heap.

"It was a real ego buster," he remarks. "I *was* the company. I had created all the things that were working well." Instead of stewing over the affront, however, within three days Yerkes simply decided to build his own solar cell company—and make it ten times bigger than Spectrolab.

His mother, two aunts, and a friend at Boeing each put up $10,000. He also sold all his Textron stock, putting $50,000 into the new partnership and keeping $50,000 to live on. Eventually, he also had to stake his entire Hughes "golden parachute" on the new venture. In launching STI, Yerkes's agenda was to make a better product at a lower price than anybody else, and become a large-scale manufacturer of a standard solar module. Wasting no time in the late summer of 1975, Yerkes soon found a suitable industrial building in Chatsworth, California, for only $12,000 a year plus a month's free rent. By shopping intensively on a shoestring budget, he obtained top-grade surplus production equipment at bargain prices from other electronics firms. He and two

J. W. (Bill) Yerkes, founder, Solar Technology, Inc., former president of ARCO
Solar and Spectrolab, later the manager
of the Boeing company's Electronics-
Prototype Lab. *Courtesy of J. W. Yerkes.*

former Spectrolab employees quickly installed everything and got it
operating.

From the start, STI's equipment was superior to Spectrolab's, and its
overhead was far lower. To keep production costs down, Yerkes took the
best production technology he had helped develop at Spectrolab and improved it, striving for simplicity. At Spectrolab, he had been involved in
developing the still-novel use of screen printing metallic contacts directly
onto solar cells. This turned out to be much faster and less expensive than
the industry's common practice of evaporating metal through a grid mask
in a vacuum.

Yerkes further simplified cell production by eliminating the diffusion
furnace (which the industry used to add dopants), by simply spinning
on his dopants and cooking them in a belt furnace. Batches of cells
could progress through the new production process in less than a week.
By contrast, "at Spectrolab people put things in trays with a tweezer,"
Yerkes explained. "A person had to stand there and feed the machine
and take each cell off [it]." At STI, whereas a human initiated each action, a machine finished it, so a person could start something and then
leave. Yerkes also aligned several machines and fashioned ramps between them to move cells efficiently from one production step to the
next.

INSPIRATION IN A JUNKYARD:
ADVENT OF THE MODERN
SOLAR MODULE

Before joining Boeing, Yerkes had worked in automotive engineering at Chrysler Corporation. He now remembered having noticed in Michigan that the windshields of rusty cars in junkyards remained in good condition though exposed to the elements. He wondered about the piece of plastic between the windshield's laminated glass sheets, and if it would be a good material for encapsulating solar cells. At Spectrolab, he had relied on silicone as a solar panel encapsulant, but silicone was messy and smelled bad. Worse, it tended to "creep out of the room and spread all over everything in the building," said Yerkes.

At the time he started STI, other solar cell companies were putting solar cells on a circuit board and pouring silicone encapsulant over them like syrup on waffles. Yerkes decided instead to make his module like a windshield. Rather than using a circuit board as a substrate, Yerkes used a tempered glass *super*strate. "The idea of gluing the cells behind the glass and letting them look through it was my idea," he said. The glass was a good insulator, so he could use aluminum or stainless steel extrusions to frame the edges of his glass modules like windows. He knew this would be a low-cost process, because many other products were made in volume this way. Today, all solar cells, even the thin-film modules, are built like this. Tempered glass is long-lasting and provides excellent protection for cells.

Yerkes combined his low-cost production process and his new module design with three-inch diameter silicon wafers that made each of his solar cells twice as powerful as Spectrolab's two-and-a-quarter-inch cells. When the equipment he needed wasn't available or was too expensive, Yerkes designed and built it himself. To make a solar-cell tester, for example, he drove two nails into a board, strung a piece of nickel-chromium wire between them, and attached the end of another sliding wire to it to create a homemade variable resistor. He then connected a solar cell to the resistor and to a rented x-y plotter. By illuminating the cell with an infrared test lamp and varying the resistance, he used the apparatus to plot the cell's current and voltage.

He etched his silicon wafers 50 at a time in a simple plastic box inside a $13.95 stainless steel pan from a restaurant supply house, warmed by a $9.95 heater from Kmart. (Texture etching makes the front of the cell

"fuzzy," so it doesn't reflect light.) Cell edges also needed etching to separate the positive and negative faces of the cell, so he made a little Teflon
fixture that enabled a dexterous person to load and etch five cells at a
time in a sink full of acid. Workers then used an ordinary kitchen sprayer
to clean off the acid, which was neutralized before disposal. Much later,
Yerkes invented a now-patented plasma etching gas furnace process that
enabled him to etch the edges of a thousand cells simultaneously at the
push of a button.

THE TACO BELL CONNECTION

STI's first order was filled by five production workers—young women
who had worked at Taco Bell and had been through its fast-food training
program. On a Monday, they would etch the cell surfaces, apply diffu-
sant, and bake the cells in a belt furnace. On Tuesday, they would screen-
print the metallic front and back contacts and fire them again in the
furnace. Then the edges were etched, and the cells were measured. On
Wednesday, they would solder strings of solar cells together to increase
voltage, and on Thursday they would laminate the circuits. They would
finish the last of the production run on Friday and test the finished modules outside in the sun at the back of their little factory.

What at the start of the week had been 1,000 blank silicon wafers and
35 pieces of glass (total value: $1,140) had now become 35 solar panels
worth over $10,000. Yerkes would type up invoices for $10,000, pack and
ship the boxes himself by 4 o'clock. Tired, but satisfied and relaxed, he
would go home while the southern California sun was still shining and reward himself with a swim in his pool and with a pleasant family evening.

A SOLID PRODUCT. A FLAMBOYANT SALES PITCH.
AND MONEY FOR EXPANSION

On the lookout for customers, Yerkes convinced a San Diego motor-
home company that a small panel on the motor-home roof would keep
the battery charged during storage. The company decided to offer the
solar panel as standard equipment on its 25-foot motor home. He then
got an important order for solar panels from Jet Propulsion Laboratory
(JPL) under a DOE block purchase program and manufactured panels to
JPL's specifications. STI's panels soon passed JPL's rigorous testing program, and sales picked up.

Yerkes's standard panels put out 100 watts. Motorola's competing panel produced only 65 watts; the next competitor's produced only 45 watts. Not only was Yerkes's module the most powerful, but due to its simplicity, it was relatively inexpensive to produce. Thus, Yerkes was able to keep STI's prices relatively low, just as he had planned, and word of mouth about the company's successful new panels traveled fast. "Customers loved them," says Yerkes. "They were reliable. They were tough. I would go to a meeting at JPL and put it between two chairs and stand up on it and jump up and down. I'd throw coffee cups at it, and they'd break." He soon had more orders than STI could handle. Within a year, he had paid off all his capital equipment and was running a bustling small business.

Yerkes, however, wanted to make a larger and better product, which required advanced manufacturing equipment. He therefore brought in a few additional investors and transformed Solar Technology International from a limited partnership into a corporation called Solar Technology, Inc. The new capital made it possible to acquire needed equipment, such as a new diffusion furnace.

He still lacked sufficient investment capital to exploit his business potential fully. Major corporate investors, however, remained leery about risking money on a small firm that was obliged to compete with companies like Exxon's Solar Power Corporation. "How could we possibly beat big outfits like that?" says Yerkes rhetorically nowadays. The multinational oil firms were also perceived by customers as more stable suppliers. Those companies were attracted at the time to a $100 million U.S. Department of Energy (DOE) PV program.

To keep up with the competition, Yerkes continually improved his product and crunched production costs. By early 1977, he was able to hire a solar-cell processing expert and a plant manager, freeing himself to go out and obtain more orders.

THE ARCO COURTSHIP

Large oil companies eager to participate in fashionable new technology were then shopping around for small solar cell companies to acquire. In mid-1977, a consultant for ARCO appraised Yerkes's operation and recommended that ARCO buy it. Although he had made a "stumble-free" start-up, based on his past ten years in the PV business, Yerkes was cautious about the future. "We had ten years to go and a lot of research to be done," he reasoned, "to make the product at significantly lower cost." He recognized that the scale of investment needed would be far beyond the resources of his small backers, even though, prompted by ARCO's interest, they were pressing more money on him and urging him to refuse ARCO in the hope that they could ultimately make a bigger profit by first expanding the company themselves, and then selling it later.

From ARCO's standpoint, solar energy appeared to be a promising new energy technology in which the company wanted to participate. Moreover, since ARCO was headquartered in Los Angeles near STI, ARCO could more easily assimilate STI than other more distant start-ups.

Yerkes projected for ARCO that, if they purchased STI and built a new factory, they soon would be able to cut the price of solar cells to only $5 per watt, an attractive price at the time. But he estimated that a million-dollar investment would be needed. So he outlined a game plan: To-gether they would grow their own silicon crystals to produce wafers that would be pure enough for solar cells but that didn't need to be as fancy and expensive as for computer chips. By being in the silicon business themselves, ARCO would insure themselves a reliable supply and could venture to produce a larger, more powerful, more economical wafer. ARCO agreed, provided that Yerkes would participate.

Yerkes's principal investor responded, "Yerkes, you'll never see a mil-lion dollars in that company." Undeterred, Yerkes returned to his negoti-ations with ARCO and suggested to them that ARCO could obtain valuable solar industry intelligence and research by acquiring STI and adding a new hundred-person solar cell research department to it.

ARCO liked that idea and Yerkes sold STI to them. The sale enabled him to recoup his Hughes severance pay five times over and brought sub-stantial returns for his investors, too. Yerkes became vice president of en-gineering and technology for the new ARCO Solar. He proceeded to hire a research director, set up a research lab, and hire the 100 people. He now estimates that probably 90 percent of ARCO's investment in ARCO Solar was defrayed by tax credits.

"They wanted ARCO Solar to grow as fast and as big as they could, be-cause of the tax implication," says Yerkes, before tax law changes oc-curred. During the next ten years, to the astonishment of Yerkes's early investors, ARCO invested over $200 million in its new acquisition.

A BUMBLING BEHEMOTH

Even though ARCO's chief executive officer (CEO) had personally gone to inspect STI's plant before their buy-out offer, Yerkes quickly concluded that ARCO didn't really understand why they were buying a

PV company, nor what the business was exactly, nor where it might go, nor where the costs were. ARCO, however, immediately leased a 20,000-square-foot building and started turning 10,000 square feet of it into the new research department. In the other half of the space, they began setting up a bigger production line. Confident of their management capabilities, ARCO then told Yerkes they would provide a plant manager to run the plant so Yerkes could concentrate on the products and planning. The company advertised the general manager job in their in-house paper, the *ARCO Spark,* and then chose an ARCO mining engineer, who had just planned the world's largest coal mine, to run the company.

"He didn't know anything about semiconductors, or glass, or bonding, or manufacturing," said Yerkes. After one year, he was replaced. His successor began trying to ease Yerkes out, bringing in a friend to be head of production. Still, Yerkes was successful in convincing ARCO management that they should build a new automated manufacturing plant in Camarillo, California, for a new, larger solar cell. He wanted the company to grow their own crystals and take other cost-cutting steps, rather than to try and increase production in their old plant with three-inch cells. His idea was to start up the new plant successfully before shutting down the existing facility. One of Yerkes's responsibilities was helping design the new plant, outfitted with "next generation" solar cell production equipment. Meanwhile, the company's solar cell yield in its old plant was steadily declining, and the new manager appeared to lack proper experience.

"One day," Yerkes said, "[the new manager] decided he couldn't wait for the Camarillo plant. He wanted to switch from three- to four-inch wafers, and he wanted to make the new product in the existing plant. He sold [management] on the idea that he'd save a lot of money by just going from three- to four-inch wafers." Yerkes knew this would be more expensive and problematic.

"The plant went down after he switched. They couldn't make anything then. So for about three months, they had no production at all, and the sales department went ballistic." About that time, *Forbes* magazine visited ARCO Solar for an article on the PV business. "The cover picture showed a pimply faced youth with a spatula at the end of a belt in that plant taking the cells and putting them onto the next belt.[1] There was a joke (in the issue) that this was ARCO bringing automation to the industry."

SWEET SUCCESS

Like the mining engineer before him, the new manager had never run a factory and apparently never had to produce below a cost ceiling. "He had dropped the price of the product quite a bit below what we were making it for on the chance that we could get higher volume," said Yerkes. After about another year of operation, the second manager also was relieved of his responsibilities. ARCO then sent a troubleshooter to the plant. He was Ron Arnault, ARCO's director of strategic planning.

After reviewing Yerkes's résumé and holding a brief discussion with him, Arnault said, "Why aren't *you* running this company? Looks to me like . . . you've got all the experience for this."

"Well," replied Yerkes, "I probably *should* be running the company—because there's certainly no one you've sent out here so far [who] has the foggiest idea how to do it, and I certainly do. I think I could straighten it out pretty easily."

Yerkes was made acting manager and told ARCO the Camarillo plant should be started immediately. In two weeks, Yerkes got the new plant running. It ran for about a month and then there were problems. During a two-week shutdown, Yerkes's team simplified the production process. The plant then resumed operation and has never shut down since. In the next six months (October 1 to March 31, 1980), production doubled every month. At the sixth month, the plant shipped a million dollars' worth of product for the first time. "That factory is still the factory that's the cash cow for the company," says Yerkes.

Six months after the plant was operational and two and a half years after ARCO bought STI from Yerkes, a ceremony was held, and Yerkes was made president of ARCO Solar. One year later, just about three and a half years from the date of ARCO's takeover, ARCO Solar was shipping three megawatts worth of solar cells and bringing in $17 million in sales, surpassing the other oil company solar subsidiaries owned by Exxon, Amoco, and Mobil. ARCO Solar became the world's number-one PV company and fulfilled Yerkes's vow to overshadow Spectrolab. The Camarillo factory finally became part of Siemens Solar Industries and in 1994 produced 13 megawatts of cells.

DEMOCRATIC DECISION MAKING

In addition to producing silicon solar cells, ARCO Solar also had a large solar cell research program on thin-film silicon, amorphous silicon, copper indium diselenide (CIS), and cadmium telluride semiconductors. The way the company made its research-and-development decisions, however, contributed to Yerkes's eventual departure.

In the early to mid-1980s, 50 of ARCO Solar's 100 researchers were working on amorphous silicon, and 45 were working on CIS, whereas only 5 were at work on cadmium telluride cells. The large number of people working on the CIS and amorphous-semiconductor technologies was a powerful constituency in the R&D area within ARCO Solar.

Sometime around 1983, a new chief engineer was hired. To cut costs, he set about trimming ARCO's three thin-film development efforts to two. ARCO Solar had been spending money on amorphous silicon for seven years, ever since buying some early R&D from inventor-entrepreneur Stanford Ovshinsky in Detroit. ARCO Solar had also been working on CIS technology for about five years and had gotten cells with up to 10 percent efficiency. Its cadmium telluride work had started only the year before. With only five people working on them, those cells had made more progress in efficiency after but a year's effort than the other technologies had made in four or five. The question was, which technologies should be retained?

Astonishingly, this issue was resolved not on the basis of a high-powered technical evaluation by the company's most qualified scientists, engineers, and consultants, but by something resembling a democratic town hall convention. "We all went," said Yerkes, "like everybody—including the janitor." Yerkes wrote a short position paper in which he plotted the efficiencies of the new thin-film cells versus the number of years in development at ARCO. On the basis of his analysis, Yerkes concluded that it seemed much easier to make cadmium telluride cells than any other thin-film cell.

But ARCO retained amorphous silicon and CIS. "The political constituency was there, with 95 people [in the company] voting for that as opposed to only 5 people voting for the other thing," said Yerkes. "So that's how we made decisions. And I could see that ARCO was going to make more political decisions, and they were going to make a bunch of mistakes."

ARCO Solar went on to develop an amorphous-silicon solar cell, building on technology acquired from Ovshinsky and ARCO's own research. Despite Yerkes's opposition, the chief engineer decided to commercialize an amorphous-silicon panel. Yerkes was concerned, because the cells were below 10 percent in efficiency and suffered efficiency losses through degradation in the field. When he raised the issue, Yerkes says the chief engineer in effect replied, "Well, we've got a fix for that."

Yerkes then urged that the panel not be sold as an ARCO Solar product to avoid damaging the existing product's reputation. His advice was rejected. "Then they had to withdraw it from the market," says Yerkes. The product was then reintroduced and was withdrawn a second time. "It was falling apart," Yerkes says.

THIN FILMS: THE CREDIBILITY GAP

Bill Yerkes today is still not yet a great believer in thin-film cells. He does not regard the degradation problem as completely solved and is concerned about the cells' lower efficiency. At efficiencies of 6 or 7 percent, he argues, the nonsemiconductor component costs are greater than those of the cells themselves. Low-efficiency cells require more glass, framing, supports, and installation work. All these area-related costs make it difficult to drive down the costs of inefficient cells.

In addition, Yerkes thinks thin films are neither cheap nor easy to manufacture. "They all say they're cheap . . . when they get to big [production] volumes," he says, "yet it takes a lot of vacuum machines and fancy things." By contrast, Yerkes sees solid silicon wafer processing today as easy, and he points out that his ARCO product required no vacuum or lithography. "And if silicon wafers are getting cheaper and cheaper . . ." (his voice trails off, so you can draw your own conclusions).

Thin films also have a credibility problem, Yerkes asserts, because of past unsubstantiated efficiency claims. "Any thin-film cell is going to have to be twice as good. It's going to have to be half the price and have higher efficiency, or no one's going to buy it, unless it's for a government demonstration."

✦ ✦ ✦

ARCO Solar's initial false start with STI, its foray into amorphous-solar panels, and its later abandonment of cadmium telluride were not the firm's only questionable solar technology decisions. ARCO had used existing solar cell technology in 1982 to build a pioneering 1-megawatt solar cell power plant for Southern California Edison (SCE) at the Lugo Substation in Victorville, California, a town in the Mojave Desert. The 50 percent federal and state tax credits then in effect made it possible for

ARCO to show a 13 percent return on this capital-intensive investment, even though the plant cost nearly $18 per watt (1992 dollars) and thus made very expensive electricity that cost well over one dollar a kilowatt-hour, for which SCE only paid an average of 4.5 cents a kilowatt-hour in 1989 and a mere 3.5 cents when the plant closed in 1992.

"That plant went in in less than a year," says Yerkes, "just like clockwork." Its completion emboldened ARCO to embark on a 6.5-megawatt solar cell plant at Carissa Plains, in California Valley, east of San Luis Obispo. The facility was to be the world's largest PV power plant and would sell its power to Pacific Gas and Electric Company (PG&E).

THE CARISSA PLAINS FIASCO

Yerkes was senior vice president then, with some R&D responsibilities. He suggested that simple reflecting mirrors could increase the power from the panel and might be less expensive than panels. With some R&D funding from the company, he designed a 1.5-kilowatt system with aluminum mirrors and a single-axis tracker. The device would track the sun each day but was much simpler than the big conventional two-axis trackers. Yerkes installed his prototype in the test yard of his original solar cell plant, connected it to data collection devices, and started taking data.

The chief engineer observed this experiment, said Yerkes, "and he thought this looked like a great idea, so he took off, unbeknownst to me, and started getting some guys to design better mirrors. . . . Before I knew it, the 7-megawatt plant up there was being planned to have mirrors, and they built a special mirror-making facility that had a big automatic rotisserie. It was like a manufacturing plant."

Unfortunately, by that time, Yerkes had enough field data from his tests to see that, after about three weeks, the mirrors got so dirty that their output was zero. To make matters worse, the mirrors still heated up the panels, so the panels were always hot, and their performance suffered.

Yerkes wrote a memo to senior management saying, in effect, "I thought this was a good idea, but it looks to me that it's not, because the mirrors are going to have to be washed a lot, and there's no water there." (A lot of labor would have been required to go around and wash them using a special truck with high-pressure hoses.) Eventually, said Yerkes, "the chief engineer verified that the mirrors get dirty and, after a while, if you don't wash them, you don't get any contribution. The array drops to half power."

But it was all too late: ARCO had a contract with Fluor, the big oil-refinery designer, to design the plant. "They had already invested a lot of money to do all the drawings and get permits and do it all in style, like they do an oil refinery. ARCO was in full swing here," said Yerkes.

By one estimate, the plant cost $65 million to build—$10,000 per kilowatt. The investment was fostered by the continued existence of the state and federal solar tax credits. "They wanted to hurry and build the plant before the credits went away," said Yerkes, "so they started installing fairly early." Very high-quality mirrors were mounted on 90-square-meter devices, twice as big as the ones that were put in on the first plant. "By the end of another two years, [the chief engineer] had realized that it was a mistake. The last 700 kilowatts of the Carissa Plains plant was put in without mirrors." These arrays were actually more efficient. "Sixty-five trackers with straight solar panels on them could do what 135 trackers would do when they were brand-new clean. They would do twice as much by the time they got dirty a few weeks later." The ARCO plant was in effect a huge but unnecessary field test that ultimately confirmed the results that Yerkes had initially obtained on his cheap, backyard prototype.

In February 1990, ARCO sold ARCO Solar, Inc., to Siemens, the German electronics giant. The new U.S. company was called Siemens Solar Industries and now produces 47 percent of all solar cells made in the United States. Siemens, however, did not purchase ARCO Solar Power Ventures, Inc., which owned the extravagant Carissa Plains plant. A group of investors known as Carissa Solar Corporation purchased ARCO Solar Power Ventures, Inc., and its Carissa Plains plant.[2]

Because of the low contract prices utilities were willing to pay for power from the plant, however, Carissa Solar Corporation decided it would be more profitable to sell the two plants for scrap than to operate them. Thus, the two facilities came to an ignominious, though for the new owners not an unprofitable, end. Disassembled, cannibalized for their panels, and sold on the wholesale secondhand market, the panels were worth $2 or $3 a watt. At those prices, the Carissa Solar Corporation probably received at least $15 million or so for an estimated $2 million in-vestment.[3] In return, the company paid decommissioning costs and is responsible for reclaiming the Carissa Plains site. Thus in its hurry to get out of the solar business, ARCO gave away assets that probably cost more than $100 million for a fiftieth or so of their cost and handed the new plants' owners a possible profit of over $10 million—a 500 percent return on their investment. Dispersed to new locations, the panels, however, will continue to produce clean electric power wherever light, load, and silicon cell chance to meet again. But the story of the ill-fated Carissa Plains plant remains a metaphor for the wastefulness and injudiciousness with which society makes many of its energy decisions today.

BEYOND ARCO SOLAR

Yerkes left ARCO Solar in 1985 and is today back at Boeing, where he enjoys speculating about the shortest technological path to the truly cost-effective solar module that will finally open up huge, multibillion dollar global markets.

His scenario is a straightforward extension of current silicon-cell processing technology. He envisions a process that would more continuously make the raw silicon for a continuous caster that would produce silicon ingots. Automated wire saws cutting in a slurry would then cut 500 disks at a time from a ten-inch-diameter silicon ingot. The final costs could be under a dollar a watt for a whole module, he projects. If balance-of-system costs added an equal cost, these units could be produced for under $2 a peak watt, a price that would make PV very widely competitive for bulk commodity power.

Some research groups have a different vision of the future for solid silicon cells. Using a technology known as edge-defined film-fed growth, pioneered by Mobil Solar, researchers today are growing a ribbonlike crystal. The crystal is then cut with laser beams rather than saws. This more advanced process saves silicon and speeds up the cutting of cells.

Whatever tomorrow's solar cell technology will be, crystalline silicon will have a critical role to play and still has a lot of competitive fight left in it. The key question now is only whether it will retain its commanding lead for a few, or many, years against its competitors: the thin films. The outcome of that contest will be determined in a major way by the success of the solar concentrator cell technology mentioned by Ishaq Shahryar.

9

※

SMART SOLAR CELLS

Electricity costing less than ten cents a kilowatt-hour can be generated using PV concentrator technology, and less than five cents a kilowatt-hour is quite possible.
— Eldon Boes and Antonio Luque

In late 1994, AMONIX, Inc., announced that it had produced a new photovoltaic concentrator cell. This device—by following the sun and swapping inexpensive plastic for more costly silicon—could meet the Department of Energy's goal of a profitably installed PV system at less than $2 a peak watt.

Known as Integrated High Concentration Photovoltaics (IHCPV), the AMONIX system uses plastic Fresnel lenses to concentrate sunlight 200–500 times onto small solar cells. The concentrators, semiconductors, and tracking devices are built in the form of an integrated array of modules anchored to a pedestal. AMONIX President Vahon Garboushian told *PV News* that IHCPV technology can meet utilities' cost goals for a commercial-scale energy system—$1.50 per watt of installed capacity ($1,500 per kilowatt)—"6 cents per kilowatt-hour of generated electricity." A 20-kilowatt AMONIX system is currently undergoing testing and utility evaluation. The module has a system energy conversion efficiency of 20.5 percent at Standard Test Conditions and 18 percent at field conditions.

THE BRIGHTNESS OF 1,000 SUNS—
OR MORE

Concentrators employ a small, flat, high-quality piece of semiconductor, usually rectangular, on which concentrated sunlight is focused by inexpensive plastic lenses. The lenses can concentrate the sunlight as much as 1,000–5,000 times its normal intensity.[1] Difficulties occur at high concentration ratios, however. (Anything 300 times and above for silicon and 1,000 times or more for gallium arsenide.)[2]

Since semiconductor material is far more expensive than simple plastic lenses, substituting lens area for semiconductor area is a cost-effective way of boosting cell output. In addition, cells equipped with lenses operate at higher efficiencies, meaning they transform a higher proportion of the incident light to electricity, which also lowers power costs.

Concentrator cells do have drawbacks, however. They are inherently more complicated than flat-plate solar cells. Since concentrator cells are able to use only direct sunlight (rather than direct and diffuse light), all but the lowest power concentrators must track the sun to remain perpendicular to the incoming direct beam radiation, the angle at which all solar panels work most efficiently. Concentrators, therefore, have to be mounted on smart, computer-controlled tracking devices that follow the sun daily across the sky and also usually adjust their angle seasonally. Trackers add some mechanical complexity and, with it, modest additional operation-and-maintenance requirements.

Because heat diminishes photovoltaic cell efficiency, and concentrated sunlight heats the cell's semiconductor, it must be cooled, either by passive cooling—the addition of a metallic heat-conducting plate or fins to the cell—or by circulating coolant. Clever designs in the future may put the energy extracted by the coolant to use. The additional complexity of concentrator cells and the requirement for direct sunlight suggests that they will be most economical for utility-scale applications in the Southwest and other Sunbelt areas of the world.

Yet in the opinion of some engineering experts, once mass produced, concentrator cells will be more economical per watt than simpler flat cells that operate on unconcentrated sunlight and require no tracking. The reduction in expensive semiconductor material requirements and the high efficiency of concentrators are expected to compensate for the additional costs of tracking.

ESTIMATING COSTS

A recent comparative engineering and economic simulation analysis of three reference 50-megawatt PV plants using concentrators and flat, thin-film modules found the thin films could produce power for as little as 10 to 11 cents a kilowatt-hour at prime sites in 1995, with concentrator costs not far behind. This detailed engineering cost study, by representatives of the Bechtel Group, Research Triangle Institute, and the Electric Power Research Institute, assumes annual solar cell manufacturing rates of 25 megawatts and 100 megawatts. Although 100 megawatts of solar cell output may seem large relative to the current U.S. cell production of 35 megawatts, it is small relative to the size of even a single conventional large (1,000-megawatt) power plant.

Engineers are currently working to reduce the cost of both concentrating and tracking devices. For example, a new experimental design for a Solarmarine Photovoltaic Power Plant proposed by Pyron Energy Products of La Jolla, California, would eliminate individual tracking support structures for cells or groups of cells, replacing them by solar concentrator cells on a flat, circular, rotating platform floating on a thin layer of water. The angular velocity of the disk is set equal to the sun's rate of progress across the sky. The entire platform would turn around a vertical axis at its center to follow the sun across the sky. The goal is an almost frictionless tracking device that would use little power and virtually support itself.

THE WORLD'S MOST
EFFICIENT CONCENTRATORS

SunPower Corporation, of Sunnyvale, California, is another solar concentrator-cell company to watch, along with AMONIX. Founded in 1988 by Dr. Richard M. Swanson, physicist Richard Crane, and engineer-entrepreneur Robert Lorenzini, SunPower develops high-efficiency solar cells and optoelectronic semiconductor devices, such as those used in barcode readers and medical instruments.

Whereas most of the firm's "bread and butter" comes from optoelectronics now, the company produces the world's most efficient commercially available solar concentrator cells, and eventually expects concentrators to be the company's principal business. Its single-crystal silicon concentrator cells are now 26 percent efficient. To achieve this record efficiency, the cell's electrical contacts are on its rear so as not to shadow the cell surface. The cell's back surface is totally covered with two polarities of metal electrodes that offer low resistance and excellent light reflectance. These thin, interdigitated aluminum electrode "fingers" make contact with positive and negative copper strips that collect the cell's power.

The 15-kilowatt Alpha Solarco tracking photovoltaic concentrator module designed for utility-scale power generation concentrates the sun 400 times onto its solar cell surfaces and provides enough power for three households. *Courtesy of Solar Energy Industries Association.*

SunPower's proprietary technology was developed over a 15-year period by Dr. Swanson and his students when Swanson was a professor of electrical engineering at Stanford University. Its trademark, Controlled Carrier Processing (CCP), refers to the precise control of the motion of electron-hole pairs in semiconductors. The Electric Power Research Institute (EPRI) and the U.S. Department of Energy (DOE) supported Swanson's research while he was at Stanford and later helped finance his solar cell research and development at SunPower.

Four innovations distinguish the company's approach to solar cell manufacturing. The first is ultra-clean chemical processing to reduce

Dr. Richard Swanson, President and Director of Technology, Sun-Power Corporation, which Professor Swanson founded in 1988 with SunPower Chairman Robert Lorenzini, founder of Siltec Corporation. *Courtesy of SunPower Corporation.*

contamination that could cause cell defects and lower production yields. Next comes fabrication of cells using single-crystal silicon wafers only 80 micrometers thick—a fifth as thick as a standard wafer. The third innovation is the CCP—precise control of charge carrier flow in the highly doped regions of the cell. The fourth breakthrough is multilevel metallization—provision of two independent current-collection paths of very low resistance to minimize intracell power losses.

GRAND PRIX SOLAR CELLS

SunPower's plain (unconcentrated) solar cells are also a premium product. Honda R&D Company used a special custom-made batch in the Dream, the solar-powered car Honda entered in the 1993 World Solar Challenge race in Australia. Because of their high cost and power, Swanson likened these cells to "Grand Prix" engines. They are processed the same way as the semiconductor portion of SunPower's concentrator cells, and thus cost about the same per unit area, but are optimized to work without plastic concentrating lenses to reduce silicon requirements, so the cells are relatively expensive. While the racing cells therefore won't satisfy the world's demand for inexpensive energy, the Honda Dream not only won the World Solar Challenge but beat the previous world record for the 1,865-mile race by more than nine hours and averaged 53 m.p.h. under solar cell power.

Swanson plans to have a concentrator cell system commercially available by 1999 for under $2 a peak watt installed, which would make his concentrator cell competitive for bulk power generation by utilities. While demand for concentrator cells is very modest today, and the company is operating only a single production shift, Swanson anticipates a steady increase in demand as costs continue to fall. He also foresees a day when large, automated, multimegawatt solar cell factories will be built on the sites of planned power plants to decrease transportation costs. Other energy experts have envisioned incorporating materials production capabilities—whole glass factories, for example—in very large-scale solar cell manufacturing plants. The production system Swanson has in mind would be integrated with highly automated construction equipment out in the field that might one day "plant" the concentrator modules and their tracking support structures very inexpensively, much as automated farm equipment now rolls over fields

Powered by SunPower Corporation's solar cells, the Honda "Dream" won the 1993 World Solar Challenge Race in Australia. *Courtesy of SunPower Corporation.*

performing mechanical operations that once took large amounts of human labor.

POCKET-SIZED POWER PLANTS

One of the most amazing things to see for anyone accustomed to large-area, conventional, flat-plate solar cells is the small size and high power of SunPower's concentrator modules. For example, SunPower today makes Thermo Photovoltaic (TPV) modules that are used in specialized gas-fired power system applications. In TPV technology, burning gas heats a filament that glows brightly at just the right frequency for optimum reception by the concentrator cells' semiconductor. Reflectors focus a beam of this light on the photoreceptive silicon cells for conversion to electricity. The cells surround what looks like a giant Coleman mantel

SunPower Corporation's 150-watt multicell photovoltaic module (*top*) for parabolic dish reflectors and the company's high-efficiency Fresnel concentrator cell (*bottom*), which produces six watts of power at 26 percent efficiency. *Courtesy of SunPower Corporation.*

lamp. Although this approach obviates the need for a heat engine and an electrical generator to transform the energy of the natural gas into electricity, TPV energy conversion naturally does not eliminate dependence on fossil fuels and would not be a renewable technology unless biogas were substituted for natural gas.

SunPower's TPV cells are contained in square modules not much bigger than a small matchbox that each produce an astounding 200 watts. These tiny modules each contain 24 small cells. A strip of just six SunPower TPV modules is thus powerful enough to power a house. (SunPower concentrator modules for more conventional outdoor PV applications use a similar technology but are optimized differently during production to maximize energy capture from concentrated sunlight.)

SunPower's compact production facilities occupy only a few medium-sized rooms in a single-story building, yet are capable of producing 10 megawatts of concentrator cells a year. For comparison, the 10-megawatt thin-film solar cell plant described in Chapter 10 occupies a building the size of a large aircraft hangar. Although semiconductor components of the concentration modules were on the market in 1995, the concentrators themselves are just beginning to be commercially available.

While SunPower currently derives only about 10 percent of its business from concentrators, Swanson sees concentrators as the company's main growth area. Demand should accelerate significantly after Sun-Power and other concentrator companies reach their low-cost targets, which could happen around the turn of the century.

10

THIN FILMS

The hurdle for introducing thin films is continually being raised by
the success of the crystalline silicon technologies.
—Ken Zweibel, thin-film solar cell expert,
National Renewable Energy Laboratory

Many solar experts believe that the photovoltaic (PV) industry one day
will be dominated not by solid single-crystal or polycrystalline solar cells
but by thin-film cells only. That's because thin films have the potential to
lower solar cell costs significantly, once certain technological hurdles are
overcome. Let's now see how thin-film technology works and take a peek
at the diverse industrial activity now under way.

SOLAR CELLS AT PENNIES A WATT?

Solar cells of amorphous silicon, the semiconductor of which most thin-
film cells are made, are only about a micrometer thick yet have great light
absorptivity. (Single-crystal cells are 200–400 times thicker.) Amazingly, a
single amorphous cell layer may be as thin as 0.3 micrometers. If they
fulfill their technological promise, amorphous silicon cells will one day
be produced for only pennies a watt in large, automated factories. Paul
Maycock, former head of the U.S. Department of Energy's (DOE) Photovoltaic Program, has estimated that the material costs for 8 percent efficient amorphous silicon modules would be only 15–30 cents a watt in a
factory producing 10 megawatts of thin-film modules per year.

Despite hundreds of millions of dollars of thin-film technology investment worldwide by governments and companies—most of it for research

and development—thin-film sales are still only about a quarter of all solar cell sales.[1] Almost all thin-film output is currently amorphous silicon, the focus of most of the world's thin-film research. Newer thin films of cadmium telluride (CdTe) and copper indium diselenide (CIS) account for only a percent or so of world thin-film production. As we shall see, however, these exciting new semiconductors may eventually wrest control of the thin-film market from amorphous silicon, if its stability and efficiency do not improve.[2]

THE DRAWBACKS TO AMORPHOUS SILICON

Amorphous silicon cells today are low in efficiency, instable, and require exacting production conditions. The very best amorphous silicon cells in 1995 deliver only about 9 percent efficiency outdoors under field conditions. Rival crystalline cells are almost twice as efficient. Average amorphous cells deliver substantially less than 9 percent. But technological improvements are occurring very rapidly in thin-film efficiency, stability, reliability, and production technology. These new developments have enhanced thin films' commercial attractiveness and raised the prospects of sharp cost reductions. If these trends continue and outpace continuing improvements in solid silicon cells and concentrators, then within five or ten years, thin films will have a large, if not a dominant position in the solar-cell market. Most solar experts expect thin-film cells' lifetime costs per kilowatt-hour eventually to fall below those of solid silicon cells (without concentrators). Thin films would then begin taking a market share from their thicker competitors and probably would become a major global energy source. (Initially they would be used in developing countries, for village-scale, decentralized applications.)

Despite these long-range prospects, current production costs for amorphous silicon modules are still roughly comparable to those of single-crystal modules—about $4 a watt. PV expert Maycock attributes these high thin-film costs to "low efficiency, poor silicon utilization [in amorphous manufacturing processes], degradation, high capital costs of pilot plants, and the cost of tin oxide," used to coat amorphous silicon cells.[3]

Major research efforts are now under way to improve the amorphous silicon cells' stability. The cells currently lose as much as 10–15 percent of their power through degradation during the first few hours of exposure to sunlight. Degradation may eventually sap 15–30 percent of an amorphous silicon module's power. Amorphous silicon cells eventually do stabilize at their reduced output, and tests now suggest that the cells will perform reliably at these restabilized values for ten years, but this still must be verified in the field. "All manufacturers are developing new structures and packages to mitigate the stability issue," Maycock reports.[4] In practice, manufacturers take degradation into account when rating their cells.[5]

The fact that thin films require little semiconductor material is important from a resource perspective, and therefore affects costs. The world solar-cell industry has grown about 100 percent in production volume since the mid-1980s to about 85 megawatts in 1994. When the industry reaches a volume of 100 to 150 megawatts of silicon cells, its demand is likely to exceed the world's scrap silicon supplies. This is the "below spec" material discarded by the computer chip and semiconductor industries and used as raw material by solid-silicon cell makers. The makers of those cells today thus pay only a fraction of the manufacturing costs of their high-grade silicon. Inevitably, they either will have to pay the full market price or build their own plants for solar-grade silicon, or utilize cell technologies that require less high-grade silicon. Prices of solar-grade silicon have already risen as much as 400 percent for some contracts just from 1993 to 1996.

During solid silicon-cell production, silicon losses in wafer sawing and processing generally ensure that a solid cell will require up to 400 times as much silicon as a thin amorphous silicon cell of equivalent planar area. Even allowing for lower thin-film efficiencies, thin films thus offer impressive savings of expensive raw material.

An entire amorphous silicon cell factory operating at a megawatt scale can run for an extended period on only a 22-pound cylinder of silicon-hydride gas (silane), rather than on thousands of pounds of solid silicon. Increases in silicon costs will therefore impact the thin-film silicon makers only slightly relative to the bulk-cell makers. And instead of having to produce silicon by crystal growth, or by casting and then cutting the solid silicon into wafers, thin films are made by well-understood and relatively inexpensive technologies originally developed for coating glass.

ROBOTIC THIN-FILM PRODUCTION

Before it was put out of production by court order following a patent infringement suit brought by Solarex Corporation, the $25 million Advanced Photovoltaic Systems, Inc. (APS), facility at Fairfield, California, was a state-of-the-art thin-film manufacturing facility. Subsequently, the APS production assets were bought by BP Solar, Inc., which intends to convert them to manufacture its cadmium telluride thin-film modules. APS was the only company making a very large-area (11-square-foot) thin-film module and was the world's largest thin-film plant producing power modules. During its short operating life, the APS plant revealed some of the generic problems on the front lines of amorphous silicon commercialization but showed how thin films will be robotically mass produced.

The APS plant was a 10-megawatt facility capable of producing 200,000 solar modules a year. Crates of tin oxide–coated float glass arrived from a glass factory at one end of the plant. (The transparent tin-oxide conductive film soon would form the amorphous silicon cell's top layer and positive electrode.) After being put on racks, the glass was never touched by human hands, until almost the end of the production line.

A machine would first pick the coated glass off its stand and transport it in quick succession to beveling, drilling, washing, drying, and scribing stations. In the scribing room, guidelines for cells of uniform voltage were outlined on the tin-oxide surface by laser beams. These cells would later be interconnected in series to produce a high-voltage, large-area module. After a wash-and-dry cycle, the glass then rolled on tracks in a batch-carrier to a preheating chamber and then into a deposition oven, where it got three thin layers of doped amorphous silicon. After heating, the panels were cooled before a back reflective aluminum coating was applied in a vacuum sputtering chamber.

A laser then scribed through both the aluminum and silicon coatings, in perfect registration with the guidelines in the tin oxide. Next, about half an inch of material was sandblasted away all around the sheet's edges insulating the cells. After testing, the panels got a glass cover and frame, which was annealed to the panel in an oven. The encapsulation provided electrical insulation and cell protection to the modules.

By contrast to these large-area, thin-film rectangular cells, solid silicon wafers generally are only a few inches in diameter, and each must be interconnected with its neighbors. Because computer-controlled lasers separate thin-film cells, the cells need not be sawed apart. Cell interconnection, too, is highly automated and is done by vapor depositions. The larger the cell area, however, the greater the challenge to keep the thin-film deposition layers uniform and to avoid any dust that could interfere with proper adherence of a layer, thereby causing electrical shorts. These tight tolerances and contamination hazards make it challenging to operate a thin-film plant.

The APS modules' final stabilized efficiency was only 5 percent—not bad in the still-experimental world of thin films, but not so good compared to 15 percent efficient single-crystal cell modules, or to concentrators that can surpass 20 percent efficiency. And out in the field—where cell performance really counts—the APS module efficiency was only 3.2 percent, after deducting normal power-system conditioning losses that occur when DC is converted to AC and transformers alter voltage.[6] (That's not quite as bad as it sounds, since in the same field, the single-crystal modules were delivering only 8 percent.)

Given a history of steady progress in improving cell efficiency, solar experts in 1994 expected that ordinary amorphous silicon cells would increase in efficiency to 10 percent and that multijunction cells, which capture light of different wavelengths, would reach 14 percent by the year 2010. Although that is about the outdoor efficiency of commercial crystalline cells today, by 2010, as single-crystal and polycrystalline cells continue to improve, they will probably still be about 50 percent more efficient than the amorphous silicon cells.

However, United Solar Systems Corporation (Uni-Solar) and Energy Conversion Devices in 1996 have already produced a multijunction amorphous silicon cell that is nearly 14.5 percent efficient, and its stabilized efficiences should be at least 10 percent. Cell efficiency in the 10 percent range will not necessarily stymie thin films, provided that manufacturers can achieve production cost savings that more than compensate. (The lower a cell's efficiency, the larger the cell area required to equal the output of higher efficiency cells; this raises the installation costs for low-efficiency cells.) According to Dan Shugar, who was a manager for APS, area-related costs are typically only 20 percent of a utility's overall photovoltaic system costs. Therefore, even a 50 percent efficiency penalty would raise total system costs only by at most 20 percent, and probably somewhat less (due to economies of scale).

Thin-film efficiency can be increased by "stacking" multiple layers of semiconductor film with different light absorptivities. This raises overall cell efficiency by utilizing more of the solar spectrum. Paul Maycock, the solar cell expert, projects that stacked thin-film cells using amorphous silicon in combination with other semiconductor materials "could be the [photovoltaic industry's] principal product by 2000 with prices less than

$2 per watt and module efficiencies of 12 percent." In search of this and lower targets, Maycock estimates that the R&D community is spending $80 million to $100 million a year on furthering amorphous silicon technology.

Even the large manufacturers of crystalline cells are actively working on thin films. Siemens, the ARCO Solar successor, was engaged in amorphous silicon research and development in the 1980s and had even built a 2-megawatt plant in Munich, Germany, in 1989, with hopes of selling photovoltaic sunroofs to the automotive industry. But before that business really developed, Siemens left the amorphous silicon field after losing a patent infringement suit by Solarex Corporation. (A Siemens' spokesman said the suit was but one factor in the company's decision to leave.) Siemens is now actively developing a competing thin-film technology based on copper indium diselenide (CIS), which offers higher efficiencies and appears to be stable. A CIS submodule manufactured by Solarex in 1995 was already at 13 percent efficiency, the best performance yet from any thin film.[7] (In 1996, not long before we went to press, the National Renewable Energy Laboratory of Golden, Colorado, broke Solarex's world record by creating a 17.7 percent efficient experimental thin-film cell made of copper indium gallium diselenide [CIGS].) Various companies—including Boeing, International Solar Energy Technology, Energy Photovoltaics, Inc. (EPV), of Princeton, New Jersey, and Solarex—are also working on CIS material, and EPV tentatively announced a 10-megawatt CIS manufacturing facility to be built in eastern Europe but by early 1996 still had not obtained full funding. The firm announced in 1995 that it would accept 1996 orders for its modules at only $3 per watt in large systems, plus installation.[8]

CONTINUOUS THIN-FILM
PRODUCTION PROCESSES

United Solar Systems Corporation (USSC) of Troy, Michigan, is a joint venture of Detroit-based Energy Conversion Devices, Inc. (ECD), and Canon, the Japanese firm, which together are commercializing a unique production process for single- or multilayered thin-film cells. ECD has been conducting semiconductor research and development for nearly 35 years and controls substantial proprietary technology for producing amorphous silicon cells on stainless steel backing. In a continuous production process, a strip of 14-inch-wide stainless steel is fed continuously from a roll through an automated machine that deposits amorphous silicon, creating a half-mile-long solar cell. The cell is then cut into different sizes for various applications. As noted in Chapter 5, ECD expects to be able to make cells for $1 a watt in 1999 using the continuous roll production process.

In 1995 USSC announced (but soon indefinitely postponed) plans to build a 10-megawatt facility to produce multijunction amorphous cells at Newport News, Virginia. Virginia said it would provide financial inducements of up to $22.5 million to solar-cell companies siting plants in the state between 1995 and 1999. When production equipment for the USSC plant was completed, however, Canon decided to use it for a plant in Japan. ECD has now opened a substitute 5-megawatt plant for USSC in Troy, Michigan. The facility produces cells of an advanced design in which a thin-cell "sandwich" of nine semiconductor layers captures light of different wavelengths to boost efficiency. One of the plant's products is a solar shingle for rooftops.

Under its flamboyant founder and chairman, Stanford R. Ovshinsky, ECD has spent $300 million or so on development of solar cells, batteries, magnets, optical memories, and other products. With no formal education beyond high school, Ovshinsky started his career as a machine builder, then worked in microbiology, and later in solid-state physics, conducting scientific research on amorphous semiconductors. In the late 1960s, he noticed that amorphous semiconductors were not just glasses, but had exciting electronic properties to which he called the scientific community's attention. Ovshinsky's critics claim that he has produced relatively few commercial products, given all the research-and-development money ECD has expended. For most of its existence, ECD has indeed delivered decades of operating losses instead of profits. Perhaps because ECD's products generally have not yet had much market impact, Ovshinsky has been disparagingly referred to as the "P. T. Barnum of photovoltaics." But the inventor has a keen scientific intuition and an undisputed talent for promoting his technological breakthroughs and for striking deals with major industrial partners. ECD work has led to flat-panel displays, a nickel metal-hydride battery, and a number of spinoff companies, such as Ovonic Synthetic Materials Company and Ovonic Display Systems, Inc.

NEWER, MORE EFFICIENT THIN FILMS

Apart from amorphous silicon, CIS, and CIGS, the other important thin-film material, cadmium telluride (CdTe), has no stability problems and offers greater efficiency than amorphous silicon. Like CIGS, it, too, has reached an efficiency of about 16 percent, but has done so more quickly and with less research money than amorphous silicon. Like amorphous silicon, CdTe could be mass produced inexpensively by vapor-deposition processes.

Various companies are either developing CdTe modules (such as BP Solar, Inc., in the United Kingdom and California), are already producing small quantities (Matsushita in Japan), or are reportedly working on manufacturing processes (Energy Photovoltaics, Inc., of Princeton, New Jersey). A new firm, Golden Photon, Inc., a Coors Corporation subsidiary, has opened a plant that will be capable of producing 2 megawatts of CdTe cells a year under normal operation and more in around-the-clock operation. They plan to sell their modules integrated in packaged systems designed for specific applications, such as water pumping and battery charging. Golden Photon is using technology obtained by absorbing a small Texas company called Photon Energy. Solar Cells, Inc., of Toledo, Ohio, has opened a CdTe cell plant of similar capacity and a 10-megawatt plant was slated to open in late 1996. Because cadmium is toxic, Golden Photon plans comprehensive cadmium recycling; the company will guarantee a return value for its modules at the end of their useful life and will incorporate recovered cadmium in new modules.[9]

AstroPower Corporation of Newark, Delaware, is using another process, epitaxial crystal growth, in the production of thin-film polycrystalline silicon cells on an inexpensive ceramic backing. The company's continuous production process eliminates the slicing and trimming costs associated with conventional silicon technologies. By 1993, AstroPower was producing 10 percent efficient cells on a pilot scale (about 200 kilowatts in 1995) and was working toward 14 percent efficient cells. Production of the firm's single-crystal (bulk) silicon cells had reached 2 megawatts a year in 1995 and was rapidly increasing. AstroPower has annual revenues of more than $10 million and sales volume of 2.5 megawatts. In late 1995, the company received an important R&D cost-sharing contract from the National Renewable Energy Laboratory to develop both the continuous production process for its thin-film solar modules and a large-area version of the module. AstroPower's ultimate goal is to produce a 32-square-foot thin-film module at only $1 a watt.[10]

One of the strengths of AstroPower's epitaxial work is that it builds on the huge scientific and engineering knowledge about silicon, offering all the advantages of more expensive bulk silicon cells, such as high efficiency, stability, and reliability. Yet the substitution of the low-cost ceramic backing for the high-cost bulk silicon gives AstroPower's cells the same critically important insensitivity to refined silicon costs as other thin films have.

BONDED BEADS AND ARTIFICIAL LEAVES

A distinctly different non-thin-film solar cell technology known as the spherical cell, which has the potential for low-cost mass production, was being pursued by Texas Instruments (TI) in partnership with Southern California Edison until 1995. TI had received significant DOE support for spherical cells. But after 20 years of R&D and millions of dollars, no investor appeared interested in the project, and TI was unwilling to continue alone.[11] In 1995, however, Ontario Hydro of Canada purchased the TI process and equipment. No production schedule has been announced, but the spherical cell may yet have its day.

TI made its spherical cells of tiny beadlike spheres using very inexpensive, metallurgical-grade silicon attached to a flexible, embossed aluminum foil backing. About 4,000 of the silicon beads per square inch were bonded into holes etched in the aluminum. The spheres themselves were then etched, coated, abraded, and covered in another foil layer. Because of the cells' flexibility and the independent electrical operation of each bead, the cells were damage-resistant, but in 1995 had demonstrated efficiencies of only about 10 percent.

Laboratory researchers are also pursuing radically new thin-film technology, which, if successful, could be a technological wild card. Their goal is to leapfrog existing technology in cost or efficiency or both. Swiss researcher Michael Grätzel at the Institut de Chimie Physique of the Ecole Polytechnique Fédéral de Lausanne is developing an electrochemical cell he calls an artificial leaf, because its operation resembles photosynthesis.

While conventional silicon cells both simultaneously absorb energy and separate charge in the same material to produce current, Professor Grätzel's "noncrystalline ceramic film" cell uses a monolayer of light-sensitive ruthenium-based dye—similar in molecular structure to chlorophyll—to absorb energy. Electrical charge is then transferred from the dye to a semiconductor "membrane" of titanium dioxide, an inexpensive and widely available pigment used in paints and paper. Titanium dioxide is simply screen-printed on the conducting oxide-coated glass during the manufacturing process. The cell contains a liquid electrolyte film enclosed between the dye-sensitized anode and a platinum-coated counterelectrode. When electrons returning from the exterior of the cell after performing work reach the counterelectrode, iodine ions in the electrolyte transport the electrons back to the ruthenium dye, completing the charge-transfer cycle.

Work on a solid-state version of the cell is also progressing, and contracts have been signed between Grätzel's school and a German consortium led by Pilkington/Flachglas, Inc., for commercialization of 100-watt modules using this technology. Consumer product applications are handled by another firm, while Asea Brown Boveri in Baden is involved in developing commercial fabrication methods for the cell. One research study has predicted an extraordinarily competitive eventual cost of only 60 cents a peak watt. Dr. Grätzel states that the cell has already achieved 12.5 percent efficiency and has been shown stable for two years of accelerated testing that corresponds to six to eight years of operation under natural conditions. Experience has shown, however, that even great technological advances in cell designs and materials must still travel the long and difficult road from the laboratory through testing, scale-up, creation of a commercial product, and on to commercial-scale manufacturing. All this invariably sounds easier to proponents of a new technology than it usually is.

SPRAYED AND STACKED POLYCRYSTALLINE

Another prominent solar cell developer, Professor Martin A. Green, Director of the University of New South Wales Center for Photovoltaic Systems (Australia), is also very optimistic about the prospects for inexpensive solar cells. The center is a leader in photovoltaic research and a holder of a world record for crystalline silicon cell efficiency. Green himself is one of the world's most respected solar scientists. Using a basic cell architecture on which he has been working for years, Green has developed a promising design for a multilayered cell with laser-grooved metallic contacts. The cell does not require high-grade silicon, because it will use very thin (3 micrometer) alternating positive and negative layers of polycrystalline doped silicon sprayed on a glass backing. (Until now, polycrystalline cells were usually made by casting and slicing solid material.) The thinness of the sprayed polycrystalline layers will minimize the chance that an electron will encounter a crystal defect before it can be collected by an electrode. Having thus reduced the electron's "travel distance," the cell designer can use cheaper, lower-grade silicon with more boundaries between crystals and more numerous defects than single-crystal material.

Green's colleagues forecast that the new cell is likely to produce power five times more economically than coal-fired plants in about 20 years and will provide cells at a dollar a watt within 10 years. Current plans, however, are for five more years of laboratory research "to refine the technology," followed by another five years for industry to begin volume production. "The potential has always been there to produce an inexpensive, clean, convenient, and inexhaustible supply of energy—and now we know how to do it," said Professor Green.

POLARIZED THIN FILMS:
POWER WITHOUT SEMICONDUCTORS

Just when you think you have heard everything in the rapidly evolving solar cell field, something strikingly new appears on the horizon. Enter nanotechnology and the work of Alvin M. Marks, an 85-year-old electrical engineer who pioneered the development of polarizing films with his brother Mortimer, patent owner of 3-D America Corporation in Los Angeles.

Alvin Marks holds hundreds of patents in optics, chemistry, and electronics. Over the years he has supplemented his knowledge of engineering with advanced studies in physical optics, polymer chemistry, quantum mechanics, and mathematics at Harvard, Massachusetts Institute of Technology (MIT), and elsewhere. His early work on polarizing films was commercialized by Marks Polarized Corporation, which at one time manufactured millions of 3-D glasses for Universal Pictures. After the boom in 3-D technology ended, Marks became interested in photosynthesis and began filing patents on light-polarizing, electrically conducting film, ordered dipoles for converting light to electricity, and a molecular-scale diode for separating electrical charge.

He now believes he has found a fundamentally new way of vastly increasing the efficiency of solar cells by developing advanced nanostructure devices for transforming light to electricity for as little as 30 cents a watt using machines similar to those that the Marks Polarized Corporation used in fabricating 3-D viewers. If his devices can be made to work at these extraordinarily low projected costs, they could not only revolutionize the solar electric industry worldwide, but could also make it possible to miniaturize computer circuitry hundredfold, produce efficient laser lights, and improve high-definition television, 3-D, and flat-panel displays.

Alvin Marks works at Advanced Research Development, Inc., of Athol, Massachusetts, where he is president, and at Phototherm, Inc., where he is chief scientist. Like Michael Grätzel, Marks hopes to make electricity from light without the use of semiconductors by mimicking photosynthesis. Plant leaves use molecular structures to capture solar energy and allow energized electrons to move in one direction. Marks's approach is to use polarizing materials containing an antenna array of linear molecular diodes. This is a radical departure from conventional solar cell technology, because it does away with silicon and other conventional semiconductor materials. The new devices are no longer constrained by semiconductor efficiency limits and can achieve theoretical efficiencies as high as 60–80 percent conversion of light to electricity.

In one of two patented methods, Marks proposes using two layers of electrically conducting polymer film known as Lumeloid, containing parallel molecular antenna arrays oriented at 90 degrees to each other. About half of the incident light is absorbed by the first polarized layer; the remainder is reflected back from the rear of the cell through the second film layer, where it is absorbed. Very high cell efficiency is theoretically possible by eliminating the transformation of light to chemical energy, which occurs in photosynthesis, and by going directly from light to electricity.

Scientist and inventor Alvin M. Marks, President of Advanced Research and Development, Inc., a firm that is developing polarizing films with submicron antennae and molecular diodes for inexpensive, high-efficiency conversion of light to electricity. *Courtesy of Alvin M. Marks.*

Marks's second technology, Lepcon, uses the same principles as Lumeloid, but with a much more durable glass substrate coated with a submicron layer of polarized metal. Marks expects to produce a photovoltaic panel using Lepcon lasting 50 years that could be manufactured in 20 seconds for 50 cents a watt. Much chemical benchwork and other research and development—probably costing tens of millions of dollars—still lie ahead before it is known whether Marks's inventions will live up to his expectations.

While the research laboratories continue creating fundamentally new thin-film technologies, hard work goes on daily in today's commercial thin-film production plants to wring costs out of existing technologies, improve their performance, and expand markets.

11

✴

PRETTY POLY

I got a lot of documents out of Amoco, which showed that they always had the intention of taking over [Solarex], and that this declaration to the bank that they had lost interest in solar is absolutely
a lie.
— the late Dr. Joseph Lindmeyer, cofounder,
Solarex Corporation

I don't think Amoco ever told the bank that they lost interest in
solar energy. . . . Either [Solarex] was going to go bankrupt . . . or
Amoco was going to buy and continue to fund it.
— Dr. John Johnson, retired Amoco
chemical engineer

Polycrystalline silicon solar cells coated with silver nitride to increase
their light absorption have a deep bluish hue with beautiful sapphire-
like highlights—a jewel-like energy generator. Also known as "semi-
crystalline," polycrystalline cells are intermediate in cost, performance,
and thickness between thin films and single-crystal silicon.

The largest producer and developer of polycrystalline cells is Solarex
Corporation of Frederick, Maryland, a former Amoco subsidiary, now a
business unit of Amoco/Enron Solar, a joint venture. In fact, since the
German conglomerate Siemens purchased ARCO Solar in 1989, Solarex
has become the largest and oldest U.S. solar cell manufacturer. Solarex
manufactures a small quantity of amorphous silicon thin-film solar cells
and is developing copper indium diselenide thin films, but most of its

modules are of cast polycrystalline silicon. The company developed this semi-crystalline silicon because of their concern over the availability and reliability of crystalline silicon supplies.

Solarex was founded in 1973 by two Hungarian expatriate scientists, Dr. Joseph Lindmeyer and Dr. Peter Varadi, both of whom had worked on space solar cells for the Communications Satellite Corporation (COMSAT). Trained in electrical engineering, Lindmeyer had emigrated from Hungary in 1956 and was looking for a new area of science and technology he could pioneer without having to spend 30 years just catching up with developments in the field. Within seven years of his arrival, he completed a Ph.D. and coauthored the world's first book on semiconductors (*Fundamentals of Semiconductor Devices*). Physical chemist Peter Varadi also left Hungary after the 1956 Hungarian Revolution and, like Lindmeyer, was eventually hired by COMSAT.

A NEW ENTERPRISE IN JEOPARDY

While at COMSAT, Lindmeyer noticed that the solar cells used in satellites were made by what he terms an "ancient technology." He then quickly developed a new device known as the "violet cell," which increased the efficiency of space cells by 50 percent. Although the violet cell was a tremendous technological improvement and was used on all satellites, Lindmeyer did not earn any extra money from it, because of his position as a salaried COMSAT employee. If he ever made a major technical breakthrough again, he resolved that he would do so in a company of his own, so he could reap the financial rewards. It happened in the early 1970s, at about the time of the energy crisis: Lindmeyer got the idea that space solar cells, like the ones he was making for COMSAT, probably could be modified to produce power on Earth. He knew, of course, that terrestrial cells would have to be more economical compared to space cells—perhaps as little as a thousandth the cost. Lindmeyer then came up with an exciting new production process that he thought would yield inexpensive cells, but he was unable to interest COMSAT's management. So at a New Year's party in 1973, Lindmeyer and Varadi decided to leave COMSAT and set up their own company to develop and sell the terrestrial cells. By February 1973, they had incorporated.

Varadi took the lead in trying to get financing and began approaching venture capital companies. "First, I would have to spell 'photovoltaics,' and it would usually go downhill from there," he said. The venture capital folks

Headquarters of Solarex Corporation, Frederick, Maryland, showing roof-integrated polycrystalline solar electric cells. *Courtesy of Solarex Corporation.*

were leery of photovoltaics and skeptical that two scientists could run a business. History would show that the investors had grounds for their concern. Within two weeks, however, Varadi gave up on the venture capitalists and started raising money from friends and acquaintances. Soon he had $250,000 in start-up funds, and the two scientists-turned-entrepreneurs quit their jobs. "Until then, we got a salary every two weeks," Varadi recalls. "Now we had to provide our own somehow. It was a very strange feeling." They launched Solarex in a small, two-story office building in Rockville, Maryland, and began confronting the formidable tasks of creating a new technology and finding markets for it, all without infringing on any of COMSAT's patents, many of which Lindmeyer had obtained.

The factory was in the building's basement; the research-and-development office was upstairs. Manufacturing began in August, and auspiciously, within just eight months, Solarex began to show a profit, according to Varadi. Right away, however, they encountered an enormous problem: COMSAT alleged patent infringement and sued them.

"COMSAT was a rich company," said Lindmeyer, "and we had no money in the bank." Moreover, since Solarex was just about to raise more money, and the suit raised doubts about the fragile start-up's future, the entire new enterprise was suddenly in jeopardy. The company had started to get contracts, but some were delayed by the uncertainty.

The suit was based on allegations by Solarex's new systems engineer, an ex-COMSAT employee, who talked to his friends at COMSAT and said that Solarex had stolen its technology from COMSAT. The engineer, however, was not involved in developing the new technology, and when it came time to give his deposition, he looked more carefully at Solarex's technology and discovered that Solarex in fact was *not* using COMSAT technology. COMSAT's technology was all for producing high-cost space cells and had little relevance to production of low-cost cells for use on Earth. The engineer retracted his allegations.

But even though the foundation of the suit had crumbled, getting it legally dismissed was another matter. Its very existence left a cloud over Solarex. So Lindmeyer managed to arrange an interview with Senator Mike Gravel of Alaska, an active supporter of renewable energy technologies, in whose state COMSAT had installations. Gravel then called up the president of COMSAT in front of Lindmeyer. "Hi Joe, this is Mike Gravel," the senator said. "What is this business of trying to run down this young company? They're working on alternate energy."

"I don't know what Dr. Charnyk [the president of COMSAT] replied," said Lindmeyer, "but Gravel said, 'Look, depositions are already taken, and there is no case.' Charnyk said something else. Then, finally the senator said, 'Joe, call off your boys!' " The case and a countersuit were both promptly dropped, and Solarex was able to get back to the business of making solar cells. Surviving the COMSAT suit, however, was but the first of many challenges.

During its early years, Solarex made solar watches and was the first to collaborate with the Japanese in making solar calculators. Then the company produced remote-area power supplies for radio repeaters. Because the terrestrial solar-cell business was in its infancy then, Solarex had to scramble to create markets for its products, and that caused complications. "One month," said company physicist John Wohlgemuth, "the manufacturing wouldn't work very well. Then the next month we wouldn't have the business."

When an order came in, the whole company would work all night and all weekend, if necessary, without any overtime pay, to assemble it. "We certainly had one big advantage," said Lindmeyer. "There was a mission. Everybody became almost religious about alternate energy. It wasn't just a job at this point." But revenues were disappointing, the founders' resources dwindled, and they worried each time payday approached about how to cover their payroll. "It's always a horrible feeling," Lindmeyer said. "You've got to have the guts to live like this."

Solarex's manufacturing technology was no more fail-safe than its finances in those early days. At one point, someone accidentally blew up one of the radio frequency furnaces in a violent hydrogen explosion that shook the whole building. Another time someone dropped an iodine solution in the basement, and fumes flooded the entire building, which had to be evacuated. On other days, brown clouds of chemical vapor poured from the facility during mishaps in the silicon-etching process. The fire department frequently responded with sirens blaring.

LARGE INVESTORS ARRIVE

Although the little company bounced from rock to shoal during the 1970s, Lindmeyer was successful in landing several million dollars in European investment capital from a French and Dutch company in 1979. Then Amoco bought a share of the company, and so did the Italian oil company ENI. At the time Amoco joined the enterprise, its participation seemed like a great coup to Lindmeyer, since Amoco immediately invested additional millions of dollars. The European partnerships had also begun flourishing, and the company was soon growing fast. Sales in 1983 were cresting at $25 million a year.

Amoco showed its enthusiasm by arranging to have one of its service stations in Chicago powered with Solarex photovoltaic panels and spent about $7 million to build Solarex a PV-powered solar-cell factory in Frederick, Maryland. Lindmeyer referred to the facility as a solar energy "breeder," since solar power was used to produce solar cells that would in turn produce more solar power. (The notion was that the plant's rooftop array would generate enough electricity to operate the plant.) Though it seemed that Solarex was truly Amoco's favored child and would flourish under its tender care, disaster was about to strike; Lindmeyer would eventually come to bitterly regret Amoco's participation. Amoco was later to conclude that the solar breeder did not work.

THE MARTINSBURG GAMBIT

With Solarex riding high, Lindmeyer and Varadi were eager to find supplies of inexpensive silicon and decided to build a plant at Martinsburg, West Virginia, that was to use a new refining technology they were developing to produce raw silicon from quartz. Here accounts differ. Peter Varadi

says that after the "large facility was set up, it turned out not to be needed."
According to Solarex physicist John Wohlgemuth, Lindmeyer raised money
for the plant based on the result of what Wohlgemuth termed "a single experiment that was later seen to be flawed and unrepeatable."

By contrast, Lindmeyer claimed that experimental results with cells
from the new silicon refining process were promising. "We made some
experiments here in Rockville, and some cells were made out of this raw
silicon. The cells were reasonable—5, 6, 7 percent efficient—which was
encouraging."

Amoco Technology Company vice president for photovoltaics John R.
Triebe says the plant never worked properly and was sized for "an absolutely enormous market, which still hasn't materialized." The plant was
a "tremendous failure," John Wohlgemuth said bluntly. "They spent $15
million on that plant, [but] it turned out that Solarex really didn't have a
process to make solar-grade silicon," said Amoco chemical engineer Dr.
John Johnson. "So that was $15 million down the drain"—money the tiny
company could ill afford to lose. Johnson also asserted that the solar
breeder project cost Solarex more than $6 million. "The concept never
really worked," he said, "[and] was simply not an economic project."

Solarex was trying simultaneously to develop four different solar technologies. "The one that was Lindmeyer's favorite never worked and
ended up having to be shut down," said Wohlgemuth. But on another oc-
casion, when business slumped, the company shut down the one cell-
production line that was profitable.

A senior vice president who had been a bastion of support for Solarex
within Amoco resigned about this time. "Everybody got into this internal
struggle at [Amoco headquarters] in Chicago," said Lindmeyer. "We
ended up on the short end." The new senior vice president was no friend
of Solarex, he intimated. "Solarex ended up as a toy," Lindmeyer said.

Things might have turned out differently. Lindmeyer had had a multi-
year $9 million DOE contract to help him develop the refining process
he needed for the Martinsburg plant. But the contract was terminated
after only a year when the Reagan administration reduced the DOE's
photovoltaic budget for fiscal 1981 in keeping with the administration's
pervasive opposition to renewable energy.

THE MARKET SOURS, AMOCO TAKES OVER

Amoco Vice President Triebe says that, at the time, because Solarex was
incurring millions of dollars in losses, and the market for PV had soured

(because solar tax credits had been withdrawn and rapid oil-price escalation failed to materialize), Amoco had to act quickly "to get the boat headed in the right direction and keep it from sinking." Lindmeyer claimed that Amoco then used chicanery to gain control over Solarex. "We had a $7 million line of credit with a Maryland bank, and they were loaning us money happily, because Amoco was behind us. Then Amoco declared to the bank that they had lost interest in solar energy. Thereupon the bank got very nervous."

"I don't think Amoco ever told the bank that they lost interest in solar energy," counters Johnson.

Amoco and Lindmeyer remained at loggerheads about what happened next. In Lindmeyer's words, "Maryland National was agreeing to these loans because of regular investment from Amoco. And our CFO [chief financial officer] at the time had not reported everything properly I guess—some minor reports were forgotten. So overnight the bank quickly checked whether we violated the loan agreement someplace, and they found something. The next morning, they called the loan, which we couldn't pay back. . . . Amoco was just standing by, so we were just instantly bankrupt."

"Finally," said Lindmeyer, "they came around and said, 'Okay, okay, we're going to loan you this money, $7 million, for three months.' And I was begging them, 'But three months,' I said, 'you know I can't raise this money in three months. Give us at least six months.' No, no, they were unwilling. So they gave it for three months and, of course, I tried, and we all tried, Peter and myself, to get GE [General Electric] and some other people involved—somebody with big enough pockets to take it over, but the three months was up. No decisions. And then they were totally in control. They said they would buy it just because they were feeling sorry, or something."

Amoco took over Solarex in 1983. Lindmeyer's wife and son both died during the takeover period. "I didn't have the energy to fully and truly fight them," he said. "I came from the funeral to a shareholders' meeting where they were to vote the issues, and I couldn't even talk."

Later, Lindmeyer sued Amoco in a compensation dispute that ended in arbitration. In the depositions, he got documents from Amoco that he claims showed that they always had the intention of taking over the company, and that "this declaration to the bank that they had lost interest in solar is absolutely a lie."

Amoco's actions, according to Lindmeyer, enabled it to buy the remainder of Solarex for much less than the company's maximum valuation. "I had written into the agreement [with Amoco] that, if they ever reached 40 percent ownership, they would have to buy out everybody else at whatever the highest price there ever was. . . . At the time of the takeover, they owned 39 percent. They calculated. There was one guy there that calculated every day." Lindmeyer died of a heart attack in 1995 at age 66, but in a conversation about a year earlier it was clear that he still believed that Amoco intentionally forced the sale just in time to prevent that clause in the agreement from being invoked.

Looking back on that period of Solarex history, John Wohlgemuth has a different view. By 1983, "Everything that got shipped out the door at the time cost more to make than they got back from it," he said. Solarex was losing millions, and investors were wary of throwing good money after bad.

While Johnson acknowledges that Amoco might have owned as much as 39 percent of Solarex at the time of the buyout, he denies Amoco intended all along to take over Solarex. "The reason it happened at the time was because . . . Solarex got into financial difficulty. They had a certain continuing cash burn [of probably a million dollars a month] and no place to get the money [besides] Amoco, because nobody else would be willing to put money into a very risky situation like that. Either the company was going to go bankrupt . . . or Amoco was going to buy and continue to fund it."

STRATEGIC ACQUISITIONS

Before Varadi and Lindmeyer lost control of Solarex, Varadi had decided to acquire RCA's amorphous silicon solar cell capability for Solarex. (RCA at the time had decided to exit the business.) RCA's capability had been based on the research of solar pioneer David Carlson and a group of about ten colleagues. Varadi saw to it that Solarex got all the RCA patents, machinery, and several key employees from the amorphous cell operation. This proved to be an important acquisition.

Because of Solarex's strong patent position as owner of the RCA technology, Solarex subsequently was able to exact royalties from the Japanese on their amorphous silicon cell sales. As mentioned earlier, Solarex also successfully barred both Siemens Solar Industries and Advanced Photovoltaic Systems, Inc., from the amorphous business and is challenging United Solar Systems Corporation, another thin-film manufacturer.

About the time of the Amoco purchase, Varadi also saw to it that Solarex absorbed Solar Power, Inc., an Exxon Corporation solar-cell venture that became available when Exxon quit the solar cell business. "That [withdrawal] was caused mostly by large government contracts, which stopped," said Varadi. The Reagan administration's assault on renewable energy had begun shortly after President Ronald Reagan took office in 1981. The repudiation of solar energy not only discouraged oil companies and others from pursuing photovoltaics, but hurt Solarex as well in the early 1980s. The DOE in that period sharply cut back its spending on solar energy and reduced purchases of solar modules for demonstration purposes. Congress also rescinded federal solar energy investment tax credits. By 1986, Solarex had shrunk from a peak of 400 employees to about 150.

POLYCRYSTALLINE: A TOUGH MOVING TARGET

Apart from establishing Solarex and its successful joint-venture agreements in Europe, Lindmeyer was proud of developing the polycrystalline solar cell. "I made the introduction [of semi-crystalline cells], and I turned it into a product before the Amoco takeover," he said. "Today they make the same thing and it's beautiful. They haven't changed the technology as far as I can tell; it's all the same." Solarex's polycrystalline modules today are sold with 20-year power-production warranties. After Lindmeyer's death, however, Solarex did change its technology, cutting module production costs in half, tripling production capacity, and raising cell efficiency to nearly 15 percent.

Although Amoco's ownership has led to the development of new solar cell applications and reliable products, Solarex in 1993 was still unprofitable ten years after Amoco acquired it. However, the situation improved greatly in 1994 and 1995: The entire Solarex output at its polycrystalline plant in Frederick, Maryland, sold out in 1994, and orders were backlogged. After embarking on a $7–$8 million expansion of the polycrystalline plant to triple production there, Solarex is planning to open a $25 million amorphous silicon/amorphous germanium cell plant near Williamsburg, Virginia, that will initially manufacture up to 5 megawatts of large-area, thin-film modules. A $100 million 50-megawatt power plant is also planned with Enron for Rajasthan, India.

Physicist Wohlgemuth stated that, while the company's crystalline silicon operations are not losing money, losses are still occurring due to Solarex's long-range research on noncrystalline thin-film cells. These amorphous silicon cells are currently only about 10 percent of Solarex's production volume, but they are a large part of the company's future hopes and plans.

Like some other manufacturers of crystalline solar cells, Solarex expects that in four or five years, it will be able to get the price of its poly-crystalline modules down to about $2 per watt, which translates to a system cost of about $3 to $3.50 per watt for a grid-connected system (which does not require storage batteries). That will roughly halve the cost of PV power relative to today's costs, and will be a tough moving target for thin-film cells to match, even after their reliability and efficiency in the field are proved.

When installed PV-system costs come down to just $2 a watt, then, says Solarex CEO Harvey Forrest, "we're all going to be selling gigawatts of power, and we'll have big successful businesses." This will take another ten or twenty years, Forrest says. Unlike Ishaq Shahryar of Solec, Dr. Forrest believes that it will be difficult to achieve total system costs of $2 a watt with crystalline technology, and he contends that amorphous modules "will cross the finish line of $2 a watt first." As we discussed earlier, some companies have since claimed they can reach the $2 a watt threshold with mass-produced silicon concentrator cells or even $1 a watt by 1999 using continuous-roll thin films.

"TURNKEY" SOLAR CELL PLANTS

The solar industry does not live by its cells alone. Every high-technology industry needs an industrial infrastructure, including specialized tools and equipment. That's primarily where Spire Corporation of Bedford, Massachusetts, fits in. Apart from being a preeminent provider of solar manufacturing equipment, products, and services for the industry, and selling optoelectronics and biomedical products, Spire is essentially an engineering R&D organization.

Its reputation in photovoltaics has been largely built on its sales of automated solar-cell manufacturing equipment—a complete "turnkey" solar-cell factory that can produce a megawatt of solar cells annually. Founded in 1969, Spire also provides customers with comprehensive training and technology-transfer services. To date, Spire has delivered over ten solar-cell production lines, and its equipment has processed millions of watts of solar cell modules. The production line includes cell testers, tabbers, connectors, solar simulators, module encapsulators, and other equipment.

Not the pet project of a large oil or aerospace company nor the brain-child of a wealthy entrepreneur, Spire built its high-technology capability slowly, growing from a one-employee company to 175 people in an era when terrestrial PV technology was still barely commercial and government support was fickle.

Spire's CEO and founder, Roger G. Little, is a physicist and triathlete who started Spire in an office above a garage in the woods of rural Massachusetts with a $15,000 loan as his initial capitalization. Originally, Little did not intend to focus on solar energy. The company initially began consulting on the effects of radiation on electronics systems. A U.S. Air Force research contract in 1970 then gave Little enough working capital to hire a couple of employees and create facilities in which to conduct radiation experiments. Gradually, as he got more contracts, he was able to expand. In the early 1970s, Little won a new Air Force contract to look at radiation effects on solar cells. That gave his researchers insights into what was wrong with existing solar cells, and it got the company involved in solar-cell fabrication, which led to a third contract to make more efficient cells. Spurred by the energy crisis, the company—still only three or four people—translated its work on space solar cells to terrestrial cells in the mid-1970s.

Though he had virtually stumbled on the solar-cell phase of his business, creating an enduring business entity to service that industry was not an easy matter. "Achieving it is very complicated, very difficult, very stressful," said Little. "It's one thing to have people give you a list of things to do; it's another to sit in a blank space in a blank volume and create your own work plans and activity right out of nothing. I can recall many days when the phone wouldn't even ring, and it was up to me to sit there and essentially make snowballs out of snowflakes."

During the 1970s, Little came up with a lot of good ideas for terrestrial solar cells, which resulted in some technical papers, patents, and a substantial amount of support from the U.S. Department of Energy and Jet Propulsion Laboratory. As attracted as he and his Spire colleagues were to the solar field, the fact that nine or so large oil companies were active in it in the late 1970s and early 1980s deterred him.

After a great deal of that thinking, Little and his colleagues decided to utilize Spire's strengths in equipment engineering and focus on providing equipment to the industry rather than to compete as a module manufacturer. "The PV graveyard is full of companies," said Little. Spire survived by choosing the equipment niche and getting very good at filling it. Despite this prudent course, the contraction of the solar industry caused by the Reagan administration's antisolar policy shift limited Spire's domestic photovoltaic equipment business and contributed to its decision to diversify into the related fields of optoelectronics and "biomaterials" (specialty biomedical coatings and products).

The job of building Spire has always been "a battle for survival," said Little, "and it continues to be so." As a triathlete, Little views running Spire as an endurance test: Spire is now entering its hundredth quarter. Both triathlons and corporate leadership require the same kind of hard work. When Little gets to his desk at Spire at 8:30 A.M., he has already risen at 5:30, run ten miles, and spent half an hour on the Stairmaster. His equipment is working hard, too, at plants around the world, producing reliable products that are helping the PV industry earn its excellent reputation for clean, dependable power.

12

SOLAR SKINS

In the 1970s, lots of people jumped on the solar bandwagon hoping to make a small fortune. Well, they were right: After 20 years,
my fortune is still small.
              —Steven J. Strong, President,
                Solar Design Associates, Inc.

Solar architect Steven J. Strong is as dedicated to photovoltaics as
Roger Little, Ishaq Shahryar, and Bill Yerkes, but Strong also has the vision of a builder and social engineer. Unsurpassed as a solar designer and
pioneer at integrating PV in building structures, Strong also has training
as an electrical and mechanical engineer, and is the founder-president
of Solar Design Associates, Inc. (SDA). Above all, Strong is driven to
demonstrate that solar technologies are safe and reliable. "They work!"
he says. "They are not pipedream." If you accept that, he adds, "then the
concept of a society based on renewable energy can't be idly dismissed as
fantasy."

AN ALL-ELECTRIC TREE HOUSE

Whereas most architects are not inventors but clever integrators of existing components and systems, Strong is willing to take new ideas from
conception through painstakingly slow development to commercial use.
SDA in the late 1970s conceived and later built integral solar roofs, for
example, using glass solar roof modules that become the building's finished weathering skin and displace conventional roofing and structure.
These roofs have since become a well-accepted innovation worldwide in

the advanced solar architect's tool kit. To build its own innovative designs, SDA has had to acquire the highest-level general contractors' license.

The fact that SDA would not just design but actually build its new systems and stand behind their work helped allay customers' fears about the new technology. SDA was obsessive about engineering integrity. "Unlike the Wright brothers, we couldn't afford any crashes of the early airplanes," said Strong. "We've never had a leaky roof. We've never had a failed system."

In addition to applying renewable energy technology to residential and commercial buildings, Strong has advised senators, governors, and presidential candidates on solar policy and has published over 100 technical papers and articles on solar energy. Yet Strong remains modest and softspoken with a wry sense of humor and a passion for his work so intense it allows him little time for sleep.

His imaginative synthesis of architecture, electronics, and solar engineering had its origins in his early childhood. As a small boy, the future architect loved to watch his artist mother create beauty from a blank canvas, and he enjoyed exploring the construction projects on which his electrician father worked. By climbing around on the scaffolding, he learned how buildings were put together, and at age seven, he built himself the neighborhood's only electrified tree house. Much later, he went to engineering school, and then secured an engineering consulting job on the Alaskan pipeline.

"I realized," said Strong, "that going to the end of the Earth to extract the last drop of fossil fuels was not the best way to deliver comfort and convenience to the customer. The customer does not want to buy oil or kilowatt-hours. The customer wants warm showers, cold drinks, and a comfortable place to live. There are much better ways to deliver [them] than mining finite resources, such as oil on the North Slope."

Strong was drawn to what he calls "the elegance" of renewable energy. "The idea of creating something which does not deplete finite resources, but instead is self-sufficient, was very compelling from the earliest time. . . . When I first learned of photovoltaics as a method of generating electricity from sunlight, it was like falling in love."

"IF I KNEW THEN WHAT I KNOW NOW . . ."

With that passionate conviction and some youthful naïveté, Strong bought a drafting table after returning from Alaska and started Solar

Design Associates in an unused bedroom of his mother's house. For the next couple of years, he was SDA's sole employee, subsidizing its development with temporary technical jobs in engineering and building design while keeping overhead to a minimum.

"Every day was a crisis of one kind or another. If I knew then what I know now about business and the odds of a small business making it, I probably never would have done it," Strong says. "I'd probably be working for some architectural or engineering firm and have a comfortable, ho-hum nine-to-five job."

Instead, Strong works 60 to 70 hours a week. "It's a hell of a hard way to make an easy living," he quips. "I'm becoming a millionaire though. I've got all the zeros. I'm just looking for the 1 in front of them." During the early days, he had some "lean periods" when he didn't take a paycheck for months, especially during the Reagan years. "I'm very bitter about what Reagan did to the solar industry," he says. "I saw a lot of my friends lose their business, go bankrupt, lose their homes. Very, very well-credentialed scientists and engineers that were on staff at SERI [the Solar Energy Research Institute] were terminated with half a day's notice, with no severance pay, and in such a forceful, disrespectful manner as to deliberately and permanently sour any desire to be involved in this industry again. It was a scorched-earth policy. It was not just a lack of enthusiasm for renewables. It was [a] deliberate, vindictive, and orchestrated campaign to snuff it out in as forceful and as vehement a manner as they could." The Reagan years were a serious setback for Strong, and the Bush administration years weren't much better. "That was a low point in my career. It was a dark decade." Strong regards just staying in business during that period as success.

HARVESTING SOLAR HEAT

When Strong started SDA in 1974, PV technology was still extremely expensive and was used mainly for satellites, so at first he focused only on designing residences that were energy-independent in their thermal energy requirements. As a labor of love, he helped a homesteading couple in central Maine build a self-sufficient, solar-heated home. It had no conventional heating system but stayed comfortable despite the harsh Maine winter climate, demonstrating that solar thermal issues "could be easily mastered." The home won Strong an American Institute of Architects design award and gradually attracted clients.

After building a few solar-heated homes, Strong started to search for ways to generate electricity as well as heat on-site. He wanted to enable people to achieve energy autonomy and sever all ties to conventional fuels. He therefore started to look for opportunities to construct PV buildings. Granite Place, an 11-story midrise apartment complex, was planned near Boston in 1978. Strong persuaded the owners to let him design a large solar thermal system plus a 5-kilowatt PV system. "Five kilowatts is nothing today," he said. "You do that on a typical house." But in 1978, no PV system of that size had ever been built outside of a government laboratory. And the panels were to be highly visible on a prominent building.

GRID-CONNECTED PV

For the first time in the world, a privately funded PV system was to be interconnected with a utility grid. "It was a very significant opportunity," said Strong, "and we charged right ahead." But in the excitement, he forgot to talk with the local utility about the plan to feed surplus electricity backward through their distribution network. "Just when we were about finished," Strong said, "someone in my office pointed out that, 'Gee, we should be talking with the utility about this, shouldn't we?'" Soon after, Strong was on the roof in the equipment penthouse, where the solar power converter was, with Peter O'Connell, the building's co-owner.

"He was asking me if the system was complete," Strong recalls, "and I said, 'Yes, the electricians have just finished the wiring this week. We're essentially ready to turn it on, but we, really, I think, have to speak with the utility, because there are issues which have never been defined.'

"He didn't think about it for a minute. He said, 'Is this the switch to turn it on?' I said, 'Yes,' and he threw it!" That was it—no discussion with the utility whatever. The system was on-line.

"I like to joke that the sun dimmed just a little bit. In reality, a cloud went by, but outdoors it did get just noticeably darker." In any case, the installation passed what Strong called "the smoke test"—nothing burst into flames. O'Connell said, "Steven, you're going to learn in this world that it's far easier to ask for forgiveness than for permission."

Because the PV installation was so visible, Strong still felt that SDA had to deal with the utility, so he asked its interconnection engineers what provisions were needed to connect SDA's system. The utility responded with a thick binder dealing with megawatt-scale engine generators. It had nothing to do with small-scale photovoltaics. "It was totally

The 270-unit Granite Place apartment complex in Quincy, Massachusetts, designed by Solar Design Associates of Harvard, Massachusetts. The rooftop solar energy system features 7,000 square feet of solar thermal collectors that provide domestic hot water. The four photovoltaic arrays in the center foreground generate electricity directly from the sun for the building and are interconnected with the utility grid. *Courtesy of Solar Design Associates.*

intimidating, and we didn't know what to do about it," said Strong, so he brought the problem back to O'Connell. "Ignore it—it's bullshit," the developer replied.

"The utility's going to find out," Strong said.

"Don't worry about it," O'Connell answered. "I'll handle it."

Shortly after, in September 1980, O'Connell and his brother invited President Jimmy Carter to come to a major public ceremony to celebrate the solar building's grand opening. Once the president accepted, the O'Connells invited the governor, the state energy secretary, the mayor, assorted dignitaries, and, of course, the utility executives.

A rostrum draped with bunting was set up outside the building, but President Carter changed his plans at the last minute. Denis Hayes, then head of the Solar Energy Research Institute, agreed to attend instead and gave a speech in which he praised the utility for allowing the locally generated PV power to be put into their grid. Then the state energy secretary rose to his feet and underscored those points. When it was the utility executives' turn, they had little choice but to praise the project, too. Interconnection suddenly was a dead issue.

Besides being the first PV project of its kind, Granite Place was also the first time that a PV system was linked to a solar thermal system. When the sun shined brightly and energy was needed to operate pumps to circulate the fluid in the solar thermal system, the PV system provided the energy. "The brighter the sun," said Strong, "the more power the pumps would get, the more fluid would flow. It created a proportional control. When the sun went away, the system shut off." Eventually this clever system became a standard design for residential solar water heaters.

THE FIRST FULL-SIZE PV HOME

Solar thermal work had been exciting and compelling for Strong, but had been relatively easy technically. Strong had now tasted the thrill of using solar electricity in a small way on a large building. Designing and building structures that generated their own electricity was now the next logical challenge for him. The DOE and its lead PV systems laboratory had decided to solicit designs in 1979 for building-integrated PV arrays. DOE's interests meshed nicely with Solar Design's. Strong won a DOE competition to design a completely integrated PV roof.

About this time, DOE also had decided to support construction of what would be the world's first full-size PV residence and announced a formal competitive solicitation in 1979. SDA won the contract, designed the house, and oversaw its construction in Carlisle, Massachusetts. The Carlisle House, as it was called, featured passive solar heating and cooling, super-insulation, a roof-integrated solar thermal system, and a 7.5-kilowatt solar cell array. It had no fossil fuel furnace; it exported surplus power to Boston Edison (the local utility); and it was the first single-family residence ever powered by a "utility-interactive" PV system.

"It was the biggest thing [in photovoltaics] at that time," Strong recalls. The Carlisle House was also beautiful. The south-facing roof, set beneath clerestory windows, was made of deep blue polycrystalline solar cells that contrasted with the warm tones of the home's wood siding and its floor-to-ceiling glass doors and windows. The exterior was suburban yet distinguished. The interior was spacious, light, and airy. *Popular Science* ran a picture of the house on its cover, and the design received laudatory worldwide publicity. More importantly, Carlisle was a physical realization of Strong's longtime dream to power a house with the sunlight that fell on its roof.

A MAJOR PV MILESTONE

A new and interesting opportunity was soon at hand: a commission to build a 3,000-square-foot, "stand-alone," all-solar electric house. The location was New York's Hudson River Valley, almost the worst imaginable place for an independent solar building, because the weather can be cloudy for five months a year. That meant the PV system would have to be larger and more expensive than in a more suitable climate. Although Strong explained this to his client, a retired U.S. Navy admiral, the admiral wanted to be near other family members in the area.

According to Strong, the admiral was a "dyed-in-the-wool, cold-steel, New England Yankee . . . the last person you'd ever expect to want an all-solar house." He had been in shipboard command for most of his professional career and was accustomed to autonomy. "He wanted his house to be autonomous," said Strong. "I believe he saw the house as his last command." This was 1982, and the project was to be the biggest all-solar home anywhere. "It was doubly difficult," said Strong, "because it was under the purview of the New York Board of Fire Underwriters," the most rigorous code authority in the country. The local electrical inspector had stopped in mid-sentence and looked at Strong when he learned that the house wasn't going to have any electrical connection to the grid and that all the electricity was going to be made by solar cells. He gazed at the roof, and after what seemed an eternity, asked, "Does your mother know what you're doing?" Shortly, he turned the entire inspection over to the state underwriting board. "We had a long wall covered with all kinds of electronic power conditioning equipment and switch gear and disconnects and power conditioners," Strong said. "Basically, we had to create this stuff and have it custom-built in 1982."

The state electrical inspectors came from Albany a couple of weeks later. "Dull gray men in dull gray suits in dull gray sedans with black wall tires, clipboards, and briefcases and all sorts of measuring equipment, cameras, and stuff," said Strong. They stayed for an entire day of inspection, poking and prodding everything. They asked "every question you could ever imagine and plenty that you couldn't imagine. . . . Nothing was UL-labeled. Nothing was code-compliant. The National Electrical Code didn't even address photovoltaics then. They had a thousand reasons to say, 'We can't approve this.' "

"At the end of the day," said Strong, "a gentleman by the name of Coombs took out his pad, signed off, and stuck an approval ticket on the switch gear. Needless to say, the preferred libation was liberally used that evening." By getting code approval from an exacting inspection agency, SDA had helped legitimize the use of all-solar technology. Solar electricity was no longer relegated to uninspected backwoods cabins; it was becoming mainstream.

A WIN-WIN PROJECT

Strong's local utility, Boston Edison, had seen all the positive publicity that SDA's Carlisle House had created. Strong then proposed to build them a PV demonstration house that would be 20 to 25 years ahead of its time. As an added incentive, he proposed that the project be the subject of the popular national PBS home-renovation television series "This Old House." Boston Edison at the time badly needed some positive publicity because of embarrassing management problems that had led to the shutdown of its Pilgrim I nuclear power plant. SDA got complete freedom of design and a generous budget to build the "Impact 2000" house.

"It was a dream commission," Strong said. "We created a house that had a solar roof integrated with photovoltaics for electricity production, solar thermal for domestic hot water, and daylighting in the sun space." The house and the PBS series about it brought the idea of energy efficiency, solar design, photovoltaics, earth-coupled geothermal heat pumps, passive solar gain, and interior mass to an audience of over nine million people in a step-by-step process that showed not only why these concepts were desirable, but how they were implemented. The series was renamed, "The All-New, This Old House." Sponsors Boston Edison and Owens Corning, the insulation manufacturer, received favorable notice. After the television series was completed, 30,000 people toured the home. Boston Edison then used it as the basis for an educational program, and finally sold the house at a profit. In addition, the utility launched a companywide effort called the Impact 2000 Program, using the house as a highly visible symbol, to promote energy efficiency and energy conservation. The entire project was a colossal success—a "made-in-heaven" public relations program for Boston Edison.

Curiously, despite the enormous broadcast exposure and a staggering flood of requests for free information, SDA did not receive any new commissions directly from viewers, although the firm's credibility was greatly enhanced. The Impact 2000 house did lead to a major new commission for SDA, however. New England Electric Company, whose subsidiary serves much of Massachusetts, saw the significant corporate benefits that

Boston Edison received by identifying itself with photovoltaics. Thus "in the grand old American spirit of one-upsmanship," said Strong, New England Electric wanted to create an entire photovoltaically powered neighborhood, not just a single home.

Along with lots of praise, Boston Edison had received some criticism that the Impact 2000 house, sited in Brookline, Massachusetts, among many multimillion dollar homes, was out of ordinary people's financial reach. The fact that the home was a custom-designed experimental prototype did not make it any less expensive. New England Electric wanted to dissociate itself from that elitist image. Strong was also eager to retrofit existing buildings, since his goal was to create a solar society, and he knew he couldn't do it just by building expensive new solar homes, or by waiting for new buildings to displace all the old ones. Thus he was eager to see what could be done with a typical blue-collar American subdivision.

The $1.25 million project converted 30 existing single-family ranch-style homes to PV in a typical working-class suburban neighborhood in Gardiner, Massachusetts. The city hall, library, community college, and even the town's Burger King also received PV systems. Because solar cells were very expensive in the 1980s, the systems were designed to provide only 30–50 percent of each home's power. With solar cells in 1996 costing only a third what they cost in 1980 and with new state-of-the-art home appliances that use far less energy than those of the 1980s, a project like the one in Gardiner could easily provide 100 percent of an ordinary home's electricity needs on an annual basis. But given the low fossil fuel prices of 1996, even these less costly and more powerful systems would still be more expensive for most existing homes than reliance on utility power, unless the homes were located beyond the utility company's lines, or needed a great deal of high cost, peak-hour power.

AN INVERTER FOR EVERY PANEL

In addition to the cost of its PV panels, the cost of a utility-interconnected PV system includes installation, power-conditioning equipment (including a DC-to-AC inverter), plus overhead and profit for the tradesman and installer. These "balance-of-system costs" (everything but the PV module) are expensive and add 100 percent to the cost of the module itself. Strong is now working to develop advanced electronic technology that will lower that cost, simplify PV utilization, and possibly make SDA a lot of money.

The "Impact 2000 House" built by Solar Design Associates in Brookline, Massachusetts, generates its heat and electricity directly from the sun without burning fossil fuels. Photovoltaic panels integrated into the south roof produce electricity, while the panels in the roof's center gather thermal energy to heat the home and its domestic water. The house includes passive solar heating and cooling, super-insulation, low-emissivity windows, and an earth-coupled geothermal heat pump, among other energy-saving features. The "Impact 2000 House" was the subject of a 26-week, nationwide PBS television series. *Courtesy of Solar Design Associates.*

Thanks to a cost-shared, cooperative research-and-development agreement with DOE, Solar Design is currently developing a revolutionary AC photovoltaic panel equipped with its own inverter. Household current and appliances are normally AC, and since solar cells produce DC, solar panel output has to be converted to AC. Either an inverter must be installed in the circuit, or DC wiring and appliances must be used. An AC panel equipped with its own miniature inverter would solve all these problems. In addition, since electricians and code officials are not accustomed to DC, and DC electrical components cost more, using AC reduces hassles and costs. Hughes Electronics, a subsidiary of General Motors, reported not long ago to their management that the AC module is the most significant development they had seen in the entire PV industry. Solarex, too, understands the potential of this new innovation. SDA formed a partnership in 1994 with Solarex (now a division of Amoco/Enron Solar), to commercialize this product. "I'm absolutely convinced that you'll see AC modules within a year or two," said Strong. "If we can't do this here in the U.S., the Japanese and the Europeans will." Since that comment, Strong's AC panels have been chosen for the roof of a new, 325-foot-long light-rail station canopy to be built by 1997 adjacent to Baltimore/Washington International Airport.

If SDA's little engineering team succeeds in making the inverter small and efficient enough so it can operate in the closed environment of a junction box and manages to bring it to market quickly enough, then, says Strong, "we've got a tiger by the tail. . . . We could do very well." The AC module is a giant step toward eliminating a host of problems and probably cutting the balance-of-systems cost in half. It will also increase system reliability and resilience. Previously, if the central power conditioner failed, the whole system was inoperable.

Whereas early inverters were the size of a small refrigerator, thanks to printed circuitry and miniaturization, inverters may be reduced to the size of a cassette tape. "It takes a long time for people to understand, and some folks still will never get it," said Strong, "big is not better. Small is beautiful. Many small pieces create an infinite level of resiliency if you take it to its extreme, and they're also much easier to make. You can mass produce something on a printed circuit card like this in a production line. Eventually it'll be on one or two application-specific integrated power chips that have logic and power switching right on an integrated circuit. That technology is available now. But it costs you half a million dollars to do a custom chip set, and nobody, regardless of whether they have the resources, would invest [in] it because the market isn't there yet."

"It's a chicken-and-egg thing, which in large part summarizes the whole photovoltaics industry. There's a huge [market] potential here, but until costs come down, you won't realize the potential, and you can't get the costs down until you have the volume that a larger market would create. You can't create the larger market until you have the costs down."

The development of the AC module and creation of a large-area module are closely interrelated. When Strong started collaborating with Solarex, their largest module was roughly three by four feet, but Strong needed one that's four by six feet to optimize the AC module. "You can make a smaller inverter, but [proportionately] the smaller you make it, the investment you have to make in the brains, the logic driver, becomes much higher . . . because the same chip can drive a 20-watt inverter or a

250-watt or even a 1,000-watt inverter." Solarex agreed to build the 24-square-foot module Strong sought.

Strong believes that the ultimate solar cell device for buildings will be a combined photovoltaic and thermal collector. "I was certain of it in 1978 when we applied to MIT for research funding and support, and we produced working prototype modules. Fifteen years ago, we showed that this concept was sound and would work, but we couldn't find an industrial partner to commercialize it." SDA is considering possible partners today, but still needs to convince the DOE of the technology's merit, so that SDA can get some DOE cost-sharing money. Meanwhile, a Swiss company known as Atlantis Energy already offers a photovoltaic roof shingle system that simultaneously generates power and utilizes the by-product heat for space heating.

Strong is sad that the United States appears to have forsaken its world leadership role in photovoltaics "and by default handed it to the Europeans and the Japanese. The Japanese have had a consistency of vision and a coincident public policy support to allow their photovoltaics program to move forward in incremental planned steps over a multiyear period with consistent funding."

Strong feels that the Japanese and Europeans are both significantly ahead of the United States in PV in a number of respects. "Not so much in technology, not just manufacturing infrastructure, but in public policy." Abroad it is common for utility customers with solar cell panels to participate in "net metering," which allows customers to receive credit on a one-for-one basis for each kilowatt of power they generate. The United States has no national net metering policy. We also lack uniform national interconnection guidelines for tying solar generating systems into local utility grids.

"That's a national policy in Japan, and it's a national policy in Switzerland and becoming so in Germany. The U.S. is hopelessly far away from a consensus on that," said Strong. "The federal government has no focus on a leadership role [nor determination] to make this happen." When a person wants to build even a simple, single-family-home solar electric system in the United States, "you have to educate the utility at your own expense. It's very painful, time-consuming, and very expensive." Those institutional barriers are enough to stop some people from building solar dwellings. Strong was part of a small group of solar activists that got net metering passed by the California legislature and now is planning a state-by-state nationwide campaign.

Steven J. Strong, President, Solar Design Associates, Inc. SDA was the 1993 winner of the Connecticut College Inherit the Earth Award honoring decisions on behalf of the environment that sustain jobs, productivity, and profits. *Courtesy of Solar Design Associates.*

As Strong looks to the future, he has high hopes for building-integrated photovoltaics, however. He believes that the technology has proven its worth across the nation and that home, commercial, and power company PV applications will become common in the twenty-first century. Thanks to many of innovations he and others have developed, we will increasingly be living and working in buildings that cleanly, safely, and unobtrusively generate more and more of their own power.

PART III

WHIRLING POWER

13

※

WIND: A CRASH COURSE

Access to capital, as everyone in the industry knows, is the secret
of success.
                —Robert Lynette, Chief Executive Officer,
                Advanced Wind Turbines, Inc.

The United States possesses an energy bonanza, yet no dictator,
sheik, or oil mogul controls its wells and spigots—or charges by the
barrel or ton for wind. America has hundreds of billions of dollars'
worth of this free, nondepletable resource. We are a Saudi Arabia of
wind.

Just with today's technology, the 48 adjoining states have enough
wind energy to produce 4.4 trillion kilowatt-hours of electricity a year—
more than one and a half times the nation's 1990 electric power generation.[1] Advanced wind technologies soon to be commercially available
will more than double the "lower 48's" capacity. These resource estimates exclude environmentally sensitive sites and the huge potential
contributions of Alaska, the Aleutian arc, Hawaii, and offshore areas.
Plus, we have economical transmission technologies that can deliver
wind power from windy regions, where it is most cost-effective, to the
rest of the nation.[2] Yet wind power currently provides less than a tenth
of 1 percent of the United States' electric power. Why, and what could
be done about it?

Before answering, let's first put wind power in its historical context and
explain how contemporary wind systems work. Then I will return to the
institutional, technological, and economic obstacles facing wind power
and how to resolve them.

EXTRATERRESTRIAL ENERGY

To many people's surprise, wind power is actually of extraterrestrial origin. Radiant energy beamed from the sun through outer space heats the Earth's surface and the atmosphere unequally. The Earth is warmest at the equator. The hot air, however, doesn't just accumulate and hover patiently there. Warmed air expands, becomes less dense, rises, and is then displaced by denser, cooler air that flows in to take its place.

Eventually, the warmed rising air cools again as it travels and becomes dense enough again to sink back to Earth. On a global scale, these mass movements create atmospheric circulation and distribute energy from the tropics toward the poles. The Earth's rotation—which produces the cooling and warming cycles of night and day and the Coriolis deflection (the bending of a moving object's straight-line trajectory)—stirs the swirling atmospheric gases.[3] Land, water, rock, soil, and plants also vary in their rates of energy absorption, reflectance, and re-radiation of solar heat back into the atmosphere. A complicated dance of matter and energy ensues. In that dance, wind is born.

Yet wind is more than the differential heating of Earth's atmosphere and more than a vehicle for transporting pollutants. Wind carries the pollen of grasses and trees; it represents fertility and the creative spirit. It has mythic significance, too. In Homer's ancient Greek epic *The Odyssey*, Lord Aeolus, God of the Wind, gave all the winds but one to Odysseus in a goatskin flask. That one wind that Aeolus left free was the one needed to propel Odysseus's ship back across the Aegean Sea to Ithaca. But Odysseus's friends opened the flask and allowed the winds to escape. Modern society, too, has allowed its winds to escape and has not taken the easy journey back to Ithaca.

A VENERABLE HISTORY

Over millennia, people have harnessed the wind in various ways. Sail power carried Egyptians along the Nile 5,000 years before Christ. Who knows when people first captured wind in bundles of reeds to nudge rude craft along shores and rivers. A powerful shove from a long ago gale may have inspired an unknown ancestor of ours to invent that first crude sail.

The Babylonian emperor Hamurabi is said to have made plans to use windmills for irrigation as early as the seventeenth century B.C. Hero of Alexandria, who lived within the first two centuries before Christ, described a small and remarkable sail-driven windmill used to inflate a pipe organ.[4] Windmills reportedly were used in Persia as early as 200 B.C., and by the seventh century A.D. windmill building was an established craft there, using bundled reeds that turned in a horizontal plane around a vertical pole. Ghengis Khan, the Mongol emperor, was so impressed by the skills of the Persian windmill builders that he had them sent to China as prisoners to teach windmill construction for irrigation. Very likely these windmills were patterned after early waterwheels or primitive devices powered by draft animals. These toiling brutes, plodding in an endless circle and yoked to a pole, turned millstones or raised water.

By the twelfth century, more sophisticated horizontal-axis windmills were found in France and England. Crusaders, after marauding in the East, may have hastened the spread of windmills as the troops drifted back home across Europe—an early example of "technology transfer." In the centuries that followed, windmills were built by the thousands in Europe to pump water, grind grain, and to pulverize chalk, lime, oil seeds, and snuff. The Dutch used them with an Archimedean screw and other devices to drain their flooded lowlands. In the United States, windmills pumped water for railroads, livestock, and crops and to produce salt from seawater. Wind power also ran sawmills in small towns.

The terms *wind turbine* and *windmill* are not synonymous. A wind turbine is a wind-driven rotary electric generator. A windmill is a wind-powered mill used for grinding, but the term is often also used to describe a wind-driven water-pumping apparatus.

By 1890, the Danes had built the first wind-operated electric generators—true wind turbines—and had several hundred operating by 1908. Small wind turbines were also used to produce electricity throughout the rural United States in the early twentieth century, charging batteries and powering early radios and simple appliances. More than six million small mechanical windmills and electric wind turbines were used on the Great Plains between 1850 and 1970.[5]

Those electric turbines were the soul of simplicity compared to wind turbines today. Wringing kilowatts out of wind now depends on computer science, aerodynamics, electrical and mechanical engineering, and materials science. Doing it cheaply and reliably has tested the mettle of the modern world's best engineers. Often a dose of junk bonds, stock manipulation, advertising hype, high finance, big government, international trade, and multinational competition is added before electricity is delivered. All this is a far cry from the bundled reeds of Babylonia.

Before wind power can be produced and sold nowadays, windy sites need to be identified and assessed; then, land leases are negotiated, power purchase and transmission agreements are concluded, and land-use permits are obtained. Turbines must be designed, manufactured, and installed, and power collection equipment, substations, and interconnection facilities must be built, tested, and maintained. At times, the business seems so complicated and challenging that it's astonishing anyone in this field can prosper, especially given vacillating government energy and tax policies and changing Public Utilities Commission regulations.

These complexities do not explain why wind provides so little of our electricity today, especially since even complex engineering systems will be readily adopted, provided they are profitable. A behind-the-scenes look reveals why wind power has taken off so slowly, and why the U.S. wind industry remains in struggle. By contrast, the international wind industry is booming; world-installed wind capacity grew 35 percent to 5,000 megawatts in 1995 and is expected to more than triple within a decade. Worldwide wind industry sales totaled $1.5 billion in 1995.

WIND ON A ROLLER COASTER

Following the energy crises of the early and mid-1970s, the late 1970s ushered in a period of millennial excitement about wind: a golden age of renewables seemed finally to have arrived. All too soon, however, a profound setback occurred. The federal tax credits that had jump-started the industry vanished prematurely. Most of the budding American wind industry promptly died. The surviving firms were gradually recovering by the late 1980s and early 1990s as best they could despite competition from low fossil-fuel prices when yet another crisis occurred: the restructuring and impending deregulation of the U.S. utility industry. In between these difficult critical periods, and even in some cases during them, wind industry pioneers have shown confidence in their technological ability to overcome all obstacles to cost-effective wind electricity.

Kenetech Windpower, Inc. (USW), is one such pioneering company. The firm pulled itself up by its bootstraps during the 1970s to become the world's largest wind turbine manufacturer in the late 1980s and early 1990s. Though not typical of the wind industry, USW in some ways epitomizes the industry's gyrating fortunes, the technological hurdles, managerial challenges, and financial obstacles that wind (and other) renewables must surmount to fulfill their potential as major sources of electrical power. The company's roller-coaster history is also a tale of technological and entrepreneurial derring-do and of a firm with a seemingly bright future that overreached itself and gambled heavily on its future prospects once too often.

After telling much of the industry's story through the saga of USW, I will summarize the activities of other leading wind energy companies, such as SeaWest Energy Corporation, FloWind Corporation, Zond Corporation, New World Power Corporation, and Cannon Energy Corporation, in addition to Enercon GmbH and other major European wind turbine companies.

THE FOUNDING FATHERS OF KENETECH

USW was founded as U.S. Windpower in 1976 by industrialists Stanley Charren and Russell Wolfe.[6] Both had engineering backgrounds and had been business partners for a long time. Their interest in wind power had been sparked (or fanned) by their acquaintance with mechanical engineering professor William Heronemus of the University of Massachusetts.

Professor Heronemus is in a sense the grandfather, or *éminence grise*, of the modern U.S. wind turbine industry. Back in 1972, the scientific establishment was promulgating the myth that no more than 1,000 megawatts of wind power—the equivalent of one large conventional energy plant—could be produced in the United States by the year 2000 and that the entire U.S. wind resource was limited to only 10,000 megawatts.[7] But Professor Heronemus correctly pointed out that the United States had a vast wind resource: "One need only go up in the sky," he said, "to find wind power of astronomical quantity."[8] He talked of the huge wind potential of the Great Plains; he proposed novel wind turbine configurations; and he described how wind power could be used to store energy in the form of hydrogen that would later produce electricity in fuel cells or from direct combustion.[9] (Both uses can power vehicles and would be entirely pollution-free, water vapor being the only effluent.)

Heronemus knew that winds offshore were much stronger than over land, and that power could be sent by cable to shore, or used on-site to manufacture hydrogen by separating it electrically from water. Today, he still stands by his calculations showing that offshore winds are sufficiently powerful to more than compensate for the costs of offshore siting and power transmission. He also still adheres firmly to the maverick view that wind could easily provide 100 percent of the nation's electricity, though of course it would be much more sensible to rely instead on a diversified mix of renewable energy sources, taking advantage of regional variations in renewable energy quality and costs.

✦ ✦ ✦

Soon after meeting Heronemus, Russell Wolfe and Stanley Charren decided to start a wind energy company. Wolfe was idealistic and wanted to provide simple, inexpensive, easy-to-build turbines to Third World countries. For technical staff, they turned to Heronemus's doctoral students Forrest S. "Woody" Stoddard, a Massachusetts Institute of Technology graduate with a helicopter-engineering degree, and Theodore S. Van Duzen, another engineer from MIT, who also ran a company that manufactured fiberglass racing boats. At the time Charren and Wolfe showed up, Heronemus and his students had been working on a wind turbine called the Wind Furnace that produced electricity as well as heat for a student-built solar home.

Wind energy in the early 1970s was not an established engineering science, and much had been forgotten from the early days of experimentation with large wind electric turbines in the 1940s. The Wind Furnace's design, therefore, was based on research Stoddard did on old British, Canadian, and Third World turbine designs. He built the Wind Furnace's blades by hand with design help from Van Duzen. Stoddard then made some basic design decisions—would the rotor be upwind or downwind of the tower, would there be three blades or two, and so on. His basic design, the turbine "architecture," set the pattern for USW's next 8,000 turbines.

In the new enterprise, Charren and Wolfe became general partners, while Stoddard and Van Duzen became limited partners. This arrangement gave Charren and Wolfe control of the company. Stoddard became the chief engineer, Van Duzen, the chief blade designer. Work began about 1977. Van Duzen built the fiberglass blades for the prototype machines. The young engineers erected and tested the machines at a small plant in the Boston area. The fundamental challenge that the new company faced was to produce a durable, cost-effective turbine. It would have to survive the stresses inflicted by sudden gusts of wind, turbulence, wind shear, and storms. It would also have to endure the baking summer sun, freezing blizzards, and the mechanical wear caused by spinning under load for thousands of hours a year.

Wolfe and Charren soon found that the new company required far more capital than they themselves could readily provide, so they turned to other entrepreneurs. About this time, Professor Heronemus introduced San Francisco businessman Alvin Duskin to engineer Woody Stoddard. Through several interesting quirks of fate, Duskin changed the course of U.S. wind energy development. In an idyllic Marin County, California, hot tub, he helped hatch a potent scheme to supercharge the nascent American wind industry.

COMMERCIALIZING WIND: A HOT IDEA

Alvin Duskin is a bold, decisive man with a tough scrappy air, and he speaks in a blunt, pithy way. He made some money as a successful garment manufacturer but sold his business and, through a mutual friend, was given a post in Washington, D.C., as a junior aide for energy legislation to Senator James Abourezk. Duskin soon started sitting in on meetings of the Senate Energy Committee. "[Abourezk] never went [to the Senate Energy Committee meetings], so he essentially voted the way I advised him to vote," said Duskin.

"Within a month of arriving in Washington," Duskin recalls, "I essentially had a vote on the Senate Energy Committee," which controlled an $11 billion DOE (Department of Energy) budget. It was an interesting turn of events for one who not long before had been an out-of-work Californian.

After advising Abourezk for a while, Duskin was soon ready for a new challenge, and an old friend suggested that he would find it interesting to write a bill and lobby it through Congress. At that point, some people at the U.S. Bureau of Reclamation called Abourezk's office. They were interested in using wind electricity to pump water and showed Duskin data indicating that they could generate more power with wind turbines along some of their High Plains coal power transmission lines than the coal plants themselves were generating. They also suggested that Duskin get in touch with Heronemus, who, they said, knew all about megawatt-scale wind power in the High Plains states.

Primed with information from Heronemus, Duskin attended a briefing by DOE for senior congressional staff about various energy technologies and how to commercialize them. "The DOE plan [for wind] was incredibly stupid," said Duskin. "They were going to put two windmills up in every state and attract popular attention to it. I said, 'This is real dumb. There are a lot of states in which there's no wind. You could put two windmills in Georgia and prove to everybody that it doesn't work. Why are you doing that? Let's put windmills where its windy—not where it's not windy.' They said, 'That's not politically realistic,' so I decided that I would write a commercialization plan for wind energy."

Written as a bill, the plan authorized the federal government to buy windmills and put them up at windy U.S. Coast Guard stations and Air Force bases to establish an industry. It passed Congress and was signed by President Jimmy Carter, but no money was ever appropriated for it.

Even before the bill was signed, Duskin says he had realized that it would not ensure the development of a cost-effective, commercial wind industry, even if the government spent millions buying turbines. Moreover "the [federal government] had botched everything to date. They had these big [research-and-development] programs with Boeing and Hamilton Standard, McDonnell Douglas, and Lockheed that didn't amount to anything. . . . I decided it wasn't really the right way to go. But I didn't know what to do."

Duskin returned to California to discuss the problem with a friend—a financial expert who worked for a company called Itel that was heading into bankruptcy. While they were soaking in the friend's Mill Valley hot tub, they came upon the idea of using tax credits to stimulate the wind industry.

A few years earlier, American farmers had harvested a bumper wheat crop so large that there weren't enough railroad cars to move it to ports to meet worldwide demand. The railroads, however, were too broke to finance the cars themselves. The problem had been solved by having the federal government provide 15 percent investment tax credits to people who bought covered-hopper railroad cars and leased them to railroads. The legislation had sent the railroads a flood of capital.

The two friends spent a day discussing how limited partnerships could be structured to take advantage of such tax benefits for commercializing wind energy. The idea eventually emerged as amendments to the Crude Oil Windfall Profits Tax Act of 1979. They provided an energy tax credit of 15 percent for both wind and solar energy investments, modeled on the existing California solar energy tax credits. A general investment tax credit of 10 percent was already in place, and in California—which became the Shangri-la of wind energy development—an additional 25 percent in state tax credits was available.

The federal and state credits combined meant that the buyer of a $100,000 wind turbine in California could put up $20,000, for example,

WIND: A CRASH COURSE ✦ **145**

as a down payment and sign a note for the balance. The tax credits would, in effect, defray the cost of the investment. Yet the investor also owned a wind turbine and had rights to the income from its electricity. If the wind turbine worked, he or she could use the revenue generated by its power sales to pay off the balance of the note. It was, indeed, too good to be true. Wind turbine technology was not yet a mature technology, but hopes of big tax write-offs were about to create a mania for turbines.

Duskin concluded that if the proposed tax credits passed, the wind industry would be launched on a flood tide of money. "At that point, I decided to get out of the Senate and look for a job, so I went to see Boeing and Lockheed and McDonnell Douglas . . . and Hamilton Standard. I visited them all at taxpayers' money. I traveled around and said, 'There's going to be this gigantic industry going. It's not going to be government, it's going to be private. There's going to be thousands of windmills out there.' "

At the time, Duskin did not know about the imminent passage of the Public Utilities Regulatory Policy Act (PURPA), which later worked synergistically with the tax credits. The tax credits enticed investors to provide capital to build wind farms comprised of tax-advantaged wind turbines. Then PURPA created a market for wind power, by obliging public utilities to purchase electric power from independent producers at the public utility's "avoided generation cost"—the marginal cost the utility would have to pay to generate or purchase an equivalent amount of power from another source.[10]

When Duskin told U.S. Windpower's Stanley Charren his thoughts, Charren was enthused by Duskin's wind industry forecasts and immediately invited him to a meeting in Boston. Duskin came and predicted an enormous market for wind power. "We've got to set up manufacturing on automotive [assembly line] scale to meet the demand," he said. Wolfe and Charren offered Duskin a job at USW.

"At the time," Duskin admits with disarming candor, "we thought that windmills would work. We didn't realize that they would all fall apart."[11] So they proceeded, knowing that if the technology worked, any smart investor was going to invest in windmills.

"I went back to Washington, went onto half-time, so I could work for Windpower," said Duskin. "My job was to push the legislation through, which we did, and eventually got the bill passed." The three entrepreneurs then decided to put USW into high gear by raising the money to build their first "wind farm" at Crotched Mountain, New Hampshire, where they planned to farm kilowatts and sell the power to the local utility.

Duskin spent the next four or five months fund-raising but had virtually no success. He then realized that he needed a superb fund-raiser and that USW needed a sizable windmill order to make itself credible. The man USW turned to was Norman Moore, who had commercialized the microwave oven at Litton Industries. Moore had obtained a Ph.D. in physics from MIT in 1941, gained microwave experience in the U.S. Navy, and had gone to work for Litton Industries in 1948, when it consisted of 57 people in a Quonset hut and was just about breaking even. In but two years, Moore was appointed president. Starting in 1963, Moore built Litton's microwave business from scratch into a half-billion-dollar-a-year division. By 1967, his Litton stock had appreciated hundreds of times. He left the company a wealthy man, to become an independent entrepreneur.

By the time Moore joined USW, the company had raised only about half a million dollars from the principals and their immediate associates. The money was largely spent. "If I remember," said Moore, "they were building a 28-foot-diameter blade machine. . . . I don't think it had ever run five minutes without something happening." One observer called that first-generation machine "a bucket of bolts on a pole." Had it worked, it would have generated only a few kilowatts.

Moore was undaunted by the fact that Wolfe and Charren did not yet have a sound, proven technology. "I liked the guys," he said, and their plan made sense to him. "It looked like fun. Starting things from scratch is what's fun." If they had had a proven technology, said Moore, "there would have been no need for me."

Wolfe and Charren hoped to raise another $500,000 with Moore's help. Moore told them that they'd never be able to produce a working machine with so little money. He also predicted that thereafter, "it'll be a hell of a lot tougher to raise the next monies, because we will have taken in two 500s and still not have a machine. This is at least a $3 million job."

Stoddard, the Heronemus graduate student who became USW's chief engineer, says that a decision to raise that money by selling stock was made behind his back. Also, to his chagrin, new managers were brought in: Norman Moore hired Dr. Herb Weiss from MIT's Lincoln Laboratory to take charge of engineering. Eager to work on solutions to the "energy crisis," Weiss went from a prestigious post and a $35 million budget to sixth man on a tiny, cash-starved company's staff.

Referring to his initial $3 million target figure, Moore remarked, "I had picked a number out of the air. I was there a few weeks and said, 'No, this is at least a $5 million job.' Turned out ultimately to be a lot more than that, but at least $5 million. So we set out to raise $5.4 million in the summer of '79, so after the accountants and lawyers got through with us, we'd still have $5 million."

To support his fund-raising, Moore produced a 200-page "redbook" describing the company and its plans. "Today," said Moore, "Gerry Alderson [Kenetech's longtime president] hopes that my 200-page redbook never sees the light of day again, because there were so many exaggerations in it."

"I still remember Stan [Charren] and Russ [Wolfe] and myself and Alvin [Duskin], maybe, sitting in some Holiday Inn somewhere, swilling coffee and guessing how much towers were going to cost, blades were going to cost, transmission lines were going to cost." Their estimates, Moore recalls, were "wildly optimistic."

"We first went to the venture capitalists. As I joked later, they all turned us down because they didn't know me—or because they did. Finally, I went to people that I had known in my business career and said, 'This is important.' "

Those people turned out to be folks like Bob Noyce, founder of Intel, and Howard Vollum, founder of Tektronics. They invested a total of $3 million, generally in units of at least $180,000 each. Actress Jane Fonda and MIT physicist Henry Kendall also were early investors. Eventually the Prince family in Chicago—through their own funds and those of the First National Bank of Chicago, in which they were the largest investors—brought the total to $5.4 million. One observer attributed the company's success in raising development money solely to Norman Moore's "silver tongue."

Moore was never able to sit back on his laurels after bringing in that first mother lode. Toward the end of his presidency at USW, after the initial $5.4 million was raised, Moore would still wake up around three in the morning at least three days a week and would usually not fall asleep again. "You were living on the edge for a long time," he recalls, "but that was fun."

While Moore was raising the USW development money, the company was operating from an unprepossessing plant in Burlington, Massachusetts. A dusty, unheated, garagelike space on the premises was used for fabrication. Wind machines were tested on a couple of towers about a hundred yards away.

Once the early major investors were committed, the founders decided to hold a board meeting. The company had not yet run its prototype machine for more than a minute or two. USW's test engineer was Louis Manfredi, another graduate of Professor Heronemus's renewable energy research program. Manfredi had intended on getting experience with the machine the day before the board meeting, but there had not been enough wind. That night, he had a nightmare. In his dream, a blade broke off the new turbine and sailed straight up in the air. As it came down, it hit and broke the second blade. That blade then smashed into the third blade, which also collapsed.

Once the board members and principal investors had arrived for the meeting, Moore gave the order to start the machine. Some of the engineers said, "Ah, hey, Norm, we haven't tested it yet." Moore replied, "The directors are here. It's their money. Let's see where we are."

Manfredi was in the control shed behind the plant, and the board members, at Moore's invitation, had all adjourned to the yard behind the plant to savor a moment of technological triumph made possible through their farsighted, courageous investment. All stood patiently waiting. Manfredi turned on the turbine. Gradually it came up to speed.

Then, suddenly, a blade snapped off at the rotor hub and shot straight up in the air. As it descended, it hit one of the two remaining blades. Within milliseconds, all three 15-foot-long fiberglass blades crashed into the ground with a soft crunching noise: *Fhro-fhro-fhruf.* Exactly as in Manfredi's dream.

Moore turned white. Skip Smith, the venture capital man from First National Bank of Chicago, put his arm around him. "We've all been in the venture capital business," he said. "You'll survive."

"None of us who were there will ever forget the sound," Moore said later.

Once the investors and board left, Moore called his troops together for a locker-room pep talk. The company, he said wryly, had merely been testing its automated deblading system.

"I had a hard time sleeping for a week or so after that," Manfredi remembers. He was just beginning to appreciate what a difficult thing getting commercial power from the wind would be. More importantly, this event showcased the haste and pressure under which the engineering was done at USW. Woody Stoddard had specifically told Charren and Wolfe that the turbines were not ready to be shown to the investors. His warnings were not heeded, and a proper balance between financial and engineering imperatives was never struck. In a sense, USW was born and baptized amid the crashing of turbine blades.

14

THE NICKEL MACHINE

Economics will win. If wind is to play a significant role, it must be
economically competitive.
>   —Dale W. Osborn, former president of
>   Kenetech Windpower, Inc.

Big dogs eat first.—Folklore

The next board meeting of the tiny start-up wind turbine company was
held a few weeks later in a bilious-green room on the second floor of the
company's plant. Through one window, you could see the prototype wind
turbine behind the building. This time, test engineer Louis Manfredi had
been running the machine before the meeting. As the board attended to
its business, you could hear a comfortable *swish, swish, swish* as the
blades went past the turbine tower. Everything seemed to be going fine.

Chief Engineer Herb Weiss was confidently telling the board about
the progress the company had made when, *CROOOMPF, FRUUUI.* A
blade separated from the rotor and flew into the ground in a graceful arc.
Then, with the now-familiar *fhro-fhro-fhruf*, the other two blades quickly
broke apart at the hub and sailed off the 75-foot tower like petals from a
daisy. Machine and the men who created it were now still. Silence filled
the room for a moment. Then, scarcely altering his expression or steady
cadence, Chief Engineer Weiss went on, "obviously, we still have some
problems to solve."

Outside, alone in the control room behind the plant, Manfredi was
aghast. For several minutes he tried desperately to get through by phone to

the company's new president, Gerald Alderson, to make him aware of the disaster. A large man who can be formidable when aroused, Alderson had come to Kenetech Windpower, Inc. (then U.S. Windpower), from Itel, a bankrupt leasing company, where he had been the company's most powerful financial officer. An examiner's report that looked into Itel's affairs after its bankruptcy accused him of violating Securities and Exchange Commission rules and suggested Itel take legal action against him and other officials. In response to Manfredi's plaintive calls, the USW operator now kept insisting that President Alderson was in a board meeting and could not be disturbed. At last, a rattled Manfredi got Alderson's secretary on the line.

"If it's about the turbine," she said, "he already knows."

"Oh good!" Manfredi exclaimed, almost dazed.

"What do you mean, 'Oh, good'?"

"Oh, good," Manfredi repeated woodenly, ". . . then I don't have to tell him."

Despite Herb Weiss's board room bravado, everyone in the company was deeply disheartened by the second turbine debacle. The failed turbines were 30-kilowatt prototypes of machines that were to be sold to investors in utility-scale power plants. The product now lay in a pile of rubble.

Woody Stoddard and Ted Van Duzen had been primarily responsible for the engineering and construction of the USW turbines to this point. Along with Russell Wolfe, they had been grooming the 30-kilowatt model for what was to become the world's first utility-scale wind farm on a ridge overgrown with blueberries at Crotched Mountain in Greenfield, New Hampshire. The turbines were to be connected to the power grid owned by the Public Service Company of New Hampshire.

The pressure the young company was under to make good on investors' money soon produced internal stress and finally a rupture. Chief Engineer Weiss regarded Stoddard's turbine engineering design as unnecessarily complex and redid it. "We can't build these," Stoddard recalls objecting. "They're going to break, they're going to be dangerous, and they're going to bilk these investors. The investors are being led down the garden path—this is fraud." He ordered their construction stopped. Management overrode his objections, and Stoddard was fired along with Van Duzen.

The wind farm at Crotched Mountain cost $1.2 million, provided by about 25 investors. The governor of New Hampshire and other dignitaries came for the grand opening ceremonies. Trouble soon followed.

"If there was a problem," said former USW President Norman Moore, "we had it. . . . Not being able to handle snow and ice . . . fatigue, blades breaking, joints breaking, everything." A friend of Moore named Andy

Anderson put his arm around Moore at one point and said, "We now know more [about] how not to build a wind farm than anyone else in the world!" During that period, said Moore, "There were many a night that I'm sure none of us—if we'd wake up at two in the morning—would go to sleep again. . . . But that doesn't mean we despaired. I don't think any of us ever thought of failure."

Time thus brought Stoddard a vindication. The Crotched Mountain experiment consisted of 20 turbines. "The 20 machines did fail," Stoddard said. "They were dangerous, and they all broke." The project was also an economic failure. The steel towers and turbines were removed and laid to rest on the edge of a parking lot in Burlington, where they were stored for years. For tax reasons, they couldn't be discarded. With the help of a bulldozer, the site was returned to something like its original condition: even the blueberries were replanted.

After Crotched Mountain, no one at USW really had any proof that the company could make a profit from wind-generated electricity. But USW's managers were determined to continue developing the technology. And that would take money to solve technological problems and to market turbines. As if the company now had found a guardian angel, the next major chunk of money for USW came almost unbidden. Angus Duffy and Skip Smith, who had put the Prince family money into USW, told Norman Moore, "You might make it through the keyhole with the money you've got, but would you take more money if we were able to raise it, no fees, just by bringing others in?"

"Twist my arm," Moore replied. Duffy and Smith then proceeded to bring in another $13 million. "Getting that money was crucial," Moore recalls.

After a year and a half with USW, Moore had realized that a stronger chief financial officer was needed for the company—an executive able to mobilize not just a few million dollars, but the hundreds of millions, or perhaps the billions, that eventually would be needed to fulfill the leadership's vision for USW. At this point, Moore yielded his post to Alderson, who subsequently became the principal architect of USW's later financial triumphs and tragedy.

THE ADOPTION OF KENETECH
AND THE SELLING OF THE 56-50

Following the Crotched Mountain struggles, the company added to its engineering staff, continued its research and development, and produced a larger turbine of 50 kilowatts with a 56-foot-diameter rotor, a more advanced version of the prototype Stoddard had condemned. USW dubbed this product the 56-50 machine. In a further confirmation of Stoddard's judgment, even these machines never worked properly for more than a few weeks at a time. Norman Moore today concedes that its early maintenance problems were severe. Nonetheless, it attracted a great deal of investment capital.

About 1981, the brokerage house of Merrill Lynch, Pierce, Fenner and Smith had asked mechanical engineer and wind energy consultant Robert Lynette to assess the technologies of the dozen or so wind turbine companies then constituting the industry and to recommend a company. Lynette, bright and enterprising, had spent 21 years in the aerospace industry, most of it working for Boeing. At the time Merrill Lynch contacted him, he was running a wind energy consulting firm and was rapidly gaining recognition as a national expert on wind turbine design, operation, and debugging. Merrill Lynch told Lynette the firm was going to sell a lot of wind-power-plant limited partnerships over the next four years—until the tax credits Duskin had engineered expired at the end of 1985. "The company who manufactures the windmills [for the partnerships] is going to be our child, if you will—we're going to adopt them—and we're going to make them the largest wind turbine company in the world," the Merrill Lynch financier reportedly told Lynette.

Even though Lynette did not believe USW had the best technology, he regarded it as fixable and was impressed by the company's management capabilities and financial resources. Therefore, he recommended that Merrill Lynch talk with USW. The other firm he recommended was ESI of Boulder, Colorado. "These guys are poorly managed; they're terribly underfunded," he told Merrill Lynch. "My guess is they won't get through, left alone. They're run by some young engineers who have no concept of business. But they have a product that's potentially ten years ahead of its time."

Merrill Lynch adopted USW and did exactly what they said they would, Lynette reported. "They sold thousands of [USW] turbines. And USW grew because of that. . . . ESI went out of business in 1985 from lack of capital, but I never forgot the machine." Indeed he did not. For later, Lynette found a way to salvage ESI's design.

With the tax credits that Duskin promoted and with Merrill Lynch hustling the capital for limited partnerships, USW was soon mass producing the trouble-prone 56-50 turbines for limited partnerships.

Could anyone have predicted then how badly the 56-50s would perform? Had Merrill Lynch's investment analysts done their "due diligence" homework properly in the early 1980s and when, together with Smith Barney, they took the company public in 1993?[1] These questions may never be satisfactorily answered. More than 700 of the 56-50s were eventually installed at investor expense, almost all in California at Altamont Pass. Although they were a major improvement over the 30-kilowatt models, the 56-50s were a technological disappointment, as the company itself was later to acknowledge.

"Most of them failed and ended up as piles of junk," said Stoddard. "[USW] would come out at night, and they would load these junk piles up in dump trucks and take them out before people could get an idea of how many turbines broke." To resolve the technological difficulties, USW hired MIT engineer Jamie Chapman in 1982.

A PROBLEM OF CREDIBILITY—
AND AN INTERESTING SOLUTION

Under Chapman's supervision, the company produced a successor to the 56-50 called the 56-100. It coupled the same basic architecture and blade diameter of the 56-50 to a 100-kilowatt generator—twice as powerful as the 56-50s. Even more importantly, the 56-100 had far fewer operational problems. Woody Stoddard bluntly summed up Chapman's contribution: "He should go down in history as the one who developed the 56-100 and saved the company's ass." As time would tell, however, the performance of this machine would also leave much to be desired.

During the development of the 56-50s, USW needed a real order. As of 1980, the company had yet to produce a single reliable commercial wind turbine model. To solve the credibility problem with investors and raise capital, Alvin Duskin called California Secretary of Resources Huey Johnson on USW's behalf. Johnson was now in charge of California's Department of Water Resources, the state's fourth largest utility. Duskin had known Johnson for years.

"We need a contract from you to build windmills," Duskin said to Johnson. "Otherwise we can't get this company going."

The people in the USW home office had just recently started to negotiate an agreement with a New Bedford utility for a 2-megawatt wind farm in Massachusetts—at best a demonstration plant. By contrast, Duskin wanted to sell a big wind farm.

"I went to Sacramento, sat down in Huey's office [and] said, 'I want a contract for 100 megawatts, and I want to sell this power to the Department of Water Resources for 5 cents a kilowatt-hour.'" According to Duskin, Johnson said, "That's terrific, that's just what we want to do. Anything you want to do, Alvin, you write it up, I'll sign it."

"We thought windmills were going to be very easy to make in those days," Duskin recalls. "Very inexpensive and very easy to make. [But] after three years of breaking them, everything changed."

As Duskin remembers, "Everybody [at U.S. Windpower] was staggered by the idea of a 100-megawatt wind farm." That would have meant a sale of a thousand turbines. The day the contract was publicly announced, and with the state legislature's public support for wind energy development, "there was a wave of resignations in the Department of Energy—people left in order to start wind companies. . . . It was amazing what that launched. Everybody started rushing into the wind business."

The state water resources contract was never consummated: USW could not provide the power profitably at the contract price. But the contract gave USW visibility and credibility and helped launch the U.S. wind energy industry. Wind companies, it seemed, were now no longer just building isolated small turbines. They were providing commercial wind power plants and were poised to enter the large and lucrative public utility market.

TAKING A STAND AT THE ALTAMONT PASS

At the time of the contract, USW had leased 5,600 acres of land in Pacheco Pass, California, and thought that was going to be its major site for limited partnerships. The next step was to verify the wind data to make sure the site would have enough wind to provide electricity at an economical price.

"Trees in Pacheco would go up like this and over," said Duskin. "Horizontal. We thought it was the windiest place in the world. We didn't know about Altamont."

USW subsequently hired a technician to actually measure the wind at Pacheco Pass with anemometers. "After about a month I met him there," said Duskin. "He looked real bad. He said, 'Alvin, the wind in Pacheco is about 40 percent less than these guys from Oregon State measured it. I don't know what they were smokin', but this is not the greatest wind site around.'" He then made a chance remark: "The wind here isn't as good as Site 300 in Altamont." Duskin subsequently finagled wind resource data about Altamont from the State Energy Commission before its formal release and, on the basis of that data, pressured U.S. Windpower to lease large portions of the Altamont site as quickly as possible.

The company subsequently leased about 32 square miles of ranch land in Altamont near Livermore, California. A few little farm ponds occupy low clefts in hills that are tawny in summer and green with pastures in the rainy season. In the heavy, low-hanging fogs of winter, it is easy to lose your way high on the winding, poorly signed dirt roads among ridges bristling with forests of turbines. One of the oak-studded grassy domes on Altamont's Walker Ranch is called Brushy Peak, in whose caves bandit Joaquin Murietta is said to have hidden the gold and money he robbed in the 1800s from stagecoaches, trains, and banks. The caves also contain ancient Native American petroglyphs.

Cattle and wildlife contentedly graze beneath Altamont's Erector-set-like turbine towers, and farmers grow wheat and oats here. The roads, tower footings, and a small substation actually occupy only a small percentage of the total land area. Golden and bald eagles, kestrels, redtails, and valley hawks patrol the skies.

USW initially installed the faulty 56-50s at Altamont but by 1983 began installing the more reliable but still-problematic 56-100s. Perversely, as the number of turbines at Altamont increased and as they remained in service longer between breakdowns, impacts on the birds of Altamont mounted.

Unfortunately, 165 raptors (birds of prey), including 27 golden eagles, were found dead at Altamont wind farms in 1993 after colliding with turbines.[2] The California Energy Commission estimates that 300 raptors a year are killed at Altamont, which is on a bird migration route and hosts a thriving small mammal prey base. (Whereas not all of the deaths occurred on USW's leaseholds, the tower design of USW's outmoded 56-100 is attractive to perching birds and may be a primary factor leading to bird deaths.) Although the bird strike-per-turbine ratio at Altamont is a little less than three strikes per year per 100 turbines, Altamont has more than 7,000 turbines.[3] USW has spent $2 million on research to solve the bird-kill problem, and has convened a scientific task force on it headed by an internationally recognized raptor expert. The company has experimented with blades painted black and white to see if they will reduce bird kills.[4] Fortunately, not all wind farms have experienced this problem.

Bird kills are not a problem unique to wind power systems. Birds of prey are maimed and killed by encounters with transmission and distribution lines that carry power from coal, oil, hydro, and nuclear power plants.[5] Perhaps because of the more serious environmental impacts of these technologies, their impacts on birds are rarely given a great deal of attention, as are the impacts of wind plants in certain locations. Tens of millions of birds are also killed every year by automobiles, hunting, buildings with plate glass windows, other tall structures, and by pollution, poisoning, and domestic cats.[6]

Thanks to the all-weather roads and microwave links that USW has built and maintains, ranchers at Altamont have better access to their property, as well as substantial land-lease and royalty arrangements compared to agricultural resources. Under contracts made in the 1980s, most wind companies paid ranchers 5 percent royalties or more for the electricity generated on their lands. Exact data about these royalties are difficult to obtain, but Randy Tinkerman, developer of the first utility scale, institutionally financed wind project in 1985, says USW provides landowners with several percent less than any other wind company. USW royalties are generally in the range of only 2 percent.

"The word around the Altamont Pass is that U.S. Windpower deals rip off the landowner," says Tinkerman. "USW Altamont landowners subsidize the high maintenance costs of USW's 56-100 turbines," Tinkerman contends. The company regarded them as a commercial success, however, because they produced about half a billion dollars in sales.

WIND WITHOUT RESPITE

USW's manufacturing plant and one of its business offices are only a short drive away from Altamont in a Livermore, California, industrial park. Plant manager Paul Smith is a tanned, crew-cut young man with a deep voice, a powerful grip, and a commanding presence. His short-sleeved knit shirt proclaims, "U.S. Windpower 56-100 Team." The numbers "98.6" on it denote the plant's near-perfect turbine "availability." In utility jargon, plant availability is simply the percentage of the year that the plant is operating or ready to do so. The 98.6 availability is impressive—far higher than that of most conventional power plants.

Power production at Altamont varies greatly from season to season. Most is generated in the summer, when the rising hot air in California's Central Valley creates a low-pressure region that sucks ocean-cooled air from the coast through Altamont's mountain passes. Sometimes the summer winds will blow at 35 m.p.h. for two or three weeks without respite.

At low wind speeds, the turbines make a gentle swishing sound. At high speeds, they produce a higher frequency sound that is virtually unnoticeable. "What you hear," said Smith, "is the howl of the wind itself."

Maintaining wind turbines is physically challenging work at tower heights varying from 60 to 140 feet. "Some of the fasteners [on the turbines] are torqued to 1,200 foot-pounds. . . . Just working in the wind is very physically demanding," manager Smith explains. "It beats on you all day. It never, ever lets up. Your eyes get affected. . . . There's times when I've come out here, and the wind is 70 miles an hour. You can literally lean your body into the wind, and it will suspend you." Because of its aerodynamic design, a heavy blade suspended in a sling on a boom can raise two or three large men off the ground while they try to grapple it down for maintenance.

Power produced by the turbines is sent to on-site substations equipped with capacitor banks that look like giant spark plugs. Switches here shunt electrical loads from one transformer and feeder line to another. This is also a dangerous place to work, since sparks, arcs, or electrical flashes could occur unexpectedly.

TURBINES THAT TALK BACK:
MULTIPATH, TWO-WAY COMMUNICATION

Wind power as we know it today would be impossible without computers and solid-state microprocessors. Some computer chips are located behind the blades and rotor; electrical boxes called tower control units at the base of each turbine tower contain solid-state relay switches, capacitors, fuses, and most importantly, a computer brain that communicates not only with its turbine-generator but also with the entire plant's control system.

The tower control unit processes signals from the anemometer on wind speed, signals from the turbine on turbine speed and torque, and pressure readings from the hydraulic system. The computer crunches all these numbers constantly during operation to issue instantaneous control orders. If the wind speed changes, the computer will signal the hydraulic system to feather or pitch the blades. If the wind direction changes, the computer will instruct the yaw motor to orient the turbine into or out of the wind. If output power does not meet voltage or synchronization criteria, corrections are made.

Circuitry in the turbine is tested automatically; if the computer detects a significant malfunction, it will shut the system down automatically and signal the company's central control room. Thanks to multipath, two-way communication, the status of any turbine anywhere in the field can be read instantly from a computerized control panel at the company's office.

Wind technology is progressing very rapidly these days. USW was able to eliminate 100,000 electronic components on 60–70 circuit boards in its tower control unit between the first and second generation of its latest turbine. It did so by using four advanced circuit boards that relied more heavily on integrated circuit technology and microprocessors.

THE MANUFACTURING CHALLENGE
THAT RAN AMOK

The USW turbines at Livermore are put together from major components, which are bought already assembled from over 100 suppliers. Workers then install operating software and test everything using unique software that simulates field conditions. A working wind turbine and power converter emerge at the end of the assembly line.

The company's production management and procurement office is decorated with photos of USW's new 33-meter variable speed machine (33M-VS), which USW began marketing in 1993 (and is now known as the KVS-33). The 33M-VS—which took four years and $70 million to develop—was both a financial gamble for the company and its greatest technological hope. Whereas conventional wind turbines like the 56-100 hold their speed constant and endure higher loads to their drive trains with increasing wind speeds, the beauty of a variable speed machine is that it can operate at a range of wind speeds while still delivering constant torque and power. Thus the fluctuating loads are more easily withstood by blades, drive train, and other parts delivering power to the generator. Since the variable speed machine encounters lower loads than its predecessors, its drive train can be made lighter and less expensively.

Variable speed machines are also able to deliver power over a wider range of wind speeds. The 33M-VS cuts in at a lower speed and continues to generate power at a higher one than previous USW machines. The 33M-VS also produces "clean power," that is, in-phase, relatively distortion-free power, which is highly compatible with utility system power. The advantages of variable speed, however, are to a significant extent counterbalanced by the increased cost of the control system and the "power conditioning" equipment—the electronic hardware and software—required to permit variable speed operation. Moreover, in an environment with relatively stable wind speeds, the variable speed capability is of little or no value. It is still too early to know for certain how the 33M-VS will perform over its entire operating lifetime, nor what its operation and maintenance costs ultimately will be, but there are reasons to suspect that the machine is seriously flawed, which would obviously have a disastrous impact on future demand. Troubling reports from the field have already begun arriving. In one very embarrassing recent incident, the company installed 90 of the 33M-VS machines at a wind project in Tarifa, Spain, only to have three of the giant blades fall off two machines during operation. All 90 machines then were shut down for safety reasons. This horrible black eye followed reports of cracked blades and damaged generators on USW turbines at wind projects in Buffalo Ridge, Minnesota, and Palm Springs, California.

These results are even more embarrassing, given that the design and development of the 33M-VS had technical and financial support from a consortium comprised of USW, the Electric Power Research Institute, Niagara Mohawk Power Corporation, and Pacific Gas & Electric Company. USW had ballyhooed the turbine during its early deployment saying it expected the 33M-VS to compete successfully against both clean coal and natural gas—and with all other utility generating alternatives. The company claimed the machine was superior to all other turbines and would produce power at 5 cents per kilowatt-hour—even before any tax credit or other incentives—while the previous generation of turbines produced power at 8–10 cents per kilowatt-hour. (Power from new natural gas plants in 1995 cost roughly 2 to 2.5 cents per kilowatt-hour.)[7]

THE KVS-33 ILLUSION

Optimism about the low cost of power from the 33M-VS had led USW managers in 1993 to expect a huge growth in turbine sales from their Livermore assembly plant, where they planned to produce 1,000 33M-VS turbines or more a year—at a time the company was building just two machines a week there. Large orders were secured quickly, and by the end of 1993, the company announced $600 million in firm orders[8] but thought that the flood of orders would only grow.[9] At one time, the firm projected selling nearly 3,200 megawatts of 33M-VS capacity by 1998—enough to produce residential electricity for more than two million people—and the company's director of research even thought that USW would ultimately sell 4,000 megawatts before having to introduce a more advanced turbine.

Kenetech's most modern commercial turbine, the 33-meter variable speed machine operating at Altamont Pass near Livermore, California. *Courtesy of Cynthia Cheak.*

Had the KVS-33 performed better and had all those orders materialized—and had USW been able to keep up with them—they might have provided the company more than $3 billion in revenue instead of hundreds of millions in losses. The company so misjudged its future, however, that it sold and installed only a few hundred megawatts of its KVS-33 before it felt compelled to propose a newer and larger turbine.

15

❖

THE CULTURE
OF SURVIVORS

The U.S. wind industry has always been plagued by a good
deal of hype.
        —Paul Gipe, principal,
        Gipe and Associates

By the early 1990s, Kenetech Windpower, Inc. (USW), was the nation's largest wind power producer and seller of utility-scale wind power plants. In addition to its 4,400 California turbines, USW had also created wind power projects in Europe, Inner Mongolia, New Zealand, Spain, Taiwan, and Ukraine. While the company continued its high-profile wind power development program in 1994 and 1995, its emergence as the dominant force in the U.S. wind industry remained controversial, and the company's self-promotional style irked some competitors. More significantly, its marketing claims for its much-vaunted variable speed turbine were challenged, and its corporate reputation was bludgeoned during 1995, as we shall see.

THE WIND POWER OPTIMISTS

During the company's halcyon days, Kenetech's public posture was to project a rosy future that obscured the tremendous risks the company and, to a lesser extent other wind companies, were facing. Glenn Ikemoto, for example, Kenetech's head of finance and marketing for European operations, was bullish on the company and the wind industry in 1993. He asserted then that 100,000 megawatts of new electrical generating capacity will be needed in the United States in the next decade and that another 100,000 megawatts will be needed in Europe. Were Kenetech to get even a small percentage of such sales, the company would earn billions. Ikemoto clearly expected that to happen. "If I could sell you a car for the same price as any other car," he said, "yet this car produced no pollution and, other than that, it was exactly the same, which car would you buy? That's the argument. . . . We are going to be the guy with the clean car that can sell it at the same price." The logic seemed impeccable, but Ikemoto evidently was unprepared for the wild cards that fate would soon deal Kenetech and the wind industry. He was not alone in his optimism.

When Kenetech's former production manager, John Lewelling, was asked in 1993 whether he feared that competitors might produce bigger, more powerful turbines, he said no, because they were already available. He was convinced that the KVS-33 was the world's most advanced wind turbine and would produce power for substantially less than everyone else's product. (What Lewelling said he *would* fear was a bigger, more powerful, *and* less expensive turbine.) Although the company did not comment publicly at the time, it reportedly was already quietly planning to build those larger machines, which it had begun to introduce by 1996.

PRIDE GOETH BEFORE THE FALL

The understandable pride that Kenetech managers like Ikemoto took in the company's accomplishments and "technologically superior" turbine at times led them to depict life at Kenetech in mythic proportions. When Ikemoto started work at USW in 1983, fresh from Stanford Business School, he managed internal financial planning by day and by night donned work clothes, took up his wrench, and assembled wind turbines. "Our basic jobs probably were good for 60 hours, and our night jobs were probably another 40," he asserted.

"We launched [USW] from zero in an industry that is full of great big players. . . . Boeing, Westinghouse, Hamilton Standard, Mitsubishi, and $500 million of government R&D, and we launched out of a garage, with a couple of visionaries, and we ground [the competitors] into the dirt. We ran past all of them," he said, ignoring the accomplishments of half a dozen international wind companies.

Ikemoto claimed the main reason for Kenetech's early success was its commitment to its mission. "We never lost track that our game is large, utility-scale, grid-connected economic, clean power. . . . We entered a market with companies as competitors that spent more capital in a day than we had for two years. In the land of these giants, [with] big footfalls crashing down all around you, that breeds this sort of pirate's mentality. You know, 'I'm going to get in there, going to raid the cupboard. I'm going to be creative, going to move faster. . . .' That's the culture of a survivor. . . . If you don't have that culture, you don't survive as a small capitalization company in [this] industry."

Although the "culture of the survivor" at first appeared to make Kenetech efficient and give it an initial edge over its competitors, it did not create a corporate "nice guy." "We fire people very quickly," Ikemoto said when Kenetech was in its heyday in 1993. "The company thinks about the people who are staying, not the people who are going. . . . We recognize that the company will evolve faster than people, so that there are people to serve at different points in time. When you don't have any resources to spare, which we don't, you know, dealing in the land of the giants, you cannot think about the people that have to be let go. . . .

"USW pushes the bounds of wind technology," Ikemoto said, "it pushes the bounds of financial technology; it pushes the bounds of management technology. It's a very innovative environment." To characterize the company's growth, Ikemoto used the image of a hockey stick with the base on the floor and the handle pointed to the right. "Are we worried about financing this company's growth trajectory as it turns the corner of the hockey stick?" he asked confidently. "No, we're not worried." Those sentiments were shared in 1993 by Kenetech's then president, Dale Osborn who showed no hint of concern about the company's finances or risk-taking. He simply regarded Kenetech as the most successful wind energy company in the world and believed the KVS-33 to be "the best wind turbine in the world."

"Many other participants in this industry view us as arrogant, aloof, unwilling to cooperate, uncompromising of views, and that's right," he said, to a visitor's surprise. "Every one of those is a fair description. . . . The arrogant part was troubling to me, in the early years. . . . But it doesn't trouble me at all now that somebody perceives our confidence . . . that we know what we're doing, we got the thing that works, we do what we say. If that's arrogance, somebody can call us arrogant. My belief is, it's supreme confidence.

"Do we have our warts?" Osborn asked. "Sure we do. Do we have problems? Sure we do. Do we tell them to the public? Never."

Kenetech was publicly dismissive of its wind industry competition. Competitors knew they could not compete with Kenetech's technology, Osborn maintained. But if Kenetech is successful, he said, there would be leftover pieces of the market for them. "Big dogs eat first," said Osborn, "and in this business, we're the big dog right now."

SELLING ABROAD

Because the U.S. wind turbine market became relatively stagnant, Kenetech actively sought international sales for the KVS-33 and formed an exclusive joint-venture partnership with Wing-Merrill Group, Ltd., of Cambridge, Massachusetts, to market and build wind plants abroad.

Kenetech representatives are particularly proud of a major foreign licensing arrangement the company made in 1993 in Ukraine. Although Kenetech had stopped marketing the economically uncompetitive 56-100 in the United States, the company arranged to have 5,000 outmoded 56-100 turbines built under a license agreement in Ukraine, to partially replace electrical power from the damaged Chernobyl nuclear power plant. The turbine factory deal was set up as a joint venture between Kenetech and Krimenergo, the Ukrainian electric utility, financed mainly by the Ukrainian Ministry of Energy. Another partner, the actual builder, was a former intercontinental ballistic missile factory.[1] Ukrainians were to build and install the machines; Kenetech was to take its royalty payment in generators, blades, and other parts.

Did the Ukrainians know they were buying and building flawed 1980s technology in the mid-1990s? The Ukrainians hoped to complete a 500-megawatt wind farm and then export 56-100s to Russia, Kazakhstan, and eastern Europe. Kenetech called it helping Ukraine "develop an industrial base." Former USW President Norman Moore, who raised USW's first millions, was enthused about the Ukrainian deal. "Perfect marriage. They need the power in Russia. They can build good generators and stuff like that. A perfect, perfect use. I was prouder of Gerry [Alderson] for that; I thought it was wonderful. That's what we should be doing." For the United States, however, Moore said he prefers deploying the newer KVS-33. "We ought to have a million and a half of them up around the U.S," he said.

HEALTHY SKEPTICISM

Although Ikemoto acknowledged that there were other variable speed turbines on the market, he called the Kenetech turbine's accurate torque control and high power quality, "Very new stuff. Very leading edge." It was the lightest in the industry, he asserted. "Nobody else even comes close."

Long before troubles with the KVS-33 became obvious at Kenetech commercial projects, some experts were wary of Kenetech's claims for the turbine. Electrical engineer and former USW Vice President Jamie Chapman cautioned that the supremacy of variable speed technology is far from certain: "[The industry doesn't] know what the best architecture is for a wind turbine," said Chapman. "We don't know whether two blades are better than three, whether variable speed is better than fixed speed. . . . If there were an obvious answer, it would spread like wildfire."

Chapman and other wind energy experts believe that variable-speed technology only provides a 10 to 15 percent performance advantage. Turbine performance, however, is only one determinant of competitiveness. Turbine cost depends heavily on the cost of financing, as well as on operation and maintenance and other costs. They can easily overwhelm a marginal performance difference.

One prominent Kenetech competitor is Enercon GmbH, which introduced a 55-kilowatt variable speed turbine in Germany years ahead of Kenetech (in 1984) and began series production of its E-40 500-kilowatt turbine in 1993. Enercon's half-megawatt machines are variable speed but require no gears and hydraulic systems, two sources of trouble and maintenance demands. The E-40 has performed with a very high 27 percent capacity factor (a measure of plant reliability) and also delivers very high-quality power to utility systems. Since the power factor and voltage can be dynamically regulated, utilities can use the plant to improve grid stability.[2] Evidently concerned about the competition, Kenetech sued Enercon for alleged patent infringement. Enercon countersued, and until the cases are resolved, it is unclear whether Kenetech has an enforceable claim to a proprietary variable speed technology.

As early as 1993, Chapman had serious reservations that the KVS-33 would not be big enough, powerful enough, and economical enough to compete with European technology. Woody Stoddard, USW's first turbine engineer, agreed. Stoddard, now a wind energy consultant, contended that although Kenetech was working on a larger machine, it would be too expensive and too late to compete effectively with Danish wind turbines. Foreign turbines, primarily Danish, captured more than half of California's market in the 1980s, snatching sales from under Kenetech's nose. He thought that could happen again in the 1990s.

"The wind industry is not just a technological ball game," Chapman says. "Technology is a wild card. It can tilt the table for a while, but in two to three years, the market will assimilate it." Because most advanced wind turbine technologies on the market today are fairly close in performance, Chapman believes that a wind company's competitive position is likely to hinge on its organizational experience, its ability to provide access to low-cost financing, or its warranties, rather than on imagined or real—but marginal—technological superiority. At the time we spoke, Kenetech's unraveling had not yet begun, nor had the reliability problems of the KVS-33 surfaced.

A PREOCCUPATION WITH FINANCE

Building a new renewable energy company can be a tightrope walk. It involves weathering technological as well as financial risks. During early 1993, only about 20 or 30 of the KVS-33 machines were operating. Kenetech's financiers were therefore wagering large sums on the KVS-33's unproven performance, and the company in turn was staking much of its fortunes on the turbine. As one investment banker who followed the company observed, "These guys have got a lot of debt. Their success is based on their ability to develop the new turbine and produce power for 5 cents a kilowatt-hour. If they can't produce power for under 6 to 7 cents, they're in trouble."[3]

OTHER PEOPLE'S MONEY

Relative to its net worth, the company borrowed at double-digit interest rates in the late 1980s and early 1990s. By 1993, the company owed nearly $230 million on tangible net worth of only $68 million and lost $7 million in the first half of that year. But Kenetech management was eager to take the company public in 1993, despite mediocre earnings: Unlike notes, money raised in a stock issue does not have to be repaid and can be used to retire a company's debt. To avoid any unfavorable publicity that might have given the stock market jitters about the new shares, Kenetech strictly controlled media access to its employees. The company then floated a hefty $96 million initial public stock offering to repay its earlier heavy debts and to proceed with KVS-33 development and marketing.

Without the stock offering, Kenetech probably would have remained capital-starved. The gamble paid off in late 1993 when Kenetech closed its public offering for $16 a share. But the company was not yet home free. Kenetech common rose to a high of more than $29 a share in March 1994, then lost almost half its value by the end of August 1994, partially on rumors about a pullout of major institutional shareholders. Speculation about the company included concern about the firmness of its contractual commitments from utilities, the reliability of its turbine technology, and market demand for wind in the coming era of utility restructuring and deregulation.[4] The financial troubles intensified in 1995. By late December the company's stock had fallen to just a shade over a dollar a share, it had laid off a substantial part of its workforce, and had taken on $56 million in bank debt, just in the first seven months of the year. Officially the company attributed its difficulties to slow or nonpayment by important customers. The industry, however, buzzed with rumors that Kenetech had overextended itself financially and was having difficulty providing wind farms as inexpensively as promised.

Earlier in its development, the risks and uncertainties the company had faced from day one seemed not to inhibit management in setting its compensation. Historically, while they grew the company on note holders' money, the Kenetech board spent millions in scarce company funds on high salaries and internal stock repurchase deals for itself and some senior managers. By 1992, then-president Gerald Alderson held nearly 80,000 shares of Kenetech common, and the company's directors and executive officers had more than 200,000 shares—nearly 15 percent of the entire company at the time. In just one transaction in June 1992, the company repurchased 7,371 shares of its stock from Alderson for $737,100.

These maneuvers enable senior company officials to profit enormously from a company—whether it is making money or not—behind a legalistic bulwark called "nonrecourse debt." Kenetech Corporation, parent company of Kenetech Windpower, Inc., borrowed money and passed it through to its subsidiary, for whose debts it is "not responsible." "Since the Notes are obligations of KENETECH only, its subsidiaries are not obligated or required to pay any amounts due pursuant to the Notes . . . ," as a company *Senior Notes Prospectus* says.[5] Thus in the event of Kenetech's insolvency, its lenders may lose everything. Board members, however, could still have made a fortune by then, thanks to the generous compensation the Kenetech board accorded itself. President Alderson, for example, received cash compensation in 1992 of more than $600,000, including accrued bonuses and certain other nonsalary compensation. The practice of providing lavish executive salaries financed with the help of heavy borrowing appears somehow unseemly and incongruous in what not long ago was a lean and virtuous looking renewable energy start-up company.

Another dark cloud over Kenetech's reputation is a charge of fraud in connection with the now-annulled bids the company submitted in the California PUC's 1993 renewable energy contract auction. Under the PUC's procedures, wind energy companies had bid for rights to provide wind power to California utilities under a complex PUC formula. In motions filed with the PUC, attorneys for FloWind Corporation have charged Kenetech and two other wind power bidders (SeaWest Energy Corporation and Zond Corporation) with deliberately submitting false bid information, cheating, and with fraud in the auction by overstating how well their projects would perform.[6]

In response to these charges, but before the Federal Energy Regulatory Commission invalidated the whole auction process, the PUC itself suspended the auction results, including huge contracts that Kenetech appeared to have won to provide 945 megawatts of wind power to Pacific Gas and Electric, San Diego Gas & Electric, and Southern California Edison.[7]

If Kenetech's turbine technology had really been the hands-down winner in the cost and performance sweepstakes, as Glenn Ikemoto, Dale Osborn, John Lewelling, and other company officials consistently maintained, why would the company have resorted to submitting questionable data about its wind plant performance? The atmosphere of secrecy in which the firm at times operated and the company's behavior before the PUC also raised questions about the senior company officials who knew of, or approved, the company's bid data. Submission of anything as important to a company as multibillion dollar auction bids must have been reviewed, one presumes, at the very highest levels of the company.

THE COLLAPSE OF KENETECH

Although various wind power companies and experts have affirmed that wind power can be generated at 5 cents a kilowatt-hour, USW had great difficulty delivering on its promise. The company in 1995 had fallen far behind on its production schedule, lost $250 million, and had to lay off large numbers of employees. As its stock nose-dived from nearly $30 in March 1994 to only $1.18 in December 1995, the company dismissed its

CEO, Gerald Alderson, and hired a corporate "turnaround specialist" to take over the company and try to save it from possible imminent bankruptcy. The effort was unsuccessful, and Kenetech Windpower, Inc., filed for bankruptcy in the spring of 1996. The promising but complicated KVS-33 had become a huge liability that was costing USW tens of millions of dollars in warrantied repair costs.

As industry leader, Kenetech's disappointing performance cannot fail to hurt the wind industry's reputation and embolden its detractors. Kenetech's financial straits have probably compromised the efforts of other legitimate wind energy companies to get large utility contracts and financing in an already difficult business environment, because of low fossil-fuel prices and excess U.S. generating capacity.

THE PRICE OF WIND POWER

Damaging as the Kenetech bankruptcy was to the wind industry—in no small measure because of its chilling effect on new investment—the fortunes of the company and the industry are not synonymous. World demand for wind technology is strong as the cost of wind power continues plummeting, and the Kenetech disaster opens the way for financially healthy competitors with sound equipment and solid reputations who will fill the demand.

At 5 cents a kilowatt-hour, wind power, even without special renewable energy tax credits, is cost-competitive with most new hydropower facilities and a range of fossil fuel technologies, including some natural gas and new combined-cycle coal plants.[8] However, the National Energy Policy Act of 1992 contains a 1.5 cent per kilowatt-hour production tax credit for wind-generated electricity, which further reduces the net cost of wind energy. While wind gets cheaper, carbon dioxide emission charges (carbon taxes) may eventually add significantly to the cost of fossil fuel power. At least one expert believes that the present value of future carbon taxes would be "comparable to the capital costs of the [fossil fuel] power plants."[9] In addition, the risk of fuel price escalation, properly accounted for, could also easily add another cent to wind's fossil competition.

Finally, if "externalities"—environmental and social damages caused by fossil fuels—are tallied and added to fossil power costs (something that some utility commissions were cautiously beginning to do in the early 1990s), these new charges would add significantly to wind's cost advantage. Conservative estimates for externalities are of a similar magnitude as generating costs and may add 2–5 cents or so to coal, 1–3 cents for gas, and 1–2 cents for combined-cycle gas. High-end estimates imply coal costs of 17 or 18 cents per kilowatt-hour. Unfortunately, in a deregulated and restructured utility environment, significant externality charges are unlikely to be widely imposed.

Despite the Kenetech fiasco, wind technology is a sound and rapidly advancing renewable energy option whose costs are falling rapidly. The California Energy Commission found recently that, in just seven years, the capital costs of wind power plants have almost been halved, while performance (measured in energy output per installed kilowatt) has almost doubled.[10] Experts with the Utility Wind Interest Group, an industry organization, expected wind to cost only 4 cents per kilowatt-hour by the end of the decade, and wind projects had already been bid at that price by 1995. Moreover, the California Energy Commission is supporting development of a turbine it believes will deliver power for an astonishingly low 3 cents per kilowatt-hour.

16

❋

ANOTHER KIND
OF WIND COMPANY

I've never been so excited about [wind] business prospects, [not]
even in the 1970s when there was this Gold Rush fever.
—Mike Bergey, cofounder,
Bergey Windpower Company

As Kenetech's star waned, other U.S. wind energy companies continued with normal operations. Their wind farm development activities illustrate both the extensive influence and presence of foreign turbine
manufacturers in the U.S. market as well as American ingenuity at work
in developing unique domestic wind machines. As we will see, some U.S.
wind farm developers that had predominantly installed turbines from
abroad are now actively marketing competitive machines of their own.
SeaWest Corporation, FloWind Corporation, Zond Corporation, New
World Power Corporation, and Cannon Energy Corporation are among
the important American wind plant energy companies offering utility
grid–connected power.

U.S. WIND COMPANIES:
JOUSTING WITH LARGE TURBINES

SeaWest, of San Diego, California, for example, operates 371.4 megawatts
of wind capacity. Unlike Kenetech, however, SeaWest is a wind farm developer and operator that installs and maintains turbines made by other
companies. Most of its capacity is in California at Tehachapi Pass, near

Bakersfield, where SeaWest operates turbines made by Mitsubishi and by various Danish manufacturers. SeaWest also manages Europe's largest wind farm in Wales.

New World Power, of Lime Rock, Connecticut, the first wind energy company to go public, is involved in solar and hydro, as well as wind farm development, and has a joint-venture agreement to establish renewable energy projects with Westinghouse Electric Corporation. New World controls the wind power assets of the Fayette Manufacturing Company, whose turbines performed very poorly, but which owned valuable wind farm property and infrastructure at Altamont. New World has already won electric utility bid solicitations for a 40-megawatt wind farm at Big Springs, Texas, for Texas Utilities and for a 17.5-megawatt wind farm for Wisconsin Public Service. This renewable energy company is negotiating wind projects in Chile, Costa Rica, Mexico, and the United Kingdom.[1]

Zond, of Tehachapi, California, is a wind farm developer and operator with 260 megawatts of capacity in California and international development interests in Chile, India, Mexico, and the United Kingdom, as well as in Hawaii. The company has mainly installed Danish turbines manufactured by Vestas Energy A/S (the world's largest wind energy company), but Zond has developed its own 40-meter-diameter 500-kilowatt machine, the Z-40, and the even larger Z-46, a 750-kilowatt turbine. By late-1996, Zond had signed one of the largest contracts in wind industry history to provide Northern States Power Company of Minneapolis, Minnesota, with 100 megawatts of capacity using the Z-46. The Z-46's rotor intercepts an area almost twice as large as that of Kenetech's KVS-33. Even the Z-40 (and the Enercon E-40 discussed earlier) sweep an area almost half again as large as does the KVS).

Cannon Energy Corporation is a smaller wind farm developer, owner, and operator with 60 megawatts of capacity in Tehachapi Pass, also consisting almost entirely of Danish turbines made by Micon, Nordtank, and Vestas. But like Zond, Cannon has now developed its own turbine, the C26/250, a 250-kilowatt machine.

FloWind Corporation of San Rafael, California, is an unusual wind turbine designer, manufacturer, owner-operator, and wind energy project developer. The company today manages about 862 wind turbines (132 megawatts of capacity) in two California wind farms at Altamont Pass and at Tehachapi Pass. Several additional projects in the United States and abroad are currently being planned. FloWind's two wind farms include

512 vertical-axis wind turbines (VAWTs), the world's largest VAWT fleet. Unlike the familiar fan-shaped, horizontal-axis wind turbine (HAWT), FloWind's VAWT resembles a giant eggbeater with two aluminum blades secured at each end to a central vertical mast, forming an ellipse. The company is now retrofitting its VAWTs with three-bladed rotors of seamless composite material. FloWind is the only U.S. manufacturer of VAWTs.

Because of their blade-support system, VAWTs require less massive blades than conventional horizontal-axis turbines, and the VAWT can use wind from any direction without needing a yaw control mechanism to orient the machine. Reduced blade mass relative to HAWT allows the turbine to start at relatively low wind speed. The VAWT also has its drive train, gearbox, and generator on the ground, making for easy inspection and maintenance.[2] FloWind's chairman, Leon Richartz, asserts that everyone will adopt VAWT technology in the long-term and that large VAWTs—which the company is developing with the help of legendary blade designer Dan Somers—will double the maximum possible power output from any wind farm.[3]

Apart from its VAWT, FloWind has installed another 350 horizontal-axis machines. These HAWTs were designed by engineer Robert Lynette and are manufactured by Advanced Wind Turbines, Inc. (AWT), which Lynette founded. AWT is jointly owned by FloWind and R. Lynette & Associates, a prominent wind energy consulting firm. AWT is known for its 275-kilowatt AWT-26 model HAWT, which FloWind calls "the most cost-effective HAWT in the world" and "the least-cost, horizontal-axis technology." FloWind projects that the AWT-26 will give customers a 15–20 percent cost advantage. At just 16,000 pounds, the AWT-26 is only 40–66 percent the weight of competing Danish and Japanese machines of similar size, and its 26-meter-diameter blades were computer-designed for maximum aerodynamic efficiency. With the AWT-26 as its calling card, FloWind won the competitive bidding for a 25-megawatt wind project in the Pacific Northwest to be developed with a consortium of eight Washington State public utility districts known as Conservation and Renewable Energy Systems (CARES). FloWind, however, has already begun marketing a 27-meter-diameter version of the unit with a lower starting speed, for producing power in less windy areas, and is also designing very large, vertical-axis wind turbines, each rated at 1 to 3 megawatts, for use in future 1,000 megawatt wind farms.

ROBERT LYNETTE AND
ADVANCED WIND TURBINES, INC.

AWT is a very different kind of energy company from Kenetech. The tone set by its founder is neither secretive, defensive, nor arrogant. Even the company's literature seems friendlier. The achievements of Robert Lynette, AWT's chief executive officer, suggests that principled and congenial wind entrepreneurs can survive today in the rough-'n'-tumble capitalistic fray of the developing wind industry.

Bob Lynette is a small man with a friendly smile who held the first meetings of his R. Lynette & Associates (RLA) consulting company in the hot tub on his farm. With support from the U.S. Department of Energy (DOE), Advanced Wind Turbines developed and now manufactures the AWT-26 turbine that FloWind markets. Like Kenetech, Lynette/FloWind has bid power contracts for as little as about 5 cents per kilowatt-hour. The recent Bonneville Power Administration contract competition that RLA won with CARES was bid at only 5.3 cents a kilowatt-hour—substantially below their nearest competitor.

Results like those did not just appear in a vacuum. Before going into business on his own, Lynette had spent most of his career in aerospace at the Boeing Company. While there, he worked on the giant MOD-2, a 2.5-megawatt turbine. Part of the Department of Energy's 1970s wind energy program, the project squandered millions of dollars on contracts to large aerospace companies for huge, cantankerous turbines. The blades of the MOD-2 were as long as a football field, and when the MOD series turbines worked at all, the behemoths were so costly that they were never brought to market. The companies that produced them were not particularly committed to renewable energy at the time and generally quit the field once their contracts ended.

Convinced he could do better, Lynette left Boeing and remortgaged his farm to start his own wind energy consulting company, which eventually became the wind industry's largest. However, for the first couple of years, Lynette recalls, "it was me on a farm with a remote phone hanging on the fence while I did the gardening." Through RLA, Lynette was retained by Merrill Lynch to survey the wind industry and recommend a company to them for investment. That led to their selection of USW, although Lynette had been more impressed with the technology of a firm called ESI in Boulder that lacked financing and management skills. "I frankly was really rooting for them," Lynette said, "because I saw the product was so far ahead, conceptually, of anything else."

Former U.S. energy secretary Hazel O'Leary and Robert Lynette, president of Advanced Wind Turbines, in front of the AWT-27 wind turbine. Lynette is also vice president of FloWind Corporation. The AWT-27, developed with Department of Energy assistance, uses advanced airfoils, controls, and monitoring electronics that permit adjustment of the machine to local wind conditions. Blades twist toward stall as wind speeds increase, thereby increasing overall energy capture. *Courtesy of FloWind Corporation.*

Lynette believed that he could readily solve the mechanical and electrical problems experienced by the old ESI Company's model ESI-80 turbine. Since ESI had gone into bankruptcy with no turbine patents, Lynette was free to use its basic turbine design. To produce the AWT-26, he adopted the ESI-80's architecture and combined it with new advanced "airfoils" (aerodynamically shaped blades) developed by DOE's National Renewable Energy Laboratory. The result was a highly competitive advanced wind turbine. "These airfoils can extract more energy out of the wind per square meter of swept area than just about any other machine operating," said Lynette, "whether they're variable pitch, variable speed, or anything else."

The AWT-26 is a lightweight downwind turbine, meaning its blades face directly opposite the wind, and it uses towers up to 160 feet high to take advantage of increased wind speed at higher altitude. Because of the simplicity of its design—two fixed-pitch blades on a teetered, free-yaw rotor—the machine offers low capital costs and low annual maintenance costs. Although the AWT-26 is a few meters smaller in diameter than the Kenetech machine, Lynette reports that the single-speed machine delivers superior performance using just two blades instead of Kenetech's three, one generator instead of Kenetech's two, and a free-yaw instead of a hydraulic-yaw drive. That means savings on machinery as well as blade materials. "It doesn't take a genius to figure out that if you can do that, you're going to have a winner. You're going to have a machine that's a hell of a lot cheaper per kilowatt-hour."

In a typical wind regime "between, say, 16 and 19 miles an hour annual average windspeeds . . . it appears to me that our machine outperforms [the KVS-33 machine], and every other machine that's in the field," Lynette says, though he is careful to add that he makes no claims to be more cost-effective than Kenetech, because he doesn't know what their machine costs. Kenetech, he believed, would face serious competition from advanced Danish, English, German, and Japanese machines. The AWT-26 will, too.

Nonetheless, Lynette's main concern about turbine competition is from huge firms in engineering, heavy equipment, electronics, and construction that are entirely outside the wind industry and have access to lower cost financing. "If the Fortune 500s get into this business, they will make U.S. Windpower [Kenetech] and all the other companies in the business look small. If they can get up on the technology and get product, they'll have no trouble penetrating the market, because of their access to capital. Access to capital, as everyone in the industry knows, is the secret of success. That, and a product." (Of course, as earlier chapters show, an effective organization and sound management are also vital.) Given Lynette's views, it was not surprising when, late in 1993, Advanced Wind Turbines, Inc., and FloWind announced a partnership with Kaiser Aerospace and Electronics Corporation's Space Product's Division to manufacture the AWT-26.[4]

FloWind and AWT's good behavior seems to be winning recognition, contracts, and financial rewards. The company in 1995 announced wind development deals worth $190 million with China and India. FloWind President Harold Koegler pointed out that the Indian and Chinese governments are promoting the use of wind to meet their massive power needs, because wind power can be installed in six months to a year, instead of the three to five years required for coal and gas power plants there. In China, the company will build turbines for projects worth $160 million in Inner Mongolia, Zhejiang, and Guangdong Provinces. Eighty-five million dollars of this business is for a 100-megawatt project in Inner Mongolia.[5] FloWind has also sold 220 turbines in India in a $30 million transaction.

Once other corporate giants like Siemens and Kaiser Aerospace also recognize that the world wind market can provide multibillion-dollar sales, they, too, will probably invade. And instead of starting from scratch in the turbine business, they will probably purchase smaller companies whose technologies they like. Then they will add high-powered engineering, vast financial resources, plus marketing and manufacturing capabilities. The heavyweights will be hard for small companies to resist and are ultimately likely to lay claim to a substantial share of the wind power market.

One such possible acquisition candidate and contemporary small player in the large turbine game is Carter Wind Turbines, Inc., founded with European capital by wind industry veteran Jay Carter, Jr. Even without the 1.5 cents a kilowatt-hour production tax credit, Carter's team claimed at an American Wind Energy Conference that its lightweight, two-bladed, downwind machine can deliver power at the highly competitive rate of 4 cents a kilowatt-hour in 1993. At that, Jamie Chapman, head of OEM Corporation (wind energy consultants), counseled, "never suspend your disbelief" about wind-turbine manufacturers' claims. As the conference hall continued to echo with the conflicting assertions of various wind companies, each claiming to have the most efficient, state-of-the-art turbine, that seemed like a good idea.

BERGEY WINDPOWER: THE TENACITY
OF A SMALL-TURBINE ENTREPRENEUR

Not everyone in the industry is given to Kenetech-style hyperbole. Bergey Windpower Company (BWC) presents a stark yet refreshing contrast to the rarified world of large turbines, fancy offices, and high finance once frequented by Kenetech. While Kenetech floated hundred-million-dollar junk-bond and stock offerings couched in inscrutable financial verbiage, Bergey Windpower made its very modest living one turbine at a time, closing deals on a handshake, a short simple contract, or a man's word.

The father-and-son company, founded by Karl and Mike Bergey, manufactures small, high-quality turbines for "stand-alone" remote power applications (not supported by a utility interconnection). BWC sells more 1- to 10-kilowatt turbines than anyone else in the world, and their turbines whirl in 47 states and more than 50 countries. For remote uses, their power is very cost-competitive with diesel and photovoltaic generators, two other stand-alone options. The BWC 1-kilowatt turbine is mainly used for residences, ranches, farms, and telecommunications power applications, such as microwave repeater stations. The 10-kilowatt unit is used not only for residential and farm utility bill reduction, but also for village electrification, community water supply, and telecommunications. You can find BWC turbines spinning at the Evangelical Baptist Mission of Timbuktu, Mali, the Australian Telecom on French Island, Australia, the Falkland Islands Development Authority, and the Desert Development Center in Cairo, Egypt.

BWC flourished in the heyday of renewable energy during the 1980s, when the renewable energy tax credits were readily available. It then narrowly escaped bankruptcy when orders suddenly melted away after the tax credits expired and oil prices collapsed in 1985. At that point, only the fittest small turbine firms with the most reliable machines survived. BWC's hard-won victories and its survival are a testimonial to its founders' perseverance, initiative, and sound business instincts.

The impetus for BWC came in 1977 after Mike Bergey, then a University of Oklahoma mechanical engineering student, helped build a wind turbine that won a national student design competition. In the flush of victory, Mike realized he had found what he wanted to do in life. The wind industry was tiny in those days, and Mike could not find a company he wanted to work for. So he convinced his father, Karl, designer of the popular Piper Cherokee airplane, to go into partnership. Karl is also an aeronautical engineer on the University of Oklahoma's engineering faculty. Along with a few family friends, he raised the $70,000 or so to start the company in 1977, building on wind energy research conducted at the university since 1970. Then, while Mike's buddies took jobs at $2,000 a month as petroleum engineers for big companies, Mike began working for his father at $400 a month, a choice he did not regret. "I was doing something I loved to do," Mike recalls.

"My father was an academic. I was a student. We had other students and our young engineers working there. We didn't have any business people. We had never manufactured anything before. There were a lot of mistakes and a lot of inefficiencies."

Mike Bergey has serious blue eyes and a prim appearance that belies a resourceful, enterprising businessman, a compulsive worker, and the driving force behind the company. Well tempered as the steel of his turbines,

Mike typically works 70 to 80 hours a week and doesn't take off weekends or vacations. By necessity, this evangelist for small wind turbines has had to be entrepreneur, engineer, designer, salesman, and marketer. Somehow he also has energy to spare for the American Wind Energy Association, of which he is a board member and past president. He has also authored more than 40 technical papers and articles on wind energy.

Early in their collaboration, Karl impressed on Mike the virtues of simplicity in design. BWC adopted direct-drive alternators to get rid of the turbine's gearbox. They used a unique passive blade pitching system Mike invented, using weights mounted forward of the blade. As the speed of rotation changes, the weights automatically alter the blade pitch to optimize performance. The turbine also has a passive overspeed protection to furl the turbine out of high winds and an inside-out alternator configuration (the outside of the alternator turns). "That allows you to put the blades right on the front and simplifies the whole structure," said Mike.

At times, he and Karl had to fight to keep BWC's doors open. In response to market demand for a turbine larger than the 1-kilowatt unit with which BWC began, the company brought out their 10-kilowatt machine in 1983. At first, demand was brisk. "We were backlogged all through 1983, just fighting to keep up," Mike remembers. Revenues peaked at about $2 million a year. Then, by mid-1984, the company's orders just dried up. To their chagrin, Karl and Mike discovered that they had simply satisfied a limited pent-up demand for their product from customers who had been waiting for something like it to reach the market. But because BWC had been expanding to keep up with the earlier order frenzy, the firm got into cash flow problems. "We had too much inventory, couldn't pay our vendors, had to lay people off. It was our first taste of real trauma as a company."

Karl came from a frugal Pennsylvania Dutch background, and not being able to pay bills hit him particularly hard. He grew tense, conveyed a sense of desperation, and was unable to enjoy his work. Mike wasn't happy either. "I came to the realization," he said, "that people had for the most part lost interest in energy and weren't really beating a path to our door, and [that] maybe it was going to be a pretty tough slog, and maybe an impossible slog. . . . I felt cheated and abandoned and just rudderless. . . . Getting past that was the most difficult thing." Low fossil-fuel prices, saturation of their market niche, and the expiration of the renewable energy tax credits clobbered BWC.

After his 1984 crisis, Mike somehow regained his faith that the market would eventually come back in the 1990s, if not sooner, and therefore he no longer felt desperate. As he told himself, "What we have to do is stay in the game, keep our technology, keep our good name, expand our lines when we can, become a better business. . . . But eventually it will come our way again."

At about this time Karl suggested that they close BWC. "If I remember correctly," says Mike, "I think we did talk with a bankruptcy attorney, whose advice was, 'File tomorrow!' " Mike, however, believed that 1985 would be a good year, because it would be the last year of the tax credits and people would be making purchases to use them before they expired. The tax credits, however, had a $10,000 limit for residential applications, and so BWC's market was only partially stimulated. Mike thought he could change that through lobbying in Washington but was disappointed. Although sales were lower in 1985, the company made a profit because of reductions in overhead. By 1986, however, BWC was still doing barely a tenth as much business as in 1984. They had to let their engineering team go, which Mike found particularly painful. "We didn't have staff, we had friends," he said. "We laid everyone off. I went without salary for two years. . . . We worked some deals with our rent. It was pretty brutal."

The company essentially had no domestic market left—the United States was awash in cheap energy—and so the Bergeys decided to concentrate on developing markets abroad, something that required a lot of initiative. "We've never just sat around waiting for the phone to ring," said Mike. Building foreign sales also took great perseverance and patience: "We didn't know beans from bananas in doing projects overseas." It took the Bergeys several years to learn about government programs that could provide assistance and to begin getting help from the U.S. Agency for International Development (AID). "All the sales were fairly difficult," said Mike. But, although sparse at first, these sales kept the manufacturing operation going.

Two projects melded together to give the shaky firm a boost. Mike had been interested for some time in entering the field of wind-electric water pumping. Although windmills can operate water pumps directly without converting the wind's mechanical energy to electricity, these mechanical water pumps require much more maintenance. The AID program in Morocco had approval for the installation of two wind-powered electric water-pumping systems. "We made a proposal saying we could do it," said Mike. "We knew [at that point] we couldn't do it. We had evidence that it could be done from USDA tests, but we had never done it ourselves. So we said we could do it; they believed us; and we went towards signing a contract. Fortunately, about the same time, the [federal] Solar Energy Research Institute put out a solicitation for research, and one of the topics was water pumping."

The Bergeys responded to the solicitation, won the award, and developed the necessary technology with the Alternative Energy Institute at West Texas State University. The pumps were developed and tested in the United States, and the first commercial deliveries went to Morocco. The $220,000 AID contract gave BWC an important new product and some new markets. "[It] really matured us as a project company. . . . We were up to then primarily a company that shipped products overseas that other people installed." The new contract required BWC to also install the product and see that it worked. "We had to suffer through the customs, duties, the local logistics, the bureaucracies, all of the protocols. . . . It was really a miserable experience, but it set us up well to go on to what we're doing now very successfully. . . . We can take on a job to install systems virtually anywhere in the world."

Another important showcase project for BWC was the installation of a six-turbine unit in Xcalac, Mexico (pronounced *eks-ca-lack*), a fishing village of 40–50 homes with white sandy beaches, crystal clear water, and big coral reefs near Belize. Xcalac, however, had no utility power, and connecting Xcalac to the nearest utility grid would have cost $3.2 million—too much for such a small Mexican village. That made Xcalac an ideal candidate for stand-alone renewable energy systems using wind and sun. Solar panels were provided by Condumex, S.A., of Mexico City. The whole project was so cost-effective at $450,000 that it resulted in savings of $2.75 million compared to the cost of extending power lines. A large market exists in developing countries for similar systems: Mexico alone has some 80,000 additional unelectrified villages.

BWC's financial fortunes are improving to the point that Mike doesn't have to work so hard—in 1992 he actually planned a vacation. "We're stable, growing profitable, and we're more secure than we have ever been since we started the company . . . ," Mike says. "I've never been so excited about business prospects, even in the 1970s when there was this Gold Rush fever." BWC now has a modest backlog of orders, including some large projects, and it appears that Mike's perseverance through 17 years of struggle has paid off. When asked why he hung on, Mike Bergey responded pragmatically, "Just from sheer will and faith that eventually we'd need to turn to wind energy."

17

EUROPE RACES AHEAD

We'd like to see Clinton say that 2 percent of our energy should be
obtained from renewables. That's a big number. If he does that,
this industry is off and running.
—Dale Osborn, former president of
Kenetech Windpower, Inc.

For the first time since the 1980s, Europe in 1995 surpassed the United
States in installed wind turbine capacity. The Europeans put in about 750
megawatts of wind capacity in 1995, according to estimates of the American Wind Energy Association, while the U.S. wind industry barely grew.
Total European installed wind capacity is now about 2,300 megawatts
versus about 1,770 megawatts for the United States. European leadership is likely to continue for the foreseeable future as Europe is expected
to add 700–900 megawatts of new wind capacity every year at least
through A.D. 2000. By contrast, the United States is expected to add only
107 megawatts in 1996 followed by additions of 100–200 megawatts
through the turn of the century. European wind companies are also leading American wind firms in some aspects of turbine technology.

Why has the United States lost its world wind power leadership to the
Europeans? The answer can be found in the "dismal science" of economics. The Europeans have simply made it more profitable to build wind
farms than we have because they are more interested in, and receptive to,
renewables. OEM Corporation principal Jamie Chapman, who studied
the overseas wind industry, concluded, "Europeans are more concerned
about carbon dioxide, greenhouse gases, pollution," as well as other consequences of nuclear and fossil power than are Americans.[1] Chapman be-

lieves that our government has yet to convince the public that there's a carbon dioxide pollution problem, much less to galvanize the nation in favor of renewables. Moreover, in contrast to the narrow spectrum of U.S. groups interested in wind, a broad cross section of European society— from Parliaments to utilities and the manufacturing sector—is involved in all phases of the European wind industry. That political constituency influences European energy policy on behalf of renewables.

THE LEAD THAT SLIPPED AWAY

Although the United States still produces about 40 percent of the world's wind electricity, our lead has been speedily evaporating—we had 95 percent of the world's installed generating capacity as recently as 1985. Today we have only a scant third of it. Our lead was visibly eroding even in the early 1990s: Europe in 1992 installed 20 times the capacity we did—250 megawatts versus our 12 megawatts (only enough power for a few thousand residences).[2]

In 1995 Germany alone added about 500 megawatts of wind capacity, five times U.S. additions. Two German states alone, Niedersachsen and Schleswig-Holstein, are each planning to install 1,000 megawatts by the years 2000 and 2010, respectively. Even France, with its heavy dependence on nuclear energy, is seeking to have 500 megawatts of wind by the year 2000. Most European Community nations plus Sweden are actively integrating significant amounts of wind generation into their utility grids. Denmark, the Netherlands, Germany, Italy, and Sweden are leading the way.[3]

Utility and government projections studied by Chapman indicate that more than 4,500 megawatts of wind power is scheduled to be in operation in Europe by 2000—an investment of well over $3.5 billion. Recent developments suggest that installed capacity might be considerably more. The European Wind Energy Association, comprised of wind industry members, projects that by the year 2030, 100,000 megawatts of wind will be installed in Europe, equal to about 10 percent of the continent's electric-generating capacity. By contrast, in 1995 U.S.-installed wind capacity was far less than a percent of our total electrical capacity.

Though European wind power growth is rapid in percentage terms, the wind power capacity installed in Europe to date is still small as a share of total capacity; even Denmark, which has Europe's largest stock, got only slightly more than 2 percent of its electricity from wind in 1991. Denmark, however, plans to obtain a modest 10 percent contribution from renewables, including wind, by 2000. The United States, with its vast wealth, technological prowess, and renewable energy resources, should set ambitious wind power capacity targets in its national energy plan, as discussed in Chapter 25. Most utility systems could accept *at least* 15 percent of their energy from wind systems, without becoming unreliable.

TRICKS OF THE TRADE: HOW EUROPEANS
NURTURE THEIR WIND INDUSTRY

We could learn much about building a vibrant wind industry by taking some tips from the Europeans. Because the European Community is putting far more research and development dollars into wind energy than is the United States, European turbine manufacturers are making great technological strides and appear to lead us in their variety of large, advanced, utility-scale machines. The Danes, Dutch, English, Germans, Italians, and others are producing very competitive advanced wind turbines and offer stiff competition to American firms for large wind farm sales.

The wind industry has been stimulated in Europe through favorable government-mandated utility power purchase rates. Wind producers in Germany, for example, are guaranteed 90 percent of the average price a utility charges residential customers within its area, and Denmark also offers favorable buyback rates to independent wind power producers. These terms guarantee that wind developers and independent wind energy producers can enjoy a fair and predictable return on their investments. This is critical in arranging low-cost, long-term financing. That, in turn, helps keep the capital costs of wind plants competitive with more established technologies.

Another important wind industry stimulant has been government cost-sharing in nations such as Denmark to reduce the initial costs of wind turbines. Denmark has stimulated its wind industry not only by attractive power purchase terms for wind power, but by subsidizing domestic purchasers of Danish turbines. Market stimulants in Denmark were coordinated with a very useful government wind turbine testing-and-certification program that set federal performance standards. To benefit from the subsidies, turbines had to meet the standards, which aided the industry and customers alike. In Europe, the subsidies typically are phased out over a ten-year period as the technology becomes more cost-competitive. This has already occurred in Denmark and the Netherlands.

The Dutch once gave direct capital cost subsidies of 40 percent to wind turbine owners who participated in the Dutch national wind development program. As of 1993, the Swedish government still provided purchasers of wind turbines a 25–35 percent purchase subsidy. Apart from their national energy development programs, a number of European nations are cooperating closely on wind technology research, development, and demonstration, with substantial support from the Commission of the European Communities.

European nations gained both operating and manufacturing experience from the substantial investment the United States made in California wind turbines during the 1980s by building and exporting many of the machines used in California. The Danes, for example, produced about 40 percent of the utility-scale machines bought for the California market of the 1980s. Other European nations and Japan seized shares of that market. European and Japanese companies then also benefited from the new respectability and credibility that the improvements of wind power plant performance in California conferred on the whole industry.

EUROPE: LAND OF THE LARGE TURBINES

Whereas very advanced European wind technology is already available, a great deal of even more extraordinary wind technology appears ready to move from demonstration sites in many countries to the commercial marketplace. European turbines come in larger sizes than U.S. machines and offer essentially all the features offered by U.S. wind technology firms and then some. For example, four Danish firms, Vestas-DWT, Micon, Nordtank, and Bonus all offer machines of 400 kilowatts or greater, as do NedWind and Windmaster of the Netherlands and Enercon GmbH, Tacke Windtechnik GmbH & Company, and Hüsumer Schiffswerft of Germany, Wind Energy Group, Ltd. (WEG) of the United Kingdom, and Riva Calzoni of Italy.[4] Villas Wind Technology of Austria produces a 600-kilowatt variable speed machine, and Vestas already has a 600-kilowatt turbine and is completing a prototype 1.5-megawatt machine, as is Nordtank Energy Group, which also has a 500-kilowatt turbine. Other Dutch, Danish, and German companies are developing machines in the 500–1,000-kilowatt range. By contrast, very few U.S. machines are larger than 300 kilowatts. Moreover, the Commission of the European Communities (CEC) is supporting 15 large wind turbine projects ranging in turbine size from 650–3,000 kilowatts.

In Europe, where undeveloped land is scarcer than in the United States and where its availability is more of a constraint in siting wind projects, utilities have shown greater interest in large turbines and in offshore wind development than have U.S. utilities. Scandinavians are already busy utilizing large turbines. The Danish utility consortium ELKRAFT, for example, operates an offshore wind power plant with 450-kilowatt turbines southwest of Copenhagen and another small wind plant using 750-kilowatt machines. The other Danish consortium, ELSAM, operates a 2-megawatt turbine on the coast of Jutland. The Swedes, meanwhile, are planning a 300-megawatt offshore wind plant using 3-megawatt Nasudden III turbines. Of the large turbines CEC has supported, several offer variable speed, and one offers adjustable speed. Blade mass is being reduced while retaining flexibility by incorporating glass and carbon fibers. Vestas' 39-meter turbine not only has lighter blades but a very light gearbox.

Outside Europe, wind energy is also making substantial headway in places as diverse as Canada and New Zealand, although they have plentiful hydro resources. Canada, for example, is building a 9-megawatt wind plant in southwestern Alberta—the nation's largest wind power plant—and Hydro-Quebec plans another 5 megawatts on the Magdalen Islands. Modest wind power plants are also being built or planned in parts of the Middle East, despite its petroleum riches.

OUR LOSS, THEIR GAIN

Spectacular improvements clearly have been made in wind electric technology over the past 15 years, since California was the world's center of wind farm development. Mechanical, aerodynamic, electronic, materials science, and computational advances are providing greater energy output per installed kilowatt, increased reliability, and lower costs, as well as larger and more efficient machines.

Instead of lagging behind Europe, we could choose to strengthen the U.S. wind industry, rather than attack its federal funding, as did an element of the Republican Congress in 1995. Apart from increasing the competitiveness of our turbine exports, the advantages of a vibrant domestic wind industry are manifold. Wind farms can be built modularly, that is, in small or large units of capacity, just as needed. They can be installed quickly, often in months, without tying up billions of dollars of investment capital and incurring huge cost overruns. Because of these economic advantages over coal, oil, or nuclear plants, and because wind turbines require no fuel, wind farm investors and utilities avoid both the financial risk of uncertain future power demand and future fuel-price increases. Nuclear power plants, by contrast, often take ten or more years to build and license.

How much of an effort should we exert to develop superior wind technology and nurture a strong wind industry? Does it make sense for the nation to begrudge a few million dollars of investment in wind technology to accelerate sustainable wind technology that can improve our economy, our quality of life, and the environment? Has the nation contemplated seriously enough, and deliberated openly enough, the true worth of obtaining power without worry about pollution, nuclear plant accidents, decommissioning, radioactive wastes, climate change, or oil supply cutoffs, Middle Eastern war, and oil price escalation?

The United States could easily make new wind capacity the most desirable generation option by guaranteeing premium buyback rates to producers of wind power and other renewable energy products. Renewables would not just be the cleanest new energy, but the most profitable. With this "market pull," the private sector, not the government, would voluntarily supply most of the capital to provide all new electrical capacity from renewable sources. Intelligently crafted economic incentives would eventually pay for themselves in avoided health and environmental damages, in foregone fossil fuel subsidies, in new domestic jobs, in export revenue. As wind farm capacity was installed and paid for, incentives would be phased out. The nation would be left with pollution-free, fuel-free, renewable wind-electric power.

Attractive as wind power is, no one energy technology need carry the burden of supplying all of our energy. Instead, we need a cost-effective blend, or smorgasbord, of energy technologies. Biomass technologies should be part of that blend, as the next two chapters demonstrate.

PART IV

BIOENERGY

18

BIOFUELS
AND PLANT POWER

The California biomass industry is fighting for its life.
—Gregg Morris, biomass and renewable energy
consultant, Future Resources Associates

Biomass has been used for heating, lighting, and cooking since our hunter-gatherer ancestors stalked wooly mammoths and saber-toothed tigers. The conversion of biomass to energy is the oldest energy-conversion process known—after photosynthesis, respiration, and digestion, that is. Biomass is not only an abundant, secure, and diverse domestic energy source that produces less environmental damage when burned than fossil fuel, but it can often be had free or salvaged inexpensively from industrial and municipal waste.

Biomass literally means *living matter*, but is used to refer both to plant and animal tissue and to substances derived from them, including animal waste and the organic (carbon-containing) portion of municipal waste.[1] Collectively, biomass is the largest source of renewable energy used in the United States after hydropower. Varied in nature, biomass can be exploited in many exciting ways on farms and by industry to produce relatively clean fuels and electricity.

BIOMASS, THE FUTURE POSSIBLE

Thanks to the possibilities inherent in biomass, a revolutionary change may be in store for American agriculture. Tomorrow's large farm might one day be an integrated agro-industrial complex, rather than just a place where food, feed, or fiber is grown. These super farms might also produce energy crops and process them profitably into industrial chemicals, non–fossil fuels, and electricity.

Through careful planning, these energy-growing and biomass-processing complexes could conduct their agricultural and industrial activities synergistically, efficiently, and ecologically. A locally owned power plant, for example, could burn woody wastes from nearby farms and return clean ash to enrich the farms' soils, thereby also solving a potential waste disposal problem. Meanwhile, the power plant could provide electricity for the farms as well as electricity and cogenerated heat for industrial processes, such as ethanol synthesis. Surplus power would go into the utility grid to earn additional revenues, in which farmers might share if they owned the plant cooperatively.

The ethanol plant, in turn, could earn the farmers money from the sale of ethanol transportation fuels, intermediate chemicals, solvents, and fuel additives, as well as from fertilizer for crops and nutritious feed for livestock, made from ethanol distillation by-products. Other co-located processing plants could boost farm revenues by synthesizing a variety of useful goods, such as pharmaceuticals or even plastics, from the farm's biomass.

Another option would be for manure from livestock to be biologically decomposed by microorganisms in the absence of oxygen to form methane in a local biogas plant. Residue from the process can be used as fertilizer. The gas produced by the anaerobic digestion can be burned for process heat or to produce electricity in a combustion turbine power plant, or compressed and upgraded for distribution through natural gas pipelines, or used to make power in a clean fuel cell. On-site uses of the power and heat tend to be most economical, especially when credits are accorded for waste manure conversion. Alternatively, the farm could also produce its own power from photovoltaic panels or from advanced wind turbines, while collecting heat for space-conditioning by means of ground-linked geothermal heat pumps. (See Chapter 20.)

Once energy crops become more cost competitive with fossil fuels, the future farm might raise fast-growing trees and towering stands of perennial native grasses for energy production. The deep-rooting grasses and trees would also shelter edible row crops from the wind and would rebuild marginal soils in which food crops could later be grown. In addition, the shelter belts would provide wildlife habitat and would protect

Wheelabrator biomass electric plant at Mt. Shasta, California. *Courtesy of Warren Gretz, National Renewable Energy Laboratory.*

streams, lakes, and groundwater by intercepting chemical runoff from cultivated fields.[2]

Today, this agro-industrial complex is still mostly fantasy, mainly because of the high costs of growing fuel on dedicated energy plantations, the low cost of competing fossil fuels, and because market networks would need to evolve for absorbing the fuels and co-products. Biomass fuel costs are currently being brought down,[3] however, and experts expect that biomass power will eventually be competitive with fossil fuel power and liquid transportation fuels.[4] Despite the costs of fuels grown solely for energy, fuels made from biomass wastes or residues are already in common commercial use and are the most promising types of biomass for additional near-term commercial development. Typical biomass feedstocks in the United States currently include wood waste (bark, chips, scraps, and sawdust), pulp and paper industry residues, agricultural residues, organic municipal waste, sewage, manure, and food processing by-products. The United States in 1996 had about 11,000 megawatts of biomass electric-generating capacity and almost as much biomass process heat. Domestically, perhaps as much as $22 billion have been invested in the U.S. biomass industry, and the industry provides 66,000 jobs. By A.D. 2020, the United States could have 30,000 megawatts of biomass electric capacity, employing 150,000 people. Overall, about one unit of energy out of every 20 we now use comes from biomass. Eighty percent of the nation's biomass power currently comes from wood, with more than half of that total generated from "black liquor," a by-product of paper pulp manufacturing.

THE BIOMASS INDUSTRY'S
FIGHT FOR SURVIVAL

Although the biomass industry has made great strides since the 1970s, doubling its contribution to U.S. primary energy supplies, it began losing momentum in the early 1990s, and little new biomass power plant development is expected at least until 2000.[5] Even waste- or residue-fueled power plants are not attractive investments under today's market condition, except when necessitated by regulatory or environmental concerns (about the impacts of competing fossil fuel technologies).[6] California's 800-megawatt biomass power industry, which was growing at 28 percent a year in the 1980s, has since lost 250 megawatts of capacity through plant shutdowns and utility buyouts of their long-term biomass power supply contracts. Most of the closures have occurred just since 1994. In Maine, where wood-to-energy plants once provided up to 25 percent of the state's electricity during the 1980s, biomass has now become one of the state's costliest power sources.[7]

Biomass-fueled electric power plants are currently shutting down in several other states besides Maine and California. Utilities in many states can obtain electricity from natural gas at a mere 2–2.5 cents per kilowatt-hour, and nonfirm bulk electricity prices in the southwestern United States were only 1.5–3 cents per kilowatt-hour in 1995.[8] Consequently, utilities are not entering into new contracts for biomass power from independent power producers, and contracts that expire are not being renewed. As biomass power plants shut down, biomass fuel providers in some areas have been left with a glut of fuel and with prices far below their cost of production.

But low gas prices could well be a medium-term phenomenon followed by significantly higher prices. Natural gas prices have historically been volatile over the past two decades, and supplies may one day again become tight. It would therefore be wise for utilities to maintain a diversified energy supply base and protect technologically sound renewable energy industries like biomass, so they will be robust and viable when needed.

In addition, emissions from gas combustion, well heads, and pipelines cause environmental impacts, and aging pipelines also present safety hazards. For all these reasons, it makes good sense to keep renewable energy industries like biomass healthy so they can continue their excellent progress in reducing their costs. The increased use of natural gas in essence reflects utilities' tendency to select new power sources using short-term cost considerations. Today's search for the lowest priced power often leads utilities to pass up environmentally friendly technologies, such as biomass, that are likely to be cost-effective in the long run.

Apart from the broad consensus among utilities and independent power producers to increase their dependence on natural gas, another major deterrent to greater biomass use is the Federal Energy Regulatory Commission's (FERC) decision—required by the Energy Policy Act of 1992—to allow interstate wholesale power "wheeling" (transmission by nonutility generators over utility power lines). The FERC's decision has led utilities with excess capacity to sell electricity at bargain basement rates over interstate power lines. The availability of this cheap power undercuts existing biomass power producers and will impair their future market prospects.[9] A third deterrent to further biomass development is that the favorable tax treatment biomass power has received in recent years is currently threatened in the mid-1990s by Republican congressional efforts to reduce the federal budget deficit and federal environmental programs.

Given these economic and political realities, let's now consider biomass resources, technologies, and their potential benefits.

BIOMASS: A BOUNTEOUS RESOURCE

As indicated, biomass can yield solid, liquid, and gaseous fuels, which can produce heat and electricity. (Biomass can also be converted to all commodity organic chemicals and to many specialty chemicals.) Almost any kind of current or future vehicle can be operated on fuel derived from biomass.[10] The U.S. Department of Energy (DOE) anticipates that liquid biomass fuels could be cost-competitive with gasoline and diesel fuels within a decade.[11] The actual date is very sensitive to oil prices, but in any case biomass fuels will displace a significant proportion of gasoline consumption by 2030. The DOE reports that the nation's biomass resource base is more than sufficient to produce enough liquid fuel for all the nation's cars, buses, and trucks. If that is true today, when the typical car gets less than 30 miles per gallon (m.p.g.) and many trucks get 5–10 m.p.g., imagine the ease with which the nation's vehicles could be biofueled once the already perceptible revolution in automotive design and propulsion occurs (discussed in Chapters 23 and 24). Efficiencies of 100 m.p.g. can be attained now, and very advanced hybrid electric vehicles that may get as much as 200 m.p.g. are in the works, making it even easier to imagine meeting the nation's greatly reduced future transportation fuel needs with biomass.

BIOMASS: FOUR BILLION PEOPLE'S
PRIMARY ENERGY SOURCE

Internationally, biomass provides about 15 percent of the world's consumed energy—almost as much as natural gas[12]—and furnishes 35–38 percent of all energy used in developing countries, mostly in rural areas, mainly in the form of fuelwood, charcoal, and dung.[13] This makes biomass the primary energy source for three-quarters of the world's population.

If a variety of significant challenges to wider use of biomass can be overcome, some optimistic studies indicate that, within 30 years, biomass energy production, mostly on plantations in the tropics, could produce even more energy than those developing nations currently consume.[14] That, however, presupposes that rural people would have an economic stake in the large energy plantations envisioned, that competitive commercial markets existed for the biomass, and that the overall economics of these projects made sense. We will discuss some problems of biomass plantations later in this chapter.

BIOMASS BENEFITS:
PRODUCTS, JOBS, AND ECONOMIC STIMULI

While reducing or eliminating costly waste-disposal problems today, the commercial use of biomass creates new jobs for tomorrow—in agribusiness, engineering, finance, and construction, to name a few industries. By stimulating local economic activity, especially in rural areas, biomass development can keep money in the community that might otherwise be sent far away to pay for imported energy. Domestically, biomass energy crops could supplement farmers' incomes, possibly replacing some farm subsidies, while decreasing oil imports, thereby reducing America's trade deficit.[15] These crops can also slow the atmospheric buildup of climate-destabilizing gases, help counter soil erosion, and provide green open space, wildlife habitat, edible crops,[16] and numerous useful by-products, ranging from composts and mulches to medicines, building materials, and feed supplements.

Because biomass-to-energy conversion facilities tend to be small and geographically dispersed, instead of concentrated in a few large, costly power stations, they can contribute to the overall reliability of a utility grid. Once decisively cost-competitive biomass conversion processes are widely available, woody biomass-to-energy power plants, ethanol producers, and biomass growers and processors all can benefit by using each others' by-products as resources in relatively closed-loop production cycles. Siting biomass producers and users in these clusters will encourage the use of local biomass for energy and will create new jobs and opportunities for manufacturing, while minimizing waste generation and transportation requirements.

AN ENVIRONMENTAL BONUS

One of the great advantages of biomass energy is that, except for the fossil fuel energy used in growing or processing biomass, the conversion of biomass to energy need not add any net carbon dioxide to the atmosphere. Plants use natural solar energy collectors—leaves—to capture energy with which to conduct photosynthesis. In photosynthesis, plants remove carbon dioxide from the air, combine it with water, and produce oxygen and carbohydrates. If biomass is produced sustainably, so the biomass harvested and converted to energy is constantly replaced through new plant growth, then the amount of carbon released during the combustion of biomass is equal to the carbon removed from the air during the photosynthesis that was responsible for its formation. Thus, no net increase in atmospheric carbon dioxide results, so the sustainable use of biomass energy does not significantly contribute to global warming.

By the same reasoning, the substitution of biomass energy for fossil fuels (or the co-firing of biomass and fossil fuels) reduces net carbon dioxide emissions. The combustion of liquid biomass fuels in place of gasoline also can reduce ozone formation and carbon monoxide production. According to some scenarios, billions of tons of carbon could be removed from the atmosphere and stored (in plant tissues) through the establishment of biomass plantations[17] and through reforestation of deforested lands.[18] Plans like this, however, face formidable economic, political, and institutional obstacles. Were the United States to adopt carbon emission taxes or some similar measure—which is currently unlikely—those revenues could do much to support biomass, other renewable energy technologies, energy efficiency, and massive reforestation.

Substituting biomass power for 1,000 megawatts of coal power capacity spares the atmosphere from the net addition of 5 million tons of carbon dioxide every year.[19] Yet, although the DOE has spent hundreds of millions of dollars a year on "clean-coal" technologies (itself an oxymoron), the DOE paradoxically has found only a few million dollars a year for cost-shared biomass commercialization projects.[20]

ENERGY FROM WASTE, AND WASTED ENERGY

Although biomass grown solely for energy on plantations or in forests is expensive, the costs and environmental impacts can be mitigated when biomass is recovered as a by-product of other economically useful activities, such as the production of food, fiber, or building materials. When this biomass is free or low cost, as is often true for agricultural or forestry wastes and processing residues, the transformation of biomass into useful energy and other products is a net plus for society and, generally, the environment. The cost may even be negative, if waste disposal costs are avoided. Sugar factories, for example, commonly burn bagasse (the residue produced by juicing sugar cane) for process heat. Similarly, wood waste is often burned (albeit inefficiently) to produce process heat, or power, or both, for lumber and paper mills. Although wood and wood waste combined is the largest source of biomass energy in the United States by an order of magnitude and, although it can still profitably be used to cogenerate electricity and process heat, much of the nation's wood waste is nonetheless landfilled by the millions of tons—at costs of up to $30 or more a ton.

Yet as landfill space grows scarcer and tipping fees increase, waste wood–to–energy conversion will become more attractive. By converting that waste on-site, industries not only can save waste disposal costs,[21] but can reduce or eliminate their electricity bills.[22] (Alternatively, wood waste can also be compressed into fireplace logs for residential use or into pellets for pellet stoves.)[23]

Landfills themselves are a major biomass energy resource. As organic material in landfills decomposes, it produces a gas, comprised mainly of equal parts methane and carbon dioxide. This combustible mixture can be recovered using established technology by drilling wells into the landfill and vacuum extracting the gas to a collection point for sale. Power and heat may also be produced on-site. Although only a small portion of the nation's landfills exploit this energy resource, they nonetheless produced a respectable 645 megawatts of electrical capacity in 1994. More projects would be in operation were it not for low fossil-fuel prices.

Even without strong financial incentives, new landfill gas capacity is increasing at 35 percent a year, mainly to comply with federal environmental and safety regulations under the Resource Conservation and Reclamation Act (RCRA) and the Clean Air Act. Provisions of these federal laws (and some state legislation) are designed for odor control, to prevent the explosion of uncollected gases, and to limit the release of methane, which contributes to global warming.[24]

BIOMASS PLANTATIONS:
THE ALLURE AND THE PROBLEMS

The agricultural sector is a vast source both of organic residues and "virgin" energy crops. As agricultural productivity continues to increase in the United States, resulting in less demand for cropland, tens of millions of acres of additional land will become available for biomass production. Fifty million acres are currently available in the United States for energy crop production, excluding lands needed for conventional agriculture and forestry.[25]

Neither is land scarce on a global scale for biomass production today. Well over a billion acres of tropical land is suitable for reforestation.[26] But many questions remain as to whether this would be the best use of the land, or whether it could be better used to produce higher value goods, such as food and fiber, or for multispecies wildlife habitat. If tropical energy plantations became profitable, intensified commercial demand for land conceivably could threaten subsistence farmers, or create additional incentives for tropical deforestation.[27]

Plantation biomass could certainly be produced on deforested, degraded, and eroding lands, as well as on surplus cropland. Government financial incentives could make tree planting profitable in these cases.[28] If perennial grasses, tree crops, or oil crops were grown in an environmentally sound manner, they could have beneficial impacts on the land, compared with erosive row-crop agriculture that requires intensive use of fertilizer, pesticides, herbicides, and water. Polyculture, the simultaneous cultivation of multiple crops, and agroforestry, the cultivation of food in conjunction with tree crops, are two methods for growing biomass that may offer economic and environmental advantages.

Perennial crops are especially desirable systems, since they require less cultivation than annuals and often protect the soil with their canopy and extensive roots, reducing erosion. The roots of perennial grasses tend to improve soil quality, and the stems and sod reduce runoff. Native perennials are apt to thrive without irrigation, and without fertilizer, herbicides, and pesticides. They require much less replanting than annuals and also provide valuable wildlife habitat.[29] If biomass is raised in annual monocultures, however, these advantages would be lost.[30]

When and if dedicated energy crops can be raised profitably, energy plantations could boost rural incomes, injecting new money into rural economies, while simultaneously rehabilitating derelict lands.

THE ENERGY YIELDS OF PLANTS

Most plants convert only about 1 percent of the sunlight received annually to chemical energy, and even the most efficient photosynthesizers under the most favorable conditions typically achieve annual efficiencies of no more than 2–3 percent. By comparison, the most efficient solid-silicon solar cells are ten times more efficient. Moreover, to get 2–3 percent efficiency from biomass, good land, abundant water, and adequate nutrients are required. Disease, pests, frost, drought, soil depletion, and soil erosion all can reduce actual yields, and some of the biomass produced probably will be in forms unusable for energy production. In addition, costly and polluting fossil fuel energy inputs are often used in biomass production, collection, storage, and energy conversion. (Agricultural equipment usually runs on diesel fuel, for example, and synthetic fertilizer is made from natural gas.)

In some situations, biomass from field and forest may be more valuable over the long term for maintaining and improving soils, or for other purposes than for energy production. The excessive removal of agricultural residues and forestry wastes can be harmful to the soil and its ecosystem. Soil fertility, organic matter content, and water-holding capacity would decline, while soil erosion and water runoff rates would increase.[31] Impoverishment of the soil and loss of biodiversity would be among the likely long-term results. Despite the environmental risks from overdoing or mismanaging biomass extraction, biomass fuels are still environmentally desirable, in general.

To succeed commercially, "grown-to-order" biomass must be produced efficiently and sustainably, yet at a high yield, and then must also be converted efficiently and economically to heat or power.[32] These are stringent requirements. So don't expect to see large biomass plantations springing up across the United States anytime soon. The profitable yet environmentally acceptable production of biomass will require attention to a complex array of factors and environmental constraints, as reviewed in detail by the National Biofuels Roundtable, a broadly representative coalition of industry, environmentalists, researchers, and government representatives.[33]

THOUGH THEY SPARE THE AIR, BIOMASS FUELS HAVE SHORTCOMINGS, TOO

Biomass fuels are far from perfect. Some cause the formation of slag and other deposits in combustion boilers that can interfere with the boiler's operation and cause troublesome shutdowns. Fast-growing plants, such as annual grasses, which are high in alkali metals (potassium and sodium) and silica, are the most likely to cause slagging.[34] Also, biomass generally is high in moisture when harvested, lowering its energy density (the energy content per unit mass). That increases transportation costs, a significant "factor of production," since biomass tends to be dispersed over large geographic areas, and energy must be expended for its collection. Finally, even dry biomass does not have as high a heating value as the fossil fuel with which it has to compete. Dried woody material, for example, has only about half the energy of bituminous coal.

BIOMASS: A SOLUTION TO POLLUTION?

Biomass has other advantages that more than compensate for its energy density: It is more volatile, having a higher hydrogen concentration, and is low in ash and sulfur relative to coal and oil. Unlike coal ash, biomass ash is relatively free of toxic heavy metals. Instead of having to be land-filled as a waste, biomass ash can be returned to the land as fertilizer.

Its low sulfur content also means that biomass—burned alone or co-fired with fossil fuels—can dramatically lower the emission of compounds that cause acid rain, such as sulfur dioxide. When wood is burned in a steam-cycle power plant, the sulfur dioxide emissions are less than 2 percent of those from a coal plant with scrubbers. Emissions of nitrogen oxides, which also cause acid rain and smog, are only a fourth as large as those from a coal plant.[35] Thus, by co-firing clean biomass with coal, a utility can lower its overall emissions. Under the Clean Air Act, the utility can then use the reduction to offset emissions at another of its plants, or it can sell the reduction as an "emissions credit" to another utility. (This is one of the incentives for a willow and coal co-firing project currently being developed by the Empire State BioPower Consortium.)[36]

Biomass has a particularly valuable role to play in waste management. The United States consumes 100 million tons of paper a year and sends about half to landfills. But if wastepaper is ground in a mill to particles and is slurried, compressed, and extruded as paper pellets or cubes, the paper could be used in industrial and institutional boilers. That would save the equivalent of 150 million barrels of crude oil or 32 million tons of coal a year.[37] What once cost millions of dollars to dispose of could instead provide heat, cooling, and electricity. Waste wood as pellets can also be used in efficient wood-pellet stoves. The use of pellets reduces particulates as well as carbon monoxide formation, compared with ordinary wood stove discharges. The biomass industry also prevents the disposal of millions of tons of agricultural waste, much of which would otherwise be placed in landfills or burned in fields. In California, for example, the biomass industry at its peak consumed ten million tons of green waste a year, a quarter of the total waste sent to state landfills.

Biomass systems may also be useful in treating wastewater. Passive wastewater treatment systems combined with biomass production are being studied at the University of Florida, supported by the Southeastern Regional Biomass Energy Program, a DOE program. Tall, fast-growing grasses, such as energycane, elephant grass, and switchgrass, are being used experimentally to take up large amounts of impure water that has received enough treatment for use on nonfood crops, but not enough for drinking. The irrigated grasses can then be harvested and burned to recover energy. The hope is that polluted water, especially from difficult-to-control nonpoint sources, can be inexpensively treated and used to water biomass farms without requiring expansion or construction of new waste-water treatment plants, which are energy intensive.[38]

Tests suggest that eucalyptus, which consumes extremely large amounts of water, can grow rapidly on urban storm-water runoff, nutrient-rich irrigation drainage water, or treated sewage effluent.[39] A eucalyptus biomass plantation could therefore serve a valuable waste management function while producing an environmentally desirable energy crop. If a diverse mix of perennial biomass crops were planted instead of just eucalyptus, significant wildlife habitat benefits also could be produced. These and other non-market benefits add value to proposed biomass projects.

Biomass systems could also help control water pollution from feed-lots, poultry, swine, and dairying operations. Hundreds of millions of tons of these problematic animal wastes are produced annually across the United States at thousands of sites. Simple anaerobic digestion of animal wastes in conjunction with aquaculture could help farmers and feedlot operators manage the huge waste volume by converting it into a variety of resources.

The anaerobic digestion produces carbon dioxide and methane, the main constituent of natural gas and an even more powerful climate desta-bilizer than carbon dioxide.[40] Releasing methane gas during natural waste decomposition pollutes the atmosphere, but capturing the methane from an anaerobic digestion lagoon makes it available for on-farm heating, dry-ing, or power production, saving fossil fuel and reducing adverse climatic effects. The anaerobic digestion process not only provides energy, but di-minishes the volume of slurried wastes and helps protect the quality of surface water.

In addition, purified but still nutrient-rich liquid effluent from the di-gester can be used to produce algae in ponds for animal feed, or for di-rect harvest by crustaceans, fish, or mollusks. The fish raised could be used as bait or processed into animal feed, or as feed for commercially farmed fish. The United States spends nearly $10 billion a year importing fish. Raising more fish domestically could decrease imports and provide domestic economic benefits. Even the digester's solid residue has uses as a fertilizer. Finally, a protein-rich bacterium, *Thiopedia*, associated with anaerobic sludge, can be dried and recovered for its protein.[41]

Anaerobic digestion of biomass is hardly a new idea, and millions of household-scale units have been built in China and India to handle ani-mal and human wastes. Biogas from these digesters can be used directly

for cooking and lighting and for generating electricity. Once traces of corrosive hydrogen sulfide gas are removed from biogas, it can also be used
to operate car and truck engines. A cubic meter of biogas (at standard
temperature and pressure) can produce 1.2 kilowatt-hours of electricity
or provide enough fuel for a three-ton truck going 1.7 miles.[42]

BIOMASS: AN ALCHEMIST'S DELIGHT

As suggested, biomass feedstock is extraordinarily versatile: It can be
turned into alcohol or liquid oil as well as gasoline, diesel fuel, hydrogen,
higher alcohols, and even biocrude (an oil similar to crude). If the feedstock is produced sustainably, these biomass fuels will add little net carbon dioxide to the atmosphere, as noted, and, having little sulfur, will not
produce sulfur-oxide precursors of acid rain.

Biodiesel can be produced from animal, vegetable, and microalgal oils
(lipids) by a relatively simple chemical process called "transesterification."[43] The resulting low-sulfur fuel can be used in a diesel engine with
only minor engine modifications.[44] Blended with ordinary diesel, biodiesel has desirable fuel attributes. Mileage is about the same as for conventional diesel, but biodiesel is biodegradable, low in toxicity, and
doesn't produce explosive vapors. Thus it may be useful for fueling heavy
equipment in mines. Problems that can occur include variations in fuel
quality, fuel filter plugging, and injector failure. Currently the DOE's National Renewable Energy Laboratory is striving toward a production cost
of only $1 a gallon for biodiesel. NREL research indicates that biodiesel
resources are adequate to provide 50 percent of the U.S. diesel fuel
demand.

A biodiesel from soybeans called "soydiesel" is currently manufactured commercially by Proctor & Gamble Company. Rape, a forage crop
of the mustard family, is another source of biodiesel. A diesel pickup
truck was driven 8,742 miles in 1994 on biodiesel from commercial
rapeseed oil mixed with ethanol and a catalyst. Horsepower was virtually
the same as for low-sulfur diesel fuel, and fuel economy lagged diesel by
only 3.7 percent. Emissions of hydrocarbons, carbon monoxide, and nitrogen oxides all were reduced. In fact, the exhaust smells like French
fries.[45] Beef tallow is currently being studied in the Midwest as another
source of biodiesel. Perhaps a car fueled with rapeseed oil running next
to a truck operating on beef tallow would together smell like a ham-
burger and fries.[46] Conventional diesel particulate emissions, however,

are no joking matter. Exposure to them has been linked with increased lung cancer risk.[47]

Apart from the combustible oils that plants, such as sunflowers, oil palm, and various aquatic species, readily provide, biomass can also be converted to fuel by a two-stage series of chemical reactions—gasification and fuel synthesis—to make gasoline, diesel, or even jet fuel. Biomass is first heated in a reactor vessel to produce crude oil vapors that are then condensed to a liquid. This bio-oil can then be burned in a conventional boiler in place of number 6 fuel oil or coal. The fuel can also be burned in stationary diesel engines for power or refined into a gasoline substitute or jet fuel. Because of the oil's high energy content, it can be economically transported,[48] in contrast to some raw biomass. The thermal production process for making bio-oil—pyrolysis/catalytic cracking—is not yet competitive with fossil fuels and is not yet in commercial use to a significant extent.

CHEAPER LIQUID BIOMASS FUELS WILL COMPETE WITH GASOLINE AND DIESEL

Biomass is also processed biochemically and thermochemically into ethanol and methanol, two other renewable, nonfossil transportation fuels that can be easily integrated into existing fuel distribution networks. Ethyl and methyl alcohol from biomass can power vehicles by direct combustion, or by operating a fuel cell. Alcohol combustion produces only carbon dioxide, water vapor, and small amounts of aldehydes, plus some unburned alcohol. Fuel cells transform chemical-to-electrical energy directly, without combustion, and can serve as a stationary power plant or supply mobile-source power. Because no combustion occurs, fuel cells are far more efficient than gasoline-powered internal combustion engines. Consequently, fuel cell vehicles need carry and consume far less fuel than conventional vehicles. Both ethanol and methanol are also water soluble and biodegradable, and ethanol is low in toxicity.

The costs of these two alcohols from biomass currently are significantly higher than gasoline or diesel fuel on an energy-equivalent basis. Ethanol has about two-thirds the energy content per gallon as gasoline, whereas methanol has half the energy. Ethanol can be burned neat (pure) in specially designed or modified engines, or it can be mixed with gasoline to produce gasohol and burned in ordinary internal combustion engines. Gasoline additives, or "oxygenates," mandated by the Environmental Protection Agency under the Clean Air Act can be made from either ethanol or methanol. Ethanol and methanol can be used as a component in ethyl tertiary butyl ether (ETBE) or methyl tertiary butyl ether (MTBE), respectively, for reformulated gasoline.

An interlaboratory governmental task force that analyzed biomass alcohol technologies in 1990 projected that ethanol might become competitive with gasoline by 1998. However, a recent study of ethanol production from cellulose in paper mill sludge suggests that the process would produce a 39 percent after-tax return on investment *right now,* with ethanol selling for $1.20 per gallon in 1995. Hundreds of millions of dollars worth of this valuable renewable fuel could thus be produced, just from paper mill wastes alone, while saving landfill capacity and avoiding waste disposal costs.[49] Other recent economic analyses of ethanol costs suggest that it will first be competitive as a fuel additive and that pure ethanol may not be competitive with gasoline until A.D. 2010–2020, depending on oil prices.

Whereas ethanol traditionally has been produced by the fermentation and distillation of grain starches and sugars to alcohol, the National Renewable Energy Laboratory (NREL) has been making great progress in developing a process for converting *any* cellulosic biomass to ethanol. Not only will this increase the yields of ethanol from grain, since the new process digests a larger fraction of the kernel, but also this Simultaneous Saccharification and Fermentation (SSF) method should be capable of producing ethanol from energy crops at 67 cents a gallon (in 1990 dollars) by the year 2010, a cost that will put ethanol in the same general price range as competitively priced gasoline. The DOE projects that ethanol could be made even more inexpensively from waste feedstocks at only 50 cents a gallon in 2005 and for just 34 cents a gallon in 2010.[50] NREL is currently working on thermochemical processes to produce methanol from biomass at only 50 cents a gallon and mixed ether at 95 cents a gallon.[51]

HOW MORE EFFICIENT BIOMASS POWER
COULD ALSO COMPETE WITH FOSSIL FUELS

Gasification of biomass to a mixture of methane and carbon dioxide, or to a mainly carbon monoxide and hydrogen mixture called syngas, may slightly improve the overall efficiency with which biomass is converted to energy. The biomass is heated in an enclosed vessel in which the supply of oxygen is limited, causing incomplete combustion (pyrolysis). The intense heat causes the biomass to decompose chemically, producing combustible vapor (containing carbon monoxide, hydrogen, carbon dioxide, water vapor, methane, and nitrogen) and leaving a solid residue. When the biogas is cleaned, it can also be used to power either a gas turbine power plant (up to several hundred kilowatts using just today's technology) or a diesel engine, or to make methanol.[52] Gasified biomass can also be used to operate a fuel cell.

Most commercial biomass-to-electricity conversion today is done relatively inefficiently by direct combustion in small steam-turbine power plants without cogeneration and therefore cannot compete with inexpensive fossil fuels, unless the biomass feedstock is free or obtainable at negligible cost and the electrical generator is close to transmission facilities. Because small steam-turbine plants often operate at only half the efficiency of a modern coal plant, their power costs are high.[53] If more efficient gas turbine conversion technologies were used, and if large amounts of plantation biomass were produced economically with valuable co-products, both fuels and electricity could be derived cost-competitively. Using that approach, the Union of Concerned Scientists has calculated that biomass could provide fully 30 percent of the electrical demand in the Midwest—at prices competitive with fossil fuels.[54]

Although biomass generally hasn't been used very efficiently, that could change. Gasified biomass could power highly efficient gas turbine plants in an integrated biomass gasifier/gas turbine power system—essentially a biomass gasifier coupled to a jet engine. Some energy analysts contend that electricity from such a system would compete favorably in price with fossil fuel power and produce far less air pollution than conventional fossil fuel plants.[55]

Cratech Corporation of Tahoka, Texas, is currently testing a demonstration pressurized gasification system capable of producing 1–20 megawatts of power under a cost-shared grant from the DOE's Western Regional Biomass Energy Program. The integrated biomass gasifier and gas-turbine combustion system is designed to make the efficiencies of larger systems available to power plants as small as a megawatt.[56]

PROTECTING THE PUBLIC INTEREST

Adverse market conditions have so crippled nascent biomass industries today that erstwhile biomass developers are dropping biomass efforts in

Kent Kaulfuss, president of Wood Industries Company (WICO), a biomass composting and fuel supply firm. Behind Kaulfuss is a crop of kenaf on WICO's 100-acre demonstration biomass farm. The kenaf can be digested to make ethanol or processed into paper and other products. *Courtesy of John Lindt,* The Valley Voice.

order to build fossil fuel projects. Some are giving up on the United States and targeting international markets, where energy costs are higher and biomass prospects are better. It would be unfortunate if the promising and multifaceted commercial biomass industry were allowed to go into a prolonged slump while other nations benefited from our biomass R&D. National and state energy and tax policies could instead easily make new biomass projects just as profitable as those using less costly fossil fuel. That would insure that we not lose the environmental and social benefits that biomass can bring. To summarize, bioenergy not only reduces pollution from heat-trapping gases, but often yields valuable coproducts, produces local economic activity, creates jobs, and frequently conserves valuable landfill space.

More biomass research and development will produce the crop yield increases and conversion process improvements now needed to enable biomass to compete effectively in a broad range of energy markets.[57] Federal and state commercialization programs also should be bolstered.

To salvage the industry in one state, the California Biomass Energy Alliance had proposed a 1.5 percent biomass energy set-aside in all future state resource allocation plans, as well as a temporary floor price of 6.5 cents per kilowatt-hour for biomass energy. That price could be subsidized by an imperceptibly small surcharge on all utility bills. New fees, however, did not pass the state's tax-averse legislature and were not included in the legislature's utility restructuring plan. The set-aside would have provided the industry with time to adjust to current low natural gas prices and could have served as a model for national legislation to help the biomass industry.[58]

Even without set-asides, some biomass industry entrepreneurs are not waiting for all the numbers to line up in the right column on their spreadsheets before boldly entering the biomass field with determination to succeed.

19

A BIOMASS BUSINESS

Though California was the nation's leader in biomass, in six
months we've lost half our capacity.
—Kent Kaulfuss, Chief Executive Officer,
Wood Industries Company

Kent Kaulfuss is a biomass pioneer, educator, entrepreneur, and "systems thinker." Until recently, the lean, sunburned outdoorsman with russet hair was in the biomass energy business. Today he's in the biomass resource recovery field, clearing land for a developer in Arizona or gathering downed trees after hurricanes or tornadoes. Kaulfuss is one of the hardy individuals who has managed to buck the unfavorable odds in the biomass industry successfully enough to wrest a living. Earl Anderson is a designer and manufacturer of innovative biomass processing equipment that makes the work of people like Kaulfuss easier and potentially more profitable. His handiwork will emerge later in this chapter.

A BIOMASS "SYSTEMS THINKER"

Kent Kaulfuss's firm, Wood Industries Company (WICO), is not only surviving a general downturn in the biomass electric power industry, but is profitable and has some appealing new business prospects. Kaulfuss expects that the ongoing restructuring of the electric utility industry coupled with new research on enzymatic decomposition of biomass to ethanol may create an economically viable future for WICO and the biomass-to-energy industry.

Kaulfuss's vision is at once integrated, technologically avant garde, and idealistic. He wants to see society resourceful rather than wasteful. Where others see a waste, Kaulfuss sees a multitude of resources, and nothing would please him more than being able to profitably use and reuse every molecule in any given biomass feedstock. He dreams of an efficient agro-industrial enterprise in which industrial and agricultural entities cooperate in mutually beneficial relationships, cyclically using and re-reusing each other's by-products. Kaulfuss today is striving to establish successful new biomass products, businesses, and waste-to-energy cycles while preventing the environmentally destructive waste of valuable biological capital.

As a systems thinker, Kaulfuss's fixation on the efficient use of natural resources extends from biomass to other resources, like land and water. Efficiency through integrated resource use is the unifying theme in all his projects: a demonstration biomass-to-energy farm, an agroscience and nature education center, a proposed biomass municipal power plant for Visalia, California, a waste-to-ethanol research study, and a possible collaboration with Amoco Oil Company. The goal of the Amoco partnership would be to resurrect defunct biomass-to-energy plants in conjunction with the construction of new ethanol production facilities in California's Central Valley, near Kaulfuss's farm.

AN ARBOREAL INSPIRATION

In the Sierra Nevada, where he worked in reforestation as a young man in the 1970s, Kaulfuss saw giant redwood logs, felled by the wind, lying in the forest. "Everything about the giant sequoia fascinated me," he said. He began asking what the wood inside those ancient downed trunks looked like. When he learned that trees that had lain on the ground a hundred years or more were still in beautiful condition inside, he began bidding for trees up to 20 feet in diameter and more than 1,000 years old. Although he was concerned about the environment, it made more sense to him to recover dormant resources that would eventually decay than to cut live green wood (even though downed trees have an important ecological function in forests).

In 1975 Kaulfuss established a small sawmill operation utilizing salvaged logs, all the while being careful to leave dead standing timber, which he knew was of significant habitat value to wildlife. From his downed lumber, he cut rustic finished products, including many redwood park signs for Sequoia National Park and Pinnacles National Monument.

During his Sawmill Era, Kaulfuss encountered the open-burning practices that were prevalent in the timber industry. Forest slash and mill wastes were dumped into huge, teepee-shaped, open-air burners and set ablaze like so much garbage—with no pollution controls. This resource waste, compounded by air pollution, bothered Kaulfuss.

When he came down from the mountains to California's Central Valley in the early 1980s, he winced to see whole old or sick orchards pushed into piles and burned, or trucked for burial in landfills. The blackened skies could be seen 20 miles away, and no energy was recovered. He wanted to put the wood to better use but knew he had to find an economic foundation to translate his ecological vision into a utilitarian reality. He soon sought out forestry equipment capable of turning trees into wood chips, with the goal of ultimately making the high-end products he knew were possible from wood, such as adhesives, polymers, alternative fuels, medicines—perhaps even toothpaste. He also began researching Finnish and Swedish biomass recovery practices.

A HUNGRY JACKRABBIT THAT EATS
100 TONS AN HOUR

It would have made Kaulfuss's life much easier back in those days if an ingenious resource recovery system known as the Jackrabbit had been available. Invented by biomass entrepreneur Earl Anderson, the $355,000, four-stage Jackrabbit system, nearly 70 feet long, with mobile chipper, hopper, shuttle, and loader, munches 7 to 50 acres of orchard prunings a day, reducing them to chips for power or ethanol production. Its engine roaring, the chipper drives between rows of trees, and, with a gapping four-foot-wide mouth of slowly turning steel rollers, it stolidly eats the rough brush piles set before it, spitting chips through a chute into a hopper car. Using advanced electronics and hydraulic controls, the chipper's engine also drives augers in the hopper to unload it into a mobile shuttle that dashes from orchard to loading station and back. In a mere minute and a half, the loader disgorges two and a half tons of olive chips into a waiting twenty-five-ton trailer. Today, efficient systems like Jackrabbit can prepare all kinds of woody wastes for ethanol or power plants at the same cost, or less, than what it cost the farmers to pile and burn their prunings.

THE BIRTH OF A BUSINESS:
CONVERTING WOOD TO ELECTRICITY

About 1983, the first wood-waste burning biomass power plants were being installed in Tulare County, California, where Kent Kaulfuss lived.

Kaulfuss rented a simple tree chipper and experimented in the forest on his salvage logging waste and other forest slash. Soon, he was supplying start-up fuel to the new biomass industry.

While Kaulfuss's long-term goal was to produce a myriad of products from wood and other biomass, he saw the sale of the waste for fuel as an immediate volume market and a solid first step toward creating more holistic, integrated biomass production cycles. And the waste fuel he provided to the power plants produced far less air pollution when burned with pollution controls than when burned in the open.

As several biomass-to-energy plants came on-line in California, and as it became obvious that more would follow, Kaulfuss convinced a group of investors to back him in Wood Industries Company (WICO), a limited partnership, and he made a bold commitment to purchase costly chipping equipment—large grinders costing up to $400,000 each. That decision was a risky "point of no return," said Kaulfuss, remembering the night the first machine arrived. "What am I doing?" he wondered. "I don't even know how to run one of those things—I don't know anything about this!"

Fortunately, the company's services were much in demand from the outset, because the agricultural and forestry industries needed to get rid of their wastes economically, and laws regulating burning were getting tougher. At the same time, the new power plants needed woody fuel, so Kaulfuss had a ready outlet for the materials.

Knowing that there was a great deal more to biomass than just combustion, WICO quickly offered an array of other products besides fuel: landscaping wood chips, compost, and soil amendments. While the company thrived during the late 1980s, Kaulfuss was concerned that hundreds of thousands of tons of waste wood and green municipal yard waste were still being placed in his county's landfills. He started monitoring the kinds of loads coming in for burial at the landfill, and, confident that these materials could be reused, he convinced the supervisors of Tulare County to contract with WICO to conduct a comprehensive county waste-management and recycling program.

Under Kaulfuss's program, people were told that for a little reasonable effort in separating wood and yard waste from their trash, their dumping fees would be reduced by half or more. WICO began collecting and processing this diverted material at county landfills and at his own recycling yards, but difficulties arose. Kaulfuss started processing waste at the landfill with the same knife grinders he had used successfully on clean wood waste. Each blade in the grinder cost $100. But mixed in with the wood at the dump were unexpected surprises—such as lengths of chain and

U-joints. Twenty chopper blades at a time would often break when they hit large chunks of metal—sometimes after only an hour's operation. Kaulfuss would then have to decide whether to shut down and take a $2,000 loss for the day or continue operating in the hopes of recovering the $2,000 before hitting another piece of scrap. After a year, Kaulfuss was involved in developing a much more durable hammer mill grinder now known as the Diamond Z.

A PILOT ENERGY PLANTATION

Today WICO is an established biomass composting and fuel supply firm that also operates a 100-acre demonstration biomass farm. The company recycles agricultural and municipal solid wastes, sewage sludge, manure, and biomass ash, which it then combines into new products and resells as soil amendments, soil-less potting and hydro-seeding mixes, poultry bedding, compost, and fiber products. Kaulfuss also evaluates high-yield indigenous energy crops he grows on the farm, including barley, sweet sorghum, kenaf, hybrid poplar, willow, and eucalyptus under a $200,000 first-stage research contract to the U.S. Department of Energy's National Renewable Energy Laboratory. By raising these plants, Kaulfuss hopes to isolate the most productive strains of indigenous crops for biofuels while optimizing production practices, minimizing costs, identifying useful co-products to supplement earnings, and incorporating the use of waste-water for irrigation and biomass soil amendments as fertilizer.

Durable, tree-free, high-quality paper can also be produced from Kaulfuss's kenaf fiber, and ethanol can be made from all the biomass crops. Kenaf fibers can also be made into structural panels, rope, twine, straw mats, and newsprint. Wood wastes can be processed into particle board, alternative adhesives, fumigants, herbicides, pesticides, preservatives, and other biochemicals. Some wood products high in natural phenols are used in orchards for natural pest control, and visitors to the farm drive on roads made of olive wood chips, which hold moisture and suppress dust.

TWO BIOMASS PLANTS THAT
CAN STOKE EACH OTHER

Kaulfuss deliberately sited WICO's farm near the county's wastewater treatment plant so that he can use treated municipal wastewater (gray water) for irrigation and sewage sludge for soil improvement. The site's proximity to the Southern Pacific Railroad, to an energy transmission right-of-way for a gas line, and to major freeways is ideal for resource transportation and marketing. WICO might also one day feed biogas from anaerobic digestion into the gas pipeline and use the energy corridor for cables carrying electricity produced from biomass.

Kaulfuss's vision extends far beyond his present operation. He is currently trying to lay the groundwork for a municipally owned biomass energy plant using a wood waste–to–electricity cycle and to integrate the production of ethanol with the biomass power generation.

Both fermentation and distillation of ethanol require heat, which could be produced from co-located biomass plants operating as "cogen-" erators" (producing power and process steam, a by-product of electricity generation). Ethanol production generates sillage wastes that are high in lignin. The wastes can be used either in animal feed, or as biomass boiler fuel for the electric power plant, or to provide heat for a plant that might produce kenaf paper. Lignin produced from cellulose during ethanol production is an excellent clean-burning energy source. Thus the two technologies—biomass to power and biomass to ethanol—feed each other. The wastewater used in ethanol production even has value as a liquid fertilizer ingredient for biomass crops.

Habitat restoration, sustainable agriculture, biodiversity, public education, and public access also are important components of Kaulfuss's project. Before starting WICO, Kaulfuss did wildlife and habitat restoration work for the U.S. Soil Conservation Service. From the perspective of land reclamation, Kaulfuss is excited by crops like kenaf that can be grown easily on set-aside or marginal land and by sorghum, which requires only two-thirds as much water as corn and tolerates poorer soil. The long roots of kenaf can help in land reclamation by bringing up nitrogen and extracting salts from the soil.[1] Kaulfuss has built ponds on the farm to benefit wildlife, conserve groundwater (and well-pumping energy), and he has begun restoring an abused creek. Eventually, he also plans to raise aquatic plants in the ponds as ethanol feedstocks.

AN UNENVIABLE BALANCE SHEET

It would be nice if Kent Kaulfuss, a fount of energy and ideas, could reap generous financial rewards for his enterprise, courage, and vision, but that day has yet to arrive. By late 1994, just as he was poised to start making large profits, with WICO's heavy equipment almost paid off and 40 employees on the payroll, utilities with access to inexpensive natural gas power began buying out their biomass power supply contracts. A number of other biomass power plants simultaneously closed as the fixed price provisions of their long-term energy supply contracts with their utility customers expired.[2] So demand for Kaulfuss's principal product virtually disappeared. With two hours' notice, he lost one contract that was worth almost $2 million a year from one facility alone. Several other facilities in a 25-mile radius of his property also closed in 1995, terminating their fuel supply contracts. With biomass-to-energy facilities shutting down, the market price of the wood chips Kaulfuss had been selling at $30–$40 a ton fell to less than $15—which would not even cover the cost of trucking the wood to the power plant, not to mention the other production costs.

"Ironically, though California was the nation's leader in biomass," said Kaulfuss, "in six months we've lost half our capacity." To stay in business, Kaulfuss had to let three-quarters of his workforce go, and he pays his bills by using his costly chipping equipment on land-clearing and salvage projects while he waits for the ethanol industry to develop and create a market for his products. He is hopeful, however. Amoco wants to build their first commercial ethanol facility somewhere in California and, if things work out, might co-locate with, and thereby resurrect, one of the biomass-to-energy plants that Kaulfuss used to fuel. Kaulfuss, however, is not waiting for Amoco to proceed, but has recently acquired his own pilot-scale fluidized-bed ethanol facility to produce alcohol and electricity. With a little bit of luck, and perhaps a more congenial regulatory environment at the state and federal levels, biomass will have a future, despite its temporary setback. The fundamental rationale for the industry is sound. Indeed, there's every reason for government to encourage it: The public benefits are overwhelming.

Through the alchemy of modern biochemistry allied with sustainable agriculture, integrated biomass systems can improve air quality, reduce global warming, save landfill space, support new industries, protect forests, produce new products, create new jobs,[3] lower consumer energy costs, increase agricultural income, reduce farm subsidies, diversify energy supplies, reduce dependence on foreign oil, improve the balance of payments, reclaim abused land, and promote rural regional development. All told, not a bad balance sheet—but only one example of the huge payoffs possible if we work with nature to create renewable energy systems.

PART V

UNDERGROUND POWER

20

GEOTHERMAL'S
ROCKY ROAD

We're literally riding around in space on a heat engine. If we can
find an economical way to tap it, we can have nonpolluting energy
from here to infinity.
　　　　　—David N. Anderson, Executive Director,
　　　　　　　Geothermal Resources Council

No active technology for home heating and air conditioning is
more efficient than the geothermal heat pump.
　　　　　—Earth Sciences and Resources Institute

Geothermal energy, the heat within the Earth, is a gigantic and virtually inexhaustible resource.[1] It is abundant, requires no fuel, and is environmentally desirable. Yet, unfortunately, the amount of energy we are
likely to derive domestically from geothermal sources for at least a
decade will probably be small relative to our total power supply, although
more than $6.5 billion has already been invested domestically in the geo-
thermal industry.

A DIVERSE RESOURCE
AT A PREDICTABLE PRICE

Commercially useful geothermal energy is found underground in natural-
steam and hot-water reservoirs at depths of several hundred to 14,000

feet.[2] By drilling wells into these reservoirs, crews can bring pressurized steam or hot water (or a mixture of both) to the surface. The energy can then be used for direct heating in homes and industries, or, if the temperature is high enough, for conversion to electricity in a power plant. These plants can provide power around the clock at a stable price, insulated from the gyrations of fossil fuel prices.

Most geothermal resources consist of liquid water at high temperature and pressure within rock pores and fissures. Geothermal energy rarely occurs in the form of dry steam: The Geysers power plant in California's Lake and Sonoma Counties is the only commercial steam field in the United States. This field alone, however, has produced over 150,000 gigawatt-hours of electricity and is expected to produce at least as much more power during its operating life. (A gigawatt is a billion watts.)

GEOTHERMAL POWER PLANTS:
THE TECHNOLOGIES OF SEPARATION

How do geothermal power plants operate and produce useful power? Power plants that receive a mixture of steam and water, instead of pure steam, must separate them to use the steam; they then reinject the water back into the ground. In both types of plants, the steam is piped against turbine blades, and as the blades rotate, they spin a generator to make electricity.

Underground hot water resources can produce power in either a *flash plant* or a *binary plant*. Hot geothermal waters that arrive above ground at more than 175 degrees Celsius (347 degrees Fahrenheit) can be flashed to steam so as to operate the turbine of a flash plant. The steam becomes the power plant's "working fluid." Cooler water of 100–175 degrees Celsius (212–347 degrees Fahrenheit) can also produce electricity but requires a binary plant.

Hot water in a binary plant is kept pressurized to keep it from boiling and is piped through a heat exchanger, where it vaporizes a working fluid with a low boiling point. The resulting hot vapor then drives a turbine-generator and afterward is condensed to a liquid. This working fluid is continually reused. The well water thus never comes into direct contact with the secondary heat-exchange fluid and is cleanly reinjected into the geothermal reservoir after its job is done.[3]

The details of all this plumbing and the thermodynamics of the heat transfer are unimportant to the layman. The beauty of the geothermal energy cycle is that it has very low air and water emissions, since it requires no combustion and geothermal fluids are reinjected into the ground after use. Geothermal plants produce no nitrogen oxides and only trace amounts of easily controlled hydrogen sulfide (H_2S), which smells like rotten eggs. (H_2S is present in most geothermal reservoirs.) In the early days of geothermal power production, the release of H_2S was a problem. Control technology has now virtually eliminated it: When needed, treatment reduces hydrogen sulfide releases to under one part per billion—far less than 1 percent of the amount of sulfur released by burning oil or coal. Even in the few geothermal plants that use hot, heavy brines, the "waste" solids that must be removed (mainly silica) can be used in building roads and flood-protection levees.

Besides producing power, geothermal energy can yield direct industrial process heat and direct heat and cooling for buildings, provided that the demand is situated near hydrothermal resources. Geothermal energy can also be used for warming fish hatcheries, greenhouses, swimming pools, and spas. Some 254 western U.S. cities are within five miles of geothermal resources exceeding 120 degrees Fahrenheit.[4] Abroad, Iceland obtains heat and hot water from geothermal sources for 95 percent of its buildings. Geothermal energy for direct heating purposes operates at efficiencies as high as 90 percent or more.[5]

Whereas geothermal energy is sufficiently concentrated near the Earth's surface in only a limited number of locations for economical direct heat extraction, geothermal heat pumps, which use electricity to "mine" geothermal heat, can be used almost anywhere in the United States. These "pumps" are actually heat exchangers that extract heat from shallow groundwater in winter and expel heat to the reservoir for cooling in the summer. Some purists argue that deep geothermal reserves are finite, because they can be depleted either by extracting heat faster than it is replenished, or by draining limited steam reserves; the near-surface geothermal sources, however, including the heat in shallow groundwater, are clearly renewable, as they are produced mainly by the warming effect of the sun, before the energy is withdrawn by geothermal heat pumps. These "ground-linked" heat pumps are so much more efficient than either electrical heating systems, or even advanced air-source heat pumps, that utilities can save 2,000–5,000 megawatts of installed generating capacity for every million geothermal heat pumps used.

Twenty U.S. cities already use geothermal district heating systems consisting of wells and insulated piping to deliver heat. Customers typically save 30–50 percent on their heating bills compared to conventional fuels. The direct use of geothermal heat currently saves the United States over 5.5 million barrels of oil-equivalent per year.[6] In addition to the absolute energy savings from geothermal heat pumps, they also reduce utility peak loads, which further benefits utilities.[7] Yet their use is only beginning to be appreciated.

The U.S. Department of Energy (DOE) and various electric utilities have recently formed an industry-led joint Geothermal Heat Pump Consortium, supported by the Edison Electric Institute, to expand the sale of geothermal heat pumps in the United States from 40,000 units annually to 400,000 by the year 2000. This would reduce greenhouse gas emissions by 1.5 million metric tons of carbon-equivalent annually.[8]

GEOTHERMAL ENERGY:
NOTHING NEW UNDER THE SUN

In addition to steam and hot water deposits, geothermal energy is found as hot dry rock, as geopressurized natural gas–laden brines, and as magma (hot molten rock). Virtually everywhere geothermal energy also exists as ambient ground heat in shallow soils and rocks just beneath the Earth's surface. Sometimes geothermal energy comes to the surface rapidly in hot springs, geysers, and fumaroles (steam vents) or in volcanic eruptions.

The Earth's primordial consolidation from a cloud of swirling gas four and a half billion years ago accounts for perhaps 60 percent or so of all deep geothermal energy, according to some scientists. The remaining 40 percent comes from the radioactive decay of unstable elements, such as uranium and thorium, in the Earth's crust.[9]

Geothermal energy has provided hot meals at scalding pools and baths in hot springs long before civilization arose. The Romans later used geothermal springs to heat their bathhouses, and the Etruscans are thought to have concentrated and dried geothermal brines for mineral salts, which they used in enameling their famous vases.[10]

Starting in the nineteenth century, geothermal heat was used for industrial drying and heating and has been used to produce electricity and mechanical power since the early 1900s. By 1944, Italy had 127 megawatts of installed geothermal electric capacity, and New Zealand installed a large plant from 1958 to 1963. Geothermal power generation made its American debut when a small electric generator was attached in 1923 to a geothermal wellhead at a place known as The Geysers in northern California.[11]

The first use of geothermal energy at Larderello, Italy, in 1904, showing an alternating motor and three-quarter horsepower dynamo. *Courtesy of Joseph W. Aidlin.*

THE GEYSERS: A GEOTHERMAL SUCCESS

Though Native Americans knew of The Geysers for thousands of years, early California settlers had no idea they existed until William Bell Elliot came upon them one day in 1847 while hunting grizzlies in California's remote Mayacamas Mountains. Seeing plumes of sulfurous steam hissing from holes in the rocky hills, he called the place the "gates of hell."[12]

Today, The Geysers geothermal area has an installed electrical capacity of 1,780 megawatts and is the largest geothermal development in the world. The power plants at The Geysers provide enough commercial power to Pacific Gas and Electric Company for about 1.3 million northern Californians.[13] The Geysers development eventually became the core of a billion-dollar geothermal company. In the process of building a geothermal industry, The Geysers' developers had to fight a series of interesting legal and regulatory battles in California that helped shape the geothermal industry worldwide.

The Geysers geothermal development began with John D. Grant, a Healdsburg, California, gravel contractor who organized The Geysers Development Company in 1921. Botanist Luther Burbank was one of the early stockholders.[14] Grant drilled some shallow wells and generated 35 kilowatts of power for a small hot-springs resort at The Geysers. To prove the field's capacity, more wells were drilled and vented to the atmosphere. These wild wells, with names like "Screaming Annie," continued to blast plumes of steam into the air for years.[15] But Grant's company was unable to obtain a power purchase contract from Pacific Gas and Electric Company (PG&E), or to generate sufficient revenue to allow further geothermal development.[16]

The Grant family in 1925 therefore sold its interest in The Geysers Development Company to Robert E. Bering, founder of Signal Oil Company, and another partner. Although the first impediments to geothermal development were technical, by the mid-1920s plentiful, cheap oil supplies were making further geothermal development unattractive to utilities.

Public interest in The Geysers was still high, however, because of extensive press coverage. Entrepreneur-developer, bond salesman, and financier Barkman (B. C.) McCabe paid a visit to The Geysers and was greatly impressed by what appeared to be a virtually free and unlimited power source. McCabe, who would one day make geothermal history, never forgot The Geysers, though he did not become involved with geothermal energy for another two decades.

During the late 1920s and the Great Depression, Robert Bering proved unable to develop The Geysers and was equally unsuccessful in getting PG&E to take its power during World War II.[17] As the years dragged on without any revenue, Bering encountered difficulties in meeting his financial commitments and, in 1953, went into a partnership with McCabe, who earlier had helped finance one of Bering's small oil companies. For his new investment, McCabe gained leases to some of the most valuable geothermal areas of The Geysers. The new venture was incorporated as Magma Power Company in December 1954, and the first well McCabe drilled that month was highly successful. Energy was there in ample quantity, but its sale was far from guaranteed.

MCCABE'S CHARISMA AND CAPITAL

Joseph W. Aidlin, a genial, young, and aggressive oil-and-gas lawyer with a handlebar mustache, provided legal services to the new firm. Aidlin, who had worked for both Bering and McCabe, eventually became both

John D. Grant at The Geysers' first geothermal electric generator, circa 1920s. *Courtesy of Joseph W. Aidlin.*

general counsel and board member to Magma Power and McCabe's right-hand man.[18] McCabe was a tall, imposing, well-read fellow with great curiosity and enthusiasm. He energetically raised capital for the new company and, in 1957, founded Thermal Power Company, which took a half-interest in The Geysers leases from Magma in return for development capital. Bering, however, did little to further Magma after its inception, according to Aidlin.

Despite the existence of commercial quantities of geothermal steam at The Geysers, PG&E, as in earlier decades, was still reluctant to contract for the steam. McCabe, in frustration, came up with a creative solution. He began negotiations with nearby municipalities to take the geothermal power and also threatened to supply it directly and competitively to industrial users. PG&E then quickly decided to take the power itself and built its first geothermal plant at The Geysers in 1960.[19] Magma was paid by PG&E according to the amount of power the plant produced. The initial rate was only two-tenths of a cent per kilowatt-hour, but it was the beginning of full-scale geothermal development at The Geysers.[20]

Barkman ("B. C.") McCabe, entrepreneur, financier, inventor, and founder of Magma Power Company. The inscription reads, "Jan. 1964. To my friend, advisor, and staunchest but constructive critic, Joe Aidlin. B. C. McCabe." *Courtesy of Joseph W. Aidlin.*

THE GEOTHERMAL LEASE DILEMMA

By the late 1950s, Magma Power had acquired thousands of acres of geothermal leaseholds outside The Geysers. As the company attempted both to expand its leaseholds and to develop these assets by drilling actively, it was hampered by a shortage of capital and by the total lack of state and federal

Joseph W. Aidlin, general counsel and board
member of Magma Power Company at the company's original well at The Geysers in 1954.
Close associate of Robert E. Bering and B. C.
McCabe, geothermal pioneers, Aidlin drafted
the world's first geothermal lease. *Courtesy of
Joseph W. Aidlin.*

laws to provide for the leasing of geothermal lands to private companies. In
addition, geothermal development was not considered eligible for the favorable tax treatment that competing fossil energy sources were enjoying.

"We had to write the law in California permitting the drilling of these
wells," said Aidlin. Basic questions had to be answered, such as, who was
in charge of regulating geothermal development, and was geothermal energy to be regulated as water, gas, or minerals? Could a geothermal developer take a depletion allowance or deduct intangible drilling costs?
Even more importantly, could a person who filed an oil and gas or mining
claim exploit geothermal energy under that claim, since no legal process
existed for filing a claim to lease a geothermal resource?

Geothermal wells at The Geysers being vented during a 28-day reservoir test to prove to Pacific Gas and Electric Company that the reservoir could support sustained power production. Open venting is no longer a common practice. *Courtesy of David N. Anderson, Geothermal Resources Council.*

Aidlin did the pioneering legal work on many of these issues, drawing up the world's first geothermal lease and handling Magma and Thermal's complex contracts, legislative work, and relationship with the Securities and Exchange Commission. Since he took most of his compensation in the form of shares, his friends teased him by saying that he dealt in hot air. Those who laughed and lived to see the profits Magma later reaped probably rued the day they themselves didn't buy shares in Aidlin and McCabe's operation. By the mid-1990s, Aidlin held nearly half a million shares of a billion-dollar geothermal company built from a $10,000 initial investment.[21]

"BIG OIL" ENTERS
THE GEOTHERMAL INDUSTRY

Independent geological consultant Dr. Carel Otte was in charge of exploration and research for the Pure Oil Company at Crystal Lake, Illinois, in the early 1960s, when—while on a vacation in Santa Barbara, California—he met a Magma Power Company director, George Rowan, who was a "promoter-entrepreneur-investor." Pure Oil was then looking for opportunities to diversify its oil business into related natural resource areas, and Rowan told Otte about Magma's geothermal development activities and introduced Otte to McCabe.[22]

When Otte investigated Magma Power, he found a small, struggling, entrepreneurial company that needed engineering, well design, and geological expertise as badly as it needed money. "They had wasted money in certain areas with inadequate knowledge of the geology," said Otte, and they especially needed help in resource evaluation. A geothermal power plant cannot be moved once it is built. The resource has to last at least 30 years. Plus, the resource and the power plant have to be well matched in size, which also requires resource assessment experience.

Dr. Otte saw an alliance with Magma as an exciting opportunity for Pure Oil and convinced the company to embark on geothermal energy development with Magma. Pure Oil therefore set up a subsidiary called Earth Energy, Inc., put Otte in charge, and through Earth Energy, entered into partnership with Magma in 1963.[23] Because Pure Oil was looking for larger resource deposits than was tiny Magma Power, Earth Energy was initially drawn to Magma's holdings in the Salton Sea area of California's Imperial Valley, where geothermal fluid temperatures as high as 700 degrees Fahrenheit (371 degrees Celsius) had been found. But when Earth Energy drilled in that area in the early 1960s, they found only very hot corrosive brines, ten times the concentration of seawater, that plated out on pipes and plugged their wells.

In search of a more manageable resource, Earth Energy began solo and joint developments with Magma in The Geysers field, which they expanded by acquiring adjoining lands. Although Pure Oil was acquired by Union Oil Company in 1965, Otte managed to convince Union to continue Earth Energy's geothermal energy development.

TAMING THE ROGUES:
THE CONTINUED SEARCH FOR LEGITIMACY

Earth Energy thus became the Union Geothermal Division, and Union in 1967 became a joint-venture partner with Magma and Thermal to pool and jointly operate all lands at The Geysers. Union acquired a 50 percent interest in the whole operation and became field operator and technical expert. "That infusion of money and expertise helped to develop and propel The Geysers," said Otte. "If you look at the growth curve from the time that we came in, it just starts skyrocketing."

Otte and his colleagues in the 1960s had become concerned about the lack of regulatory oversight for the geothermal industry. Unscrupulous or incompetent operators could just poke wells in the ground and walk away, leaving hot water and drilling mud oozing from the ground.

Otte therefore worked for legislation that would give the state's Division of Oil and Gas jurisdiction over geothermal resources. As a result of his efforts, and those of Aidlin and others, the Geothermal Resources Act became part of California's Resources Code in 1968.[24] The new law served as a point of departure for similar legislation not only in other western states but also in foreign lands where Union proceeded with geothermal development.[25]

Union helped McCabe make the case to Congress that geothermal energy was a finite resource and therefore should be eligible for depletion allowances and intangible drilling write-offs. Dr. Otte testified before Congress to win the right to lease government land for geothermal development. Although leasing legislation was pocket-vetoed by President Lyndon Johnson in the 1960s, President Richard Nixon finally signed the Geothermal Steam Act of 1970. The law established geothermal leasing rights and allowed people who had prior legitimate oil, gas, or mineral claims to convert them to geothermal leases.

Otte was also instrumental in gaining geothermal developers the right to lease state lands for geothermal purposes. The state had extensive landholdings in The Geysers area that were crucial to the commercial success of The Geysers' field. Had geothermal developers been unable to lease them, The Geysers could not have been properly developed.[26] The importance of The Geysers' success to the geothermal industry can hardly be underestimated. "Here was a successful project that was expanding," said Otte. Many oil companies now imitated Union by entering the geothermal industry, hoping to repeat Union's success.

THE SALTON SEA CHALLENGE

With its legal and tax status clarified in the 1970s, and with improvements in geothermal technology, the industry was poised for expansion. Union now returned to face the challenge of geothermal development in the Salton Sea, where problems of corrosion and scaling had proven insurmountable in the early 1960s. Union already knew that the Salton Sea resource base was huge and that the company had no exploration risk, so Otte focused Union Oil's attention on solving the technical problems.

To deal with the corrosion, Union's Geothermal Division built many of its pipes out of titanium. But even gold pipes would not prevent scaling—the precipitation of heavy carbonates and silicates. "The wells were plugging up, and each time they plugged up, you had to re-drill a well, and it cost you another million dollars. . . . I knew that if we could solve the scaling problem," said Dr. Otte, "we would have an immense resource."

Magma Power was also wrestling with scaling and came up with a crystallization and clarification technique for handling heavy brines. Union developed a parallel technology called acid modification. By altering the acidity of the brine, Union managed to prevent the heavy scale from coming out of solution and simply reinjected it into the ground along with the brine. By contrast, the crystallizer-clarifier technology insured that the precipitation would occur in a location where it could be handled without plugging up the system. From the wellhead, as much of the brines as possible were allowed to shoot into the crystallizer and solidify. The fluid was then treated in a clarifier, where the remaining solids were precipitated. Although Otte's division developed the acid-modification treatment, Union also contributed to crystallizer-clarifier technology.

Union's technology worked so well that no matter how briny and corrosive the geothermal fluid was, acid modification removed all the dissolved solids. The clean steam that reached the power plant was so pure it did not corrode the turbine. A measure of Union's commitment to geothermal power is that, whereas Otte had started as a one-man department with a borrowed secretary, at the peak of Union's involvement in geothermal energy he had more than 1,000 employees in his Geothermal Division.

FOREIGN DEVELOPMENT OPPORTUNITIES

The technologies pioneered by Magma and UNOCAL (which was the name Union's operating division began using in 1985) now are available for use in the developing world. UNOCAL, for example, has pioneered geothermal development in the Philippines at Tiwi and Makiling-Banahao near Manila.[27] Foreign countries will add 4,000–6,000 megawatts of geothermal capacity at a cost of $10–$15 billion over the next 20 years according to U.S. Department of Energy estimates.[28] Obviously, the sale of geothermal equipment and development of geothermal

Carel Otte, former president of UNO-CAL's Geothermal Division. *Courtesy of Carel Otte.*

leases abroad can be lucrative for American and other companies. The U.S. industry today holds about $6 billion worth of international contracts, even as the moribund domestic geothermal industry diminishes.[29] Most of the international activity is now in Indonesia and the Philippines.

Half of all underindustrialized countries have geothermal resources today, notes Dr. Michael Wright of the Utah Research Institute. Geothermal power can enable them to avoid costly energy imports that use up their hard currency. Building a geothermal plant now, in place of a fossil-fired unit, can avoid 30–40 years of pollution. Whereas geothermal energy is at best a minor energy source in most industrialized nations, for small developing nations with large geothermal potential and little installed electrical capacity, geothermal power could provide a significant portion of their electrical needs within the next few years. The need is there and the technology is available. Generally, low-cost investment capital remains the main obstacle.

Although geothermal energy does have environmental impacts, they are for the most part relatively minor and manageable. Because of the absence of combustion, geothermal plants do not make any significant contribution to acid rain or global warming, and because they require no fuel, they do not require mining. Geothermal plants, on the average, occupy only about a tenth of an acre (400 square meters).[30] By contrast, the mining required to fuel a coal or nuclear power station for its 30-year life often disturbs large areas of land and requires the processing of huge volumes of material.

Whereas some dissolved carbon dioxide and other gases (such as hydrogen sulfide) are typically present in geothermal fluids, none are released if the liquids are not permitted to flash to steam and are reinjected into the ground, as occurs in a binary-phase geothermal plant. In a flashed-steam plant, in which steam and water are separated, carbon dioxide releases are less than a thousandth that of coal, oil, or even natural gas plants (see Figure 1), due to the low concentration of carbon dioxide in solution in geothermal waters. Even these tiny releases ultimately produce no net increase in atmospheric carbon dioxide, because geothermal reservoirs naturally and constantly emit carbon dioxide anyway. The gas continually makes its way to the Earth's surface through the Earth's upper layers, which are quite porous to gases.[31]

Geothermal brines (which may be so thick in certain instances they resemble "rock soup") contain other trace gases and some dissolved toxic elements; however, most of the material unsuitable for atmospheric release can be treated or reinjected into the geothermal reservoir on-site. Water contamination by geothermal brines is an environmental concern, and for that reason, brines typically are reinjected into geothermal reservoirs. To prevent leakage into aquifers during extraction or reinjection of fluids, wells are lined with steel casing pipe that is cemented into surrounding rock.[32] In the United States the problem of waste heat from the plant is solved by cooling towers, without dumping heated water into streams or lakes.

Other environmental impacts associated with geothermal energy are those of exploration, drilling, well maintenance, power plant and cooling tower operation (such as noise and vapor plumes), and wastes—airborne and liquid. Although air pollution can be controlled, trace quantities of toxins, such as arsenic, mercury, and radon, at barely detectable levels, may be released, and control system residues require disposal. All told, however, the environmental impacts of geothermal plants are far less than for fossil units.[33,34]

AN UNDERUTILIZED RESOURCE

Geothermal energy use has been slight nationwide, due in part to relatively low levels of federal research-and-development support and to

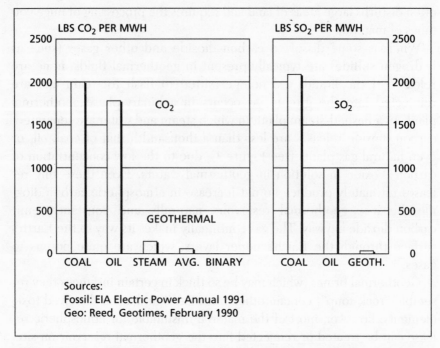

Figure 1. Pounds of carbon dioxide and sulfur dioxide produced per megawatt-hour by coal and oil plants as compared with geothermal plants. *Courtesy of Geothermal Division. U.S. Department of Energy; Energy Information Administration's Electric Power Annual, 1991; and* Geotimes, *February 1990.*

low-cost fossil fuels.[35] We had only 2,725 megawatts of installed geothermal power capacity in 1995, and these power plants produced just a few tenths of a percent of the nation's power. Geothermal power regionally, however, provides 8 percent of California's electricity and 12 percent of northern Nevada's.

Globally, geothermal energy exists in vast quantities, including many locations throughout the United States, especially in the West and abroad.[36] However, only a small part of the resource base can be developed economically because of current technological limitations. If one day the limitations are overcome, America would have an energy supply far beyond its needs.

Although known to be extremely large, the magnitude of the geothermal resource base is not well quantified.[37] By one estimate, geothermal resources exceed the world's entire coal, oil, and natural gas resources by 30 times.[38] Another estimate found the heat content of the Earth's crust (above-surface temperatures) is hundreds of thousands of times U.S. annual energy demand.[39] (The crustal volume assessed was to a depth of ten kilometers.) Worldwide hot dry rock geothermal resources are enormous, at more than 10 million quads. (The United States uses but 80 quads of primary energy every year. A quad is a million billion British Thermal Units.) For comparison, economically recoverable hydrothermal reserves today are estimated by the DOE to be 250 quads. In addition, some leading geothermal experts believe there is a tremendous undiscovered geothermal resource. But little exploration for it is going on now, because of fossil fuel competition.[40]

The U.S. Energy Information Administration (EIA) currently estimates that probable developed geothermal electrical capacity may reach 8,500 megawatts by 2010. With major increases in federal funding and R&D, the EIA has estimated that 17,000 megawatts of geothermal capacity could be attained by then.[41] Even more optimistic is Dr. Ted Mock, former head of the DOE's Geothermal Division, who believes that we could produce 40,000 megawatts of geothermal electricity in the United States using existing "binary" geothermal technology at just 4.5–6.5 cents a kilowatt-hour.[42]

Domestic geothermal power production has grown more than 400 percent from 1980 to almost 18 billion kilowatt-hours a year by 1993. Simultaneously, the cost of geothermal electricity from domestic binary power plants has dropped from 12–6 cents a kilowatt-hour between 1980 and 1993. Most U.S. geothermal systems today produce electricity at costs ranging from only 3–7.5 cents per kilowatt-hour.[43]

Along with competition from low-cost natural gas power, the main impediment to construction of additional geothermal capacity today is slack electrical demand. The current cost of geothermal power from new plants compares favorably with nuclear and oil, but is more expensive than natural gas and hydro. Therefore, few companies are building new geothermal plants in the United States now, nor are customers willing to buy geothermal power that is not competitively priced.[44]

Whereas we can ultimately hope to exploit only a small fraction of the world's incomprehensibly vast geothermal resources in an economical way, the Massachusetts Institute of Technology's Energy Laboratory has nonetheless very optimistically estimated that as much as 1.9 million megawatts of geothermal electricity could be available from hot dry rock at competitive prices by the year 2015, assuming expected advances over current technology.[45]

THE RENEWABILITY ISSUE

Hot dry rock is indeed liberally distributed throughout much of the western United States, as well as in half a dozen "arcs of fire" around the world where the Earth's crust is tortured by tectonic forces, resulting in fractures that bring heat from the Earth's depths closer to the surface. Currently, however, hot dry rock technology is still in the research-and-development stage. Cost estimates and even talk of commercialization therefore may be premature.

As to the sustainability of geothermal power, deep geothermal energy is not renewable in the same sense that solar and wind energy are constantly replenished by the sun. Geothermal reservoirs *do* lose heat and pressure when energy is extracted from geothermal zones, and they may also lose water, if it is not reinjected.[46] However, over the very long term (100,000 years, for example), deep geothermal resources will be replenished from heat radiated outward by the Earth's core. Thus, geothermal resources ultimately are virtually inexhaustible.

Some depletion of geothermal reserves has already occurred at The Geysers, which peaked in output a few years ago at about 2,000 megawatts. Depletion occurs most rapidly if a reservoir's heat-bearing fluid is removed and is not sufficiently replaced by reinjection. Any resource will in time become depleted if energy is extracted faster than it can be replenished. Some high-quality energy is degraded during power production, however, leading to an eventual reduction in the field's capacity, even with reinjection.

In practice, properly managed geothermal fields provide energy for decades or longer, and when energy production falls because of water loss, the field can sometimes be revitalized by reinjecting additional water, including low-value wastewater. In cases of mild depletion, capping a reservoir and allowing heat to reaccumulate may be enough to restore a field.

POWER ON DEMAND (24 HOURS
A DAY)—AT A REASONABLE PRICE

Fortunately, although new sources of geothermal power are generally not as inexpensive as natural gas power and hydropower, geothermal energy is not exorbitant and can provide baseload as well as peaking power. This is an important feature relative to other nonbaseload renewables. Geothermal power can be produced for about the same price as some conventional power sources, such as large new coal power plants, but with a great deal less environmental havoc, as noted. Yet with all these attractive features and the potential for expanded capacity, geothermal energy has not been a major national research priority.

More intensive research and development since the opening of the first U.S. geothermal plant in 1960 might well have resulted in costs far lower than today's. Costs can still be reduced by improvements in drilling technology, corrosion control, and development of new materials (such as high-temperature elastomers for seals, cements for well completions, and polymers) for handling geothermal fluids in drilling and in energy conversion.[47] Absent that research, the geothermal industry today faces an uncertain future, due to increased competition and possible decreases in government support in conjunction with possible impending deregulation of the utility industry.

EXPLOITING HOT ROCKS: THE ENGINEERING AND ECONOMIC CHALLENGES

The enormous magnitude of hot dry rock resources has been a source of great excitement to some geothermal enthusiasts. But although geothermal energy per se is widely distributed, it is rare to find high-quality steam and hot-water deposits near enough to the Earth's surface to be profitably exploited for electrical production with current technology. Although hot rock deposits are much more pervasive than these wet resources, hot rock deposits are generally deeper and therefore much costlier to drill. Costs also vary with the geothermal reservoir's temperature.

Unfortunately, hot dry rock by definition lacks water to convey heat to the surface. Water must be injected at high pressure through a well to fracture the rock and thereby increase its surface area for more efficient heat transfer. Another well is drilled to recover injected water. Water is then circulated through the fractured rock to absorb heat. The heated water is collected at the field's production well, through which it rises to the surface, where it is passed through a heat exchanger (an apparatus not unlike a radiator) for power production or direct uses of the heat.

At Los Alamos National Laboratory's (LANL) Fenton Hill test facility for hot dry rock technology in New Mexico, water was injected into an artificially engineered reservoir about 10,000–12,000 feet underground. The water went down one well at 21 degrees Celsius (70 degrees Fahren-

The generator of the Magma Power Company's modern Leathers geothermal power plant near Bradley, California. *Courtesy of Warren Gretz, National Renewable Energy Laboratory.*

heit) and came up the production well at 180 degrees Celsius (360 degrees Fahrenheit). Before its closure in 1996, the Fenton Hill plant produced thermal energy at the rate of 4 megawatts, which would be equivalent to about half a megawatt of electricity in a geothermal power plant. If we could have drilled a second production well, said Dr. David Duchane, LANL Hot Dry Rock Program manager, the output of our facility could probably have been increased "by about four or five times, making it a very small but 'on-the-fringe-of-commercialization' facility."[48]

In contrast, Dr. Wright, the geologist and geophysicist at the University of Utah, estimates that the cost of electricity from hot dry rock using current technology would be about 10–15 cents a kilowatt-hour. "There's no way it could be economically competitive today," he says, "[given that] conventional geothermal power can't compete with natural gas." Hot dry rock, says Wright, is a concept, not a commercial technology. Many of the basic engineering questions remain about the life span of the hot dry rock reservoir (locally) and about fluid recovery. Yet the DOE's hot dry rock research program did demonstrate steady production of geoheated water with only minor water losses and zero emissions (except waste heat).[49]

From 1994 to 1995, the DOE tried to form a 50-50 government-industry partnership to cost-share a demonstration hot dry rock geothermal plant; however, the DOE did not find a suitable industry partner. Dr. Duchane of Los Alamos nonetheless believes it is very important that the plant be built to provide actual economic data about hot dry rock plants, as opposed to theoretical cost studies. "What we've got to get is a track record," he says. Hot dry rock technology today, he acknowledges, is as immature as electronics was in the days of crystal radio sets, but he believes a lot of efficiency improvements are possible, which would lower costs. Data from a demonstration plant would show that hot dry rock plants could be operated "at a competitive rate," Duchane contends. This, he says, would set the stage for fully commercialized hot dry rock power. However, the Geothermal Energy Association, an industry group over which Dr. Wright presides, recommended to the DOE in 1995 that DOE pursue hot dry rock research but not attempt construction of a demonstration plant yet. Given the current state of hot dry rock technology and the scarcity of R&D funds, the industry representatives felt that a demonstration plant would be costly and premature.

Although the hot dry rock concept was initially developed by U.S. scientists in the 1970s, other nations are now pursuing it more vigorously than the United States. "One thing that worries me is that the Japanese might be overtaking us in geothermal right now," says Dr. Ted Mock, who led the DOE's geothermal program from 1981 to 1995.

The Japanese have a strong incentive to produce geothermal power, because they have hot dry rock reservoirs only 3,000 to 5,000 feet below the Earth's surface. Dr. Wright believes that since hot dry rock development will necessarily be based on its component technologies—drilling, fracture stimulation, and energy conversion—we should work on these technologies separately. This will aid the industry in the short term in hydrothermal resource development while automatically developing the components of hot dry rock technology. "Industry does not believe that we will lose out in future hot dry rock development if we pursue this course," said Dr. Wright.

U.S. Geothermal Corporation of Dallas is currently studying some unique drilling methods that Dr. Duchane thinks might reduce drilling costs by as much as two-thirds. That kind of cost reduction would be very significant, since drilling costs currently appear to be 50–60 percent of overall hot dry rock geothermal costs. And with decreased drilling costs, Dr. Duchane maintains that hot dry rock could even be "very practical" in the East, where hot rock is two to three times as deep as in the West. "The hot dry rock is there," he notes."[50]

THE TANTALIZING PROSPECT: ELECTRICITY, HEAT, AND MOBILITY FROM A SINGLE SOURCE

The implications of successful hot dry rock development are exciting. "If we get this technology developed and proven," says Dr. Duchane, "I can foresee the day in the not-too-distant future when a hot dry rock system might be drilled in the basement of a large skyscraper or in part of a shopping center." Hot water would be brought up, used first to generate electricity, and then to heat the entire facility. Afterward it would be pumped back underground. "You would have a totally in-house electricity-and-heating system [that] would take up very little space." The only surface facilities would be two wellheads and a small generator. The energy supply would be 5,000 to 10,000 feet underground, or maybe 15,000 feet underground. These self-contained facilities would enable people to be energy-independent and could also be used to charge electric vehicles.

In addition to hot dry rock, magma energy at temperatures often above 538 degrees Celsius (1,000 degrees Fahrenheit) may also be commercially exploited some day, though magma research and development are still in early stages. To recover magma energy, water would be injected into a magma chamber, which would solidify and fracture in the well vicinity. Heat would then be recovered for surface power production.

To those who say that America has no alternative to fossil fuels and nuclear power, the many possibilities inherent in geothermal energy should raise additional questions, particularly since no single renewable energy source need substitute for fossil fuels all by itself. Clearly, geothermal power and heat pumps, along with a diverse array of other renewable technologies, can be important and environmentally acceptable energy supply contributors.

PART VI

KNOWLEDGE
AND POWER

21

※

EFFICIENCY:
THE SLEEPING GIANT

Improving our energy efficiency means making ourselves wealthier.
—Christine Ervin, Assistant Secretary,
Energy Efficiency & Renewable Energy,
U.S. Department of Energy

A saved watt (which we may call a "negawatt") is just like a generated watt, only cheaper, cleaner, safer, and faster to produce.
—Amory B. Lovins, Vice President and Director
of Research, Rocky Mountain Institute

No discussion of energy policy options would be complete without including energy efficiency, a measure of work or services derived per unit
of energy. Efficiency is unquestionably the best energy resource covered
in this book, because it is cost-effective, profitable, safe, clean, and dependable. The greater the efficiency, the less energy required. Its use en-
tails no sacrifices or lifestyle changes, as we will see, yet public health and
safety are improved.

While energy efficiency may sound like a "free lunch," there is no
catch. Prudent investments in greater efficiency really do enhance economic well-being, services, and satisfaction, while saving vast sums of
money and protecting our natural resources and health. With the nation
spending about $500 billion on energy every year, even a 1 percent net
energy savings due to efficiency is worth nearly $5 billion a year. And the

tremendous savings possible with more energy efficiency could make vast financial resources available for investment in renewable energy systems, in creating new jobs, and in other worthy causes. (See Table 1 for additional benefits.)

On top of the direct benefits—savings on energy, power plants, and equipment—substantial indirect benefits also occur in the form of induced economic activity, measured as a multiple of the initial investment. The economic multiplier for energy efficiency investments is typically $2.32—more than 50 percent greater than for buying oil or natural gas.[1] Moreover, enhanced economic productivity per unit of energy makes the economy more competitive and increases tax revenues through increased profits. Greater competitiveness opens new export markets for American goods and for new industries based on the manufacture and sale of efficiency technologies themselves. Meanwhile, greater energy self-reliance and reduced dependence on foreign oil improves national security, promotes economic stability, and helps our balance of payments. Finally, since fewer power plants are required by more efficient industries, less fuel is consumed, less pollution is therefore created, and so less money needs to be spent to control and clean up pollution.

Renewable energy systems tend to have high initial (capital) costs and low operating costs. Because energy efficiency reduces our peak need for energy and the size of energy systems required, the high first-cost barriers blocking many renewables would fall. This is another important connection between renewable energy and energy efficiency.

Fundamentally, energy efficiency enables us to do more with less, through better technology and smarter use of existing energy systems. Efficiency can be increased in three basic ways: by wringing more work from each unit of energy as we convert it to more useful forms (say from fuels to electricity), by plugging energy leaks, and by strategic improvements in energy system controls. Better controls allow more selective energy use—when and where we need it, in the amount needed. Controls include simple light and occupancy sensors for lighting, heating, and cooling services.

Because a kilowatt can be saved for a fraction of the cost of producing a new one, saving energy actually pays, often with far better internal rates of return than typical financial investments.[2] But unlike ordinary investments, the payback is tax-free, as it comes in the form of savings, rather than as taxable income. Efficiency opportunities are often created by market failures, such as those that occur when developers or building owners make financially expedient energy decisions that tenants or lessees then have to live with and pay for to their disadvantage.

THE KEY TO MULTITRILLION DOLLAR SAVINGS

To some, saving energy may sound dull and unrewarding—an unwanted trip to the attic to inspect the insulation, or worse, a crawl through a dank, musty basement to check on greasy motors and moldy fan ducts. But when efficiency is understood as the key to a new and scarcely tapped, *multitrillion* dollar energy resource, it suddenly becomes more intriguing.

By taking full advantage of all the energy efficiency technology now available, the United States in the long term could save more than $300 billion a year in energy costs.[3] Not only could we somehow find a way to use the extra $3 trillion a decade, but—by forgoing that energy waste— we could simultaneously eliminate billions of tons of air pollution and make our economy vastly more competitive.[4]

Energy savings in U.S. buildings alone, where we now consume 40 percent of all our energy, could amount to more than $100 billion a year.[5] Savings from making electric motors and lighting systems more efficient could yield another $75 billion annually. Our electricity use could be reduced by three-quarters—with no loss in energy services. Huge improvements in energy efficiency are also possible in the transportation sector. Vehicle fuel efficiency, for example, can be increased tenfold.

Even though formidable market failures and institutional obstacles stand in the way, we have begun cashing in on energy efficiency already: Increased energy efficiency currently saves the U.S. economy more than $110 billion every year over what we would be spending had energy efficiency not begun increasing after the 1973 oil embargo.[6] By 1990, the reduction in actual electricity use per aggregate U.S. economic output (GNP), compared to electricity consumption forecasts based on GNP growth, was equal to the output of 350 large (1,000-megawatt) baseload power plants.[7] The environmental benefits of increased efficiency have also been substantial. For example, from 1973 to 1986, avoided increases in U.S. electricity consumption from all causes averted the production of more than 880 million tons of carbon dioxide.

Sometimes, surprisingly modest changes in energy use can produce vast and painless economic benefits. For example, when the U.S. Department of Energy (DOE) simply changed to energy efficient lighting in its headquarters, it began enjoying savings of $300,000 a year. The investment cost to DOE and taxpayers was zero, since the changeover was financed by a utility and an energy service company.[8]

In the private sector, Boeing Corporation cut its lighting electricity use by up to 90 percent in some plants with a 53 percent overall return on investment and bonus improvements in lighting quality and safety, thanks to the U.S. Environmental Protection Agency's "Green Lights" program, which we will discuss shortly.[9] In the utility sector, Pennsylvania Power & Light got a 900 percent return on its lighting upgrade, primarily through dramatic improvements in worker productivity. Lockheed Martin Corporation is enjoying half a million dollars a year in energy savings from a single energy-efficient office building.[10] Energy savings in new buildings, even using 1980s technology, can be nothing short of spectacular. Completed in 1987, the Nederlandsche Middenstandsbank (NMB) south of Amsterdam (the Netherlands) uses only a fifth the energy per square foot of an adjacent bank of comparable cost.[11] The efficient building saves NMB $2.5 million a year in energy costs. In the public sector, a multistage retrofit of lights, windows, and HVAC systems at the Government Center in Massachusetts cut the energy bill at the four-building complex by $2.5 million a year.[12] (HVAC stands for heating, ventilating, and air conditioning.)

Dr. David B. Goldstein, senior scientist of the Natural Resources Defense Council, and Dr. Arthur Rosenfeld of the U.S. Department of Energy's Office of Building Technologies have each made extraordinary contributions to the field of energy efficiency by uncovering vast opportunities for energy savings.

HOUSEHOLD APPLIANCES AND
BUILDINGS: THE MEGA-SAVINGS

Physicist David B. Goldstein has studied the energy efficiency of appliances and buildings for 20 years as a staff scientist both at the Natural Resources Defense Council (NRDC) and the Lawrence Berkeley Laboratory (LBL). His work illustrates how meticulous attention both to the nature of end-use energy demand and to the engineering details of energy-using devices can yield multibillion dollar dividends.

Before the 1970s, when Goldstein and others at Lawrence Berkeley Laboratory began studying appliance efficiency in great detail, few people realized what a big difference energy efficient buildings and appli-

DIRECT ECONOMIC BENEFITS
1. Reduces customers' bills
2. Reduces the need for power plant construction
3. Reduces the need for transmission construction
4. Reduces the need for distribution upgrades
5. Reduces the threat of future fuel price volatility
6. Opens up opportunities to sulfur rights as a tradable commodity

INDIRECT ECONOMIC BENEFITS
7. Stimulates economic development by engaging multipliers
8. Creates durable jobs that benefit the local economy
9. Increases competitiveness of local business and industries
10. Energy-efficient technologies reduce maintenance and equipment replacement costs
11. Many retrofits result in the HVAC bonus

ENVIRONMENTAL BENEFITS
12. Mitigates the threat of global climate change
13. Reduces emissions that cause acid rain
14. Reduces the rate of stratospheric ozone depletion
15. Reduces the threat of nuclear accidents and proliferation
16. Minimizes pollution at mines and power plants
17. Minimizes the threat of electromagnetic fields from power lines and home wiring

SOCIETAL BENEFITS
18. Enhances national security by easing dependence on foreign energy resources
19. Increases the value of real estate in soft markets
20. Increases the comfort and quality of work spaces, which can increase productivity
21. Many electric and gas efficiency measures also save water
22. DSM programs address the regressive nature of low income people's energy use
23. Energy-efficiency programs can make housing more affordable
24. Utility programs create market transformations with long-term results

Table 1. The Wider Benefits of Energy Efficiency. HVAC stands for heating, ventilating, and air-conditioning. DSM stands for demand-side management. Reprinted by permission of IRT Environment, Inc., and The Results Center, P.O. Box 2239, Basalt, CO 81612.[13]

ances could make to the nation's energy demand, or to the utilities that had to meet that demand, or to the consumers who had to pay for it. But in the 1970s, Goldstein and colleagues crunched some numbers and proved that huge energy savings were readily attainable by making buildings and large domestic appliances energy efficient. He and his associates forecast the impact of statewide energy efficiency standards on future state energy demand. Coming at the time when California's major utilities were proposing to build 17 new power plants, Goldstein's work demonstrated clearly—to many people's astonishment—that *no* new power plants were needed.[14]

It is hard today to appreciate how radical that conclusion sounded in 1976 when he presented it to the state's energy commission, which stopped the plants. In fact, it so surprised the Environmental Defense Fund's (EDF) energy researchers, with whom Goldstein was working as a consultant, that they were reluctant to make the claim public, thinking there might be a flaw in their analysis. Then, when it all checked out, they feared they might be dismissed as outside the framework of responsible debate. Release it they did, however, and their seminal work saved utility ratepayers $50 billion in California alone that would have been spent building the unneeded power plants.[15] Ratepayers in California saved additional billions through the avoided payments for unneeded delivered energy, plus the avoided plant operation and maintenance costs and the avoided environmental damages.

California instituted statewide energy efficiency standards, and time vindicated Goldstein and EDF's erstwhile radical conclusions, for although none of the plants were built in California, no power shortages ever resulted. By contrast, in states where no one defended ratepayers effectively, dozens of large unneeded power plants were begun—and subsequently canceled—costing ratepayers billions. The total savings that ensued from Goldstein and EDF's work nationally are even greater than the $50 billion saved outright in California, since the California Energy Commission's stand against energy waste inspired many states, such as Florida, Minnesota, and New York, to set their own efficiency standards and save billions, too.

While some researchers might have rested on these ample laurels, Goldstein's spectacular success at age 25 seemed merely to whet his appetite for new challenges and greater results. During the 1980s, Goldstein and colleagues at the Natural Resources Defense Council (NRDC), whose staff he joined in 1980, became active nationally, first in litigating and then in negotiating stringent appliance energy-efficiency standards with the refrigeration and air-conditioning industries.

Codified into law in the National Appliance Energy Conservation Act of 1987, the standards also covered furnaces and water heaters. With subsequent modifications (in 1988, 1989, and 1991), the standards by the year 2015 will avoid the need for 30,500 megawatts of peak power capacity. That is equivalent to the output of more than 40 large baseload power plants operating at a generous 75 percent capacity.[16] Including the effects of the broadening and tightening of efficiency standards that occurred in the National Energy Policy Act of 1992, today's appliance standards will save consumers $132 billion in net savings (over first-cost investments) throughout the life of the affected products.[17]

Although Dr. Goldstein was justifiably pleased with the savings from the new standards, he believed that more could be done and that the potential for realizing energy efficiency had barely been tapped in the appliance market, not to mention other energy sectors.

DR. GOLDSTEIN'S GOLDEN CARROT

In 1987, an international agreement was reached in Montreal, Canada, on the need to protect stratospheric ozone from destruction by the release of chlorofluorocarbon (CFC) refrigerants and foaming agents. The 1987 Montreal Protocol on Substances that Deplete the Ozone Layer initially called for a 50 percent reduction in production of CFCs and indicated that further reductions in CFC production might follow. This created uncertainty in the appliance industry, whose representatives realized that it would have to phase out CFCs, but didn't know exactly when.

This quandary gave Dr. Goldstein an idea. Could he, in providing motivation for manufacturers to solve the CFC problem, also in some way encourage them to make their refrigerators highly energy efficient? But how? Appliance manufacturers were reluctant to exceed the national efficiency standards, because consumers had a tendency to walk past energy efficient appliances that had higher first costs, even when those appliances paid back the extra initial investment many times over in energy savings during the appliance's life.

Given this economically inefficient market behavior, Goldstein—with the aid of the U.S. EPA—established the nonprofit Consortium for Energy Efficiency (CEE) and helped launch a national energy efficient refrigerator competition nicknamed the Golden Carrot Award. The nonprofit Super Efficient Refrigerator Program (SERP) was set up to manage the contest. SERP's broad goal was to advance the technology of superefficient refrigerator/freezers and bring them to market *without* banned CFCs in the mid-1990s—years ahead of when they would otherwise have been developed. SERP set an ambitious goal; it stipulated that the winner would have to design, develop, and distribute more than 250,000 new, superefficient CFC-free units between 1994 and 1997. They would have to be 25 percent more efficient than the 1993 federal appliance efficiency standards.

By passing its hat to its 24 utility members, SERP raised a $30 million Golden Carrot prize for the refrigerator manufacturer who produced and marketed the nation's most efficient refrigerator. Thanks to changes in utility regulations that the NRDC had advocated, utilities were willing to contribute, since in many states they were authorized to earn a higher rate of return on efficiency investments than for new energy supplies.

The contest results were very gratifying. Through the use of superefficient compressors, better insulation, better sealing, and smarter controls, the 1995 version of the winning superefficient refrigerator model produced by Whirlpool Corporation uses 560 kilowatt-hours a year, which is 40 percent less energy than the 935 kilowatt-hours required by the 1993 federal energy efficiency standards. And while it costs the same as similarly sized units, the superefficient model saves customers up to $600 in electricity over its working life. The Whirlpool refrigerator will also save on the order of 7.5 million tons of carbon dioxide emissions from power plants every decade and should thereby avoid a quarter to a half a billion dollars in carbon cleanup costs over the program's life.[18] There is still room for improvement, however. The world's most efficient refrigerator, a much smaller Swedish unit, uses only 220 kilowatt-hours a year!

THE $5 BILLION WINDOW
AND THE $6 BILLION LIGHT

The man whose work at the Lawrence Berkeley Laboratory got David B. Goldstein enthused about energy efficiency in the 1970s is Dr. Arthur H. Rosenfeld, recipient of the prestigious Sadi Carnot Prize from the U.S. Department of Energy for lifetime achievement in the fields of energy conservation and renewable energy. A professor of physics at the University of California in the 1970s and staff scientist at the Lawrence Berkeley National Laboratory, Dr. Rosenfeld has contributed to the field of energy effi-

Dr. David B. Goldstein, senior scientist of the Natural Resources Defense Council, winner of the American Council for an Energy Efficient Economy's 1994 "Champion of Energy Efficiency" award for conceiving and promoting the Super Efficient Refrigerator Program and for creating the National Appliance Energy Conservation Act of 1987. *Courtesy of Dr. David B. Goldstein.*

ciency through his work as founder and director of the laboratory's Center for Building Science.

Rosenfeld's career was auspicious from the start: He studied under Nobel laureate Dr. Enrico Fermi and had the honor to coauthor *Nuclear Physics* with Fermi and two other scientists. Much of Rosenfeld's career until 1973 was spent doing research on particle physics, uses of computers, and arms control, before he switched to energy utilization. Currently a senior advisor to the U.S. Department of Energy (DOE), Rosenfeld thinks he and his LBL colleagues have lately found what may be the single most cost-effective approach to the improvement of urban air quality. The method hinges on mitigating what Rosenfeld calls "urban heat islands."

Researchers at Rosenfeld's Center for Building Science have been responsible for developing low-emissivity ("heat mirror," or "low-E") windows and have played a catalytic role in developing high-frequency electronic "ballasts" for fluorescent lamps. Ballasts control the current and help start fluorescent bulbs. Electronic ballasts not only save energy but eliminate bulb flicker and hum. Sales of electronic ballasts and "low-E" windows have together saved energy users $5 billion already.[19] Once electronic ballasts saturate the market, Rosenfeld calculates they will save $6 billion every year. These are truly spectacular payoffs for a $3 million federal research-and-development investment in electronic ballasts and should give pause to critics of federal energy R&D spending.

In addition, energy-saving compact fluorescents—whose commercialization was greatly accelerated by electronic ballasts—will save consumers and business $7 billion a year. (Compact fluorescent fixture improvements made by the Center for Building Science have increased light output by 25 percent and have been widely adopted by manufacturers.)

Despite the progress already made in lighting and other energy-use technologies, much new technology is still being developed at LBL. Prototype sulfur lamps twice as efficient as the most efficient fluorescents, and with superior light similar to sunlight, are now under development in partnership with industry. A single bulb can deliver as much light as 175 full-sized fluorescent lamps.[20]

The "low-E" windows the Center pioneered through two decades of such industry partnership will save more than $5 billion annually.[21] Low-E coatings allow the passage of visible light but reflect heat in the form of infrared radiation. LBL research showed that by using inert gas to fill the gap between panes of double-glazed windows, manufacturers could produce windows for cold climates that greatly reduce winter heating requirements by trapping heat inside buildings.

By also using low-E coatings to prevent heat from entering through windows, buildings can stay cool in warm climates, reducing air-conditioning needs. The next step beyond simple low-E windows will be dynamic "smart windows," which are now under development. These low-E windows will use a small light-induced current to control their own optical properties electronically. The same window will be able to exclude heat in summer as well as trap it inside the building when needed in winter. Thin-film solar cells can be integrated into the window glass to provide current with which to control transparency without external wiring. A polymer window film coated with electrochromic materials and thin-film photovoltaic cells to control the window's transparency has already been developed by researcher David Benson at the National Renewable Energy Laboratory. The film could be glued right to the window exterior.

Rosenfeld has also plied his formidable computer skills in the quest for building energy savings, and he wrote the first version of a series of computer programs now known as "DOE-2," for analyzing building energy use and for designing and rating efficient buildings. The program, which was developed by three national laboratories (funded by DOE) and an outside consulting company, can simulate all building types in all climates. Its use has saved energy users $1.9 billion cumulatively nationwide through 1993, and it currently saves $1 billion a year in energy costs through the implementation of California's building energy standards that were developed using DOE-2.[22]

With billions of dollars in cumulative energy savings from research activities such as those just discussed, it is virtually incomprehensible that anyone would want to hamper these extraordinary efforts. Yet in 1995, it was fashionable among certain budget cutters in the Republican Congress to call for dismantling the Energy Department and eviscerating its cost-effective energy research. Some legislators seemed bent on depriving the nation of advances in energy efficiency and renewables.

SMART ROOFS AND URBAN HEAT ISLANDS

These days, Dr. Rosenfeld has found another big source of potential energy savings for utilities and ratepayers. He and other LBL researchers have determined that simply by cooling hot urban surfaces—by increasing their albedo (reflectance) and by shading them with trees—$10 billion a year in energy and equipment costs can be saved, 27 million metric tons of carbon dioxide emissions can be eliminated, and urban smog can be reduced.[23] How so?

Modern cities are about 2.5 degrees Celsius (5 degrees Fahrenheit) hotter than surrounding rural areas. That is in large part because natural surfaces and vegetation have been replaced with dark, dry, unshaded surfaces, especially asphalt roads and asphalt shingle roofs. This not only affects our comfort indoors and urban energy use, but also affects climate by warming the air over urban areas.

Surfaces that are dark in color and therefore highly heat-absorptive get a great deal hotter—up to 50 degrees Celsius (122 degrees Fahrenheit)—than their surroundings. Less absorptive light-colored surfaces, which reflect most of the light reaching them, remain only moderately warmer than the ambient temperature. Researchers at the Center for Building Science wondered what affect painting roofs white and shading buildings with trees would have on building air-conditioning loads. They also wondered whether, by changing the reflectivity of road surfaces at the same time, they could cool the whole urban area enough to affect urban smog formation, since heat accelerates the chemical reactions that create smog.

Using actual experiments to measure the effects of building roof coatings and shade trees on air-conditioning loads of buildings, plus computerized energy-use simulations, the LBL scientists found that light roofs and trees reduced summer air-conditioning loads by 40 percent in houses in Sacramento.[24] Trees alone, which cool by shading and evapotranspiration, decreased the air-conditioning loads for about 1 percent the avoided capital cost of power plants and air-conditioning equipment.

Next, the researchers used a computer land-use database to simulate a doubling in the average reflectivity of developed portions of the Los Angeles Basin. This modification could be accomplished by lightening roof coatings and road surfaces. By making these changes gradually, during periodic routine roof maintenance, and by retopping roads during regularly scheduled resurfacing, the changes could be very inexpensive but just as effective.

A moderate change in the average reflectivity of the land area could lower peak summertime temperatures in Los Angeles by 2–4 degrees Celsius. The researchers calculated that the resulting reduction in smog due to the temperature decrease would be equivalent to removing ten million cars from Los Angeles roads! Because the city-scale temperature reduction would be reinforced by lower interior cooling loads, due to cooler roofs and shade trees, the researchers estimated that summertime urban cooling energy could be reduced by 20 percent.

To reduce the problem of glare from white surfaces, Rosenfeld proposes that only flat roofs not visible to passersby would be painted white, with light-beige reserved for sloped rooftops. Roads, he explains, need only be lightened to the color of weathered concrete. If the urban heat islands of other cities behave as these findings suggest, then, when fully implemented within 20 years, a nationwide urban heat island mitigation program could produce multibillion dollar energy savings across the country.

Further study still needs to be done, however, on the offsetting effects that increased reflectance will have on winter heating needs of cooler cities. In the future, "smart roofs" (analogous to smart windows, discussed earlier) might be developed to adjust their reflectance seasonally, so as to counteract this potential problem.

PROOF OF CONCEPT: PG&E'S ACT2
PROGRAM (A HARD ACT2 FOLLOW)

Thanks in no small measure to the work of scientists such as Arthur Rosenfeld, David B. Goldstein, and Amory Lovins, major utilities, such as the Pacific Gas and Electric Company (PG&E), established large and significant energy efficiency programs. Utilities in New England and California began making major investments in efficiency once they were allowed to not only secure a rate of return for providing energy services (instead of just supplying kilowatts), but were permitted to earn greater profits from the services than from energy sales.[25]

PG&E has invested $1.069 billion in Customer Energy Efficiency (CEE) programs since 1990. The programs have been cost-effective, resulting in benefits of $1.795 billion for participating customers alone, plus additional benefits to the company, by lowering its future cost of doing business and by generating shareholder profits, all while saving enough electricity to power 345,000 homes and enough natural gas to heat 210,000 homes annually. For 1994, the company's $186 million investment in CEE was expected to save customers $57 million a year on their energy bills; PG&E's future costs will be lowered by $172 million (present value); and PG&E's shareholders will earn $15.7 million, to be recorded over ten years.[26] One of PG&E's most exciting CEE efforts is ACT2, the Advanced Customer Technology Test for Maximum Energy Efficiency, which began in 1990 and will continue through 1997.[27] The ACT2 program's goal is to demonstrate comprehensively and scientifically in real buildings how much energy can be cost-effectively saved through the use of optimized integrated systems of modern, high-efficiency technologies. ACT2's project team designs, installs, monitors, and then evaluates these technology systems at pilot demonstration sites with the intent of capturing the synergistic savings possible through the interaction of the component technologies.

Prior to ACT2, PG&E and most utilities relied on individual efficiency measures for energy savings rather than taking advantage of their combined effects, which can exceed the sum of the individual savings. For example, by using natural daylighting, artificial lighting can be reduced, internal heat gain can be minimized, and air-conditioning loads can be decreased.

ACT2 identified sites in the residential, commercial, and agricultural sectors and then tested energy efficiency systems in both new and modi-

fied residential and commercial buildings. PG&E started with a commercial office building in San Ramon, California, occupied by its R&D
division. After reducing heat gain with low-E windows and lighting
efficiency improvements, the project team installed high-efficiency air-
conditioning and digital energy-management controls for the heating
and cooling system. Building electrical demand promptly dropped by 56
percent, largely due to the air-conditioning improvements. The achievement is especially significant, since air-conditioning is the largest load in
most office buildings.

To demonstrate energy efficiency measures in new residential construction, ACT2 designed and commissioned a single-story, four-bedroom,
two-bathroom house in Davis, California. The goal was to maximize energy savings consistent with the area's $250,000 mean home price. The
ACT2 design team simulated the energy performance and cost of six design alternatives before calculating cost/benefit ratios and selecting a design utilizing a package of 20 energy efficiency measures. Although the
house is located in California's Central Valley, where daytime summer
temperatures are often in the high nineties, the home's passive solar orien-
tation, high-efficiency external envelope, windows, doors, and appliances
make it so resistive to heat gain that it remains comfortable with *no* air-
conditioning. The final design of the Davis house reduces total energy
consumption by 62 percent, saving 100 percent of the air-conditioning
energy and 78 percent of heating energy use. A new engineered wall-
framing system was custom designed for the house, which also has a combined refrigerator–water heater, and radiant floor heating. The new wall
system uses oriented-strand structural wood studs (stronger but smaller
than ordinary 2-by-4-inch studs) and rigid foam insulation panels along
with stucco and drywall. Because of the new wall's reduced wood mass
and improved insulation, it has an 85 percent higher thermal resistance
yet is stronger than conventional walls.

Another ACT2 single-family residence in the Stanford Ranch development near Rocklin, California, presented an even greater challenge,
since, unlike the Davis site, it doesn't naturally cool off much at night.
Designers chose a night-operated direct evaporative cooler with subfloor
radiant cooling that operates efficiently at night and delivers passive cooling the next day. A single gas-fired heater provides both space heat and
hot water. Other features are similar to the Davis house. The 1,700-
square-foot Stanford Ranch home's integrated design and appliances cut
air-conditioning load by 90 percent, lighting and refrigeration by 76 per-

cent, and heating by 72 percent, yet the house is designed to sell for about $172,000, the mean price in the area.

ACT2's new commercial office building for the California Automobile Association in Antioch, California, may be the most energy-efficient building in the nation and may well be ACT2's most impressive achievement. It has state-of-the-art energy efficiency equipment of all kinds and an overall energy-efficient design. Features include light-colored walls, a white roof, and 29 skylights with prismatic lenses and louvers operated by light sensors to control illumination, diffuse sunlight, and reduce glare. Light-level sensors also turn on high-efficiency, dimmable T-8 fluorescent lights; occupancy sensors turn lights off when unneeded. Heat gain is further minimized by double-pane, air-filled windows containing a thin mylar "heat mirror" that rejects most radiant heat, and by wide roof overhangs that shade south and west windows from the summer sun. Water for washing is heated by point-of-use water heaters that deliver warm water to low-flow faucets.

The net effect of these features is to reduce projected building energy use by 72 percent over California's Title 24 building energy standards— saving nearly three-quarters of total demand compared with standards that themselves restrict buildings to about a third of the electricity and gas used by comparable older buildings. PG&E estimates that the cumulative energy savings will be worth more than a quarter of a million dollars to the occupant-owner.

Very clearly ACT2 showed that the energy savings projected by energy scientists Goldstein, Lovins, Rosenfeld, and others are real and achievable at competitive costs using carefully selected and, for the most part, off-the-shelf equipment. The ACT2 accomplishments also reveal the value of an integrated design process that carefully considers the interaction of all building systems, instead of fragmenting responsibilities for design among a series of subcontractors who work in relative isolation from each other. The latter miss the valuable energy saving "synergies" that integrated design offers.

22

MINIATURIZING ENERGY DEMANDS

A building engineer puts lines down on paper, it gets built maybe
three years or five years later, and he's off on another disaster. He
never really goes back to see the first disaster.
—Lee Eng Lock, founder,
Supersymmetry Pte., Ltd.

Lee Eng Lock is a master engineer who uses elegant engineering to
shrink industrial energy demand for corporations, large and small. His
achievements at once illustrate the elegant art of energy efficiency and
the "nuts and bolts" of slashing energy consumption. But does industry
appreciate the profit-enhancing potential in his work? Not quite. Not yet.

Lee has a worldwide reputation for making large commercial and industrial buildings energy efficient. He hates waste, scorns the corruption that
is endemic in some of the countries where his international consulting firm
works, and likes to see things run well. His passion is to displace kilowatt-
hours with elegant engineering reflecting his dedication to craft and mastery of fluid flow and thermodynamics. Using intuition and fingertip feel,
Lee can walk through a commercial or industrial building, touch a motor,
glance at lighting and ductwork, and then provide a three-page summary
about what needs to be looked at right away. Lee's work illustrates the
tremendous potential for saving energy in buildings and rests on a solid
foundation of sophisticated data analysis and meticulous engineering.

Lee's *modus operandi* is to accurately measure the performance of
buildings and their energy systems so as to precisely match equipment to
actual loads. "Without actually knowing data, you tend to oversize to have
a safety factor," says Lee. Once the equipment is correctly sized, it also

must operate harmoniously. Lee's job then is not unlike that of a conductor tuning an orchestra, although the orchestra consists of motors, fans, pumps, and valves, not strings, woodwinds, and percussion.

MORE DATA THAN A MOON SHOT

Sizing and tuning all the interactive machinery routinely requires collecting and analyzing mountains of data from monitoring equipment that samples building conditions and complex equipment behavior minute by minute for periods of up to two years or even more. Grappling with all that data daunted some of his competitors, but challenged Lee.

Born into a poor Singapore family of Chinese ancestry, Lee managed to get to Surrey University in England, where he studied naval architecture before gravitating to the energy services business and eventually starting his own consulting firm, Supersymmetry Pte. Ltd., based in Singapore. Lee thinks like a master boat-builder and frequently uses boat-building analogies. He likens his work to the simplification and minimization of designing a sailboat. "You have to make the hull look good. You have to have the right sail area. It's a combination of things. . . . Craftsmanship and attention to detail; good materials, common sense."

At Supersymmetry, Lee honed his instrumentation and measurement skills in designing and building multimillion dollar dust-free "clean rooms" for computer and other semiconductor manufacturers. To build and control the clean rooms, Lee had to measure very accurately the temperature, volume, and pressure of airflow through the rooms and the concentrations of airborne dust particles. These variables had to be monitored continuously to make sure they were within specifications; otherwise, semiconductor chip yields would fall dramatically, causing millions of dollars in losses per hour. No expense was therefore spared on clean-room instrumentation.

These instruments generated more data by far than a moon shot or the *Encyclopedia Britannica.* Somehow, Lee had to figure out how to review and analyze the gigabyte-scale data—far more than normal spreadsheets could handle. So Lee went first to Silicon Graphics for the best and fastest hardware he could get. Then he hired software developers to create powerful new data-imaging methods for him.

DECIPHERING THE DATA

Once Lee examined his data sets with these new tools, he discovered that the energy-management systems for the clean rooms were operating way outside their design ranges. He therefore began replacing the oversized equipment and installed new controls. When he did so, he found that not only did energy consumption go down, but the clean-room conditions became stable, which was exactly what the semiconductor industry needed.

Using a combination of his new state-of-the-art data-imaging software and his computerized building-information systems coupled with the best energy-efficient technology, Lee then began saving major corporations in the Far East and the United States millions of dollars on their overall energy bills in factories and commercial buildings. At one Compaq Computer Corporation building in Houston, Lee cut cooling energy use by 90 percent and saved Compaq about $5 million a year in overall electricity costs plus additional capital equipment savings, according to Compaq's former facilities manager Ron Perkins (now president of Supersymmetry USA).

After redesigning the energy system for an AT&T plant in Singapore, which then used a tenth of the energy per square foot of a comparable AT&T plant in Louisiana, Lee got an assignment to check the energy design of an AT&T telephone manufacturing plant to be built in Thailand. Through his design improvements, the plant's chiller (central air conditioner) was downsized from 1,200 to 400 tons of capacity. That step saved AT&T several million dollars in capital, since smaller chillers meant a cascade of savings in smaller pumps, cooling towers, piping, and transformers.[1] "We redesigned the whole system," Lee said, "and took out all the surplus pumps and control valves and silly piping layouts." The building's electrical consumption promptly shrank by half and saved the company about a quarter of a million dollars a year in energy costs. Instead of costing the company money, energy efficiency had a "negative payback" period.

The experience in Thailand was far from unique. Lee was also able to save Siemens millions of dollars in energy capital costs for a semiconductor plant it built in Singapore. "It's very common in engineers to oversize and overkill like hell. Probably it's that you don't have to do good engineering," Lee remarked. He likes to tell of the palatial Asian Development Bank (ADB) building, which received a coveted international award for energy efficiency from the American Society of Heating, Refrigerating, and Airconditioning Engineers (ASHRAE). At 1.7 million square feet, the ADB building was one of the largest in Asia. But on the request of the International Institute for Energy Conservation, Supersymmetry did a mechanical audit of the building. The audit showed that the ADB could have realized another 58 percent in energy savings with a three-year pay-back period on a $5 million annual energy budget. "People have good intentions, and they have the money," said Lee, "but it [energy efficiency] doesn't happen, and you see this repeatedly."

MARKET BARRIERS TO
ENERGY COMMON SENSE

Lee recently identified examples of unexploited opportunities for energy efficiency in a typical commercial building—a large warehouse-style discount electronics mart in Palo Alto, California. Walking through the bustling aisles, Lee pointed to the lighting: Just changing bulbs, he noted, would reduce lighting demand by nearly 40 percent; substituting electronic for magnetic ballasts would cut ballast energy losses by more than 80 percent. Motion sensors connected to light dimmers would cut lighting energy by another half whenever aisles were unoccupied. For the ceiling, Lee suggested more insulation and heat-reflective white paint for the roof, and high-efficiency rooftop coolers with controls that would alert operators when maintenance is needed.

The megastore's aisles were noticeably drafty. "Right now, the fans are single speed, right?" said Lee. "They're on at a speed set to cool the building on the hottest day of the year, and now it's raining and cool." Variable speed fans would provide significant savings. "You could do with half the air," Lee commented. Optimizing air-duct size and reducing the number of bends and sharp "takeoffs" generally reduces fan loads by half to three-quarters.

How difficult would it be to get the building manager to adopt the seemingly simple changes that Lee was suggesting? Harder than you might think. Lee usually finds building engineers, chief engineers, or facilities managers profoundly uninterested in the opportunities he offers, due to a lack of personal incentives. Moreover, even when motivated, "building operators are not CEOs," he said. "Few people listen to them." Likewise, Ron Perkins, president of Supersymmetry USA, has found it difficult to interest U.S. companies in saving building energy.

Their behavior, which causes market failures, might generously be called "bounded rationality." Since energy costs are only a small percentage of overall operating costs for many large firms, high-level company executives tend to ignore energy expenditures to focus on larger cost centers when striving to reduce spending. They ignore the fact that energy savings can be important contributors to a company's bottom line, especially when business is bad.

As Ron Perkins has pointed out, if a company spends 1 percent of its revenues on energy and enjoys profits of 5 percent, then a 50 percent savings in annual energy costs can mean a 10 percent increase in profits. Despite this important payoff and the fact that Supersymmetry can easily save customers half of their energy use in buildings, demand for Supersymmetry USA's services is feeble, due to the market's failure to perform efficiently. "The economy just beats a company like us up," said Perkins. "When we go in and try to make changes or sell our services to building owners, we have to do battle with their design firm. And the design firm always wins."

PERVERSE INCENTIVES

Lee and Perkins have both learned through hard experience that, whereas saving energy cost-effectively is a relatively straightforward process, people sometimes feel threatened by the process or lack motivation to participate. Design consultants, mechanical engineers, and equipment suppliers are not happy to be told that their design "rules of thumb" lead to oversized and hard-to-control equipment. Consultants hired on the basis of a low bid may have no time to ask questions and do more than recycle old design work. Usually they are given no incentive to do an energy-efficient design. More precise engineering takes more work and time, which all too often might be unappreciated, or might come out of the consultant's profit.

"In engineering, you have to do things elegantly," said Lee, "which sometimes conflicts with doing it for the fees. In a fee, you are driven to jack up capital costs. If you make it elegant and cheap and better, you're working against your own financial interests." Amory Lovins concurs: "As long as we reward design professionals for what they spend, not what they save, we can expect to get nutty results."

The equipment contractor also has a perverse incentive to sell the biggest cooling or heating unit he can to realize the greatest profit and perform the additional work the larger system requires. "Maybe he has to upgrade the power supply, and so on," said Lee. "That temptation is also present for the mechanical engineer. Put in more stuff, and get more fees for the stuff."

The solution is that private incentives must be brought in line with public interests. Engineers must be paid partly according to how energy efficient their buildings are. "They should be paid for results, not the process," Lee advises. The city of Oakland, California, is developing a compensation system along these lines, and Ontario Hydro has one. Lovins recommends direct utility rebates to building designers and developers. Perverse incentives, explains Lovins, have suppressed real engineering and have cost the nation up to a trillion dollars in unnecessary energy equipment and power supplies.[2] "The result is a $12 trillion stock of obsolete buildings in this country," he says.[3]

The implications for the owner-builder of an individual office building, Lovins has shown, are that the present value of the total energy savings realized by making the building energy efficient can actually pay for the building![4] The building would also have a competitive market advantage because of its lower operating costs and would be more comfortable, because of the modern lighting and space conditioning controls. Workers tend to be more productive in those environments, and a percentage increase in worker productivity is generally worth a hundred times an equivalent percentage increase in energy savings, because of the magnitude of investment in workers' salaries compared to energy bills.

CHILLER ENVY

Fees and money are not always the cause of resistance to energy efficiency. Obstacles come in unexpected forms. At a Compaq plant in Singapore some years ago, Lee found that the facilities manager was opposed to reducing oversized cooling equipment for psychological reasons. "There's this very macho thing [among facilities managers]: 'I have a bigger chiller than you,' " Lee said. "The previous facilities manager, who was a Singapore Chinese guy, wanted to have a bigger chiller than his friends. It's unbelievable. People are willing to spend millions to say, 'I have a bigger plant than you!' " Moreover, the original consultant's design for the plant had called for three 800-ton chillers. "In Chinese," said Lee, "eight is a lucky number, and so, three 800s!"

Lee has observed a more general problem that impedes the engineering of energy-efficient buildings. "A building engineer puts lines down on paper," said Lee, "it gets built maybe three years or five years later, and he's off on another disaster. He never really goes back to see the first disaster." Not only is the engineer not paid to go back and learn from his earlier work, but until recently, no mechanisms existed to track energy-system performance and archive years of data so it could be analyzed later. Supersymmetry is working on the development of advanced software to make that possible.

Lee is also working on an automated building diagnostics project with a consortium of universities funded by the California Institute for Energy Efficiency to collect accurate building energy-use data, analyze it, and make it available on the Internet. Probably no one will be more interested in the avalanches of data the project's information systems will produce than Lee himself. For mechanical engineer Lee Eng Lock, learning how to manage energy elegantly is a lifetime challenge.

ENERGY EFFICIENCY AND THE DEVELOPING WORLD

Apart from its enormous domestic value in the United States and other heavily industrialized nations, energy efficiency has an immensely important role to play in the ongoing struggles of the world's developing nations to free themselves from poverty. A developing nation's economy that attempts to keep up with unrestrained energy demand growth by continuing to build new power plants will consume an enormous amount of capital but ultimately will fall behind demand.[5] The World Bank estimates that developing countries are attempting to spend $1 trillion on new power supplies in the 1990s.[6] This $100-billion-a-year effort will divert scarce capital from other urgent social and industrial needs. The World Bank recognized that the developing nations will have difficulty finding this huge amount of capital and will therefore have to borrow heavily from abroad.[7] Servicing that debt will then be a drag on economic expansion. In addition to the price the developing nations will have to pay for failing to use energy efficiency, that policy will eventually undermine the climate-protection efforts of the United States and other developed countries.

A quick look at some energy-consumption patterns and trends should make this clear. As of 1989, the primary energy consumption of the 246 million people in the United States was equal to the combined energy consumption of all of the 3.5 billion people of China, Asia, Latin America, the Mideast, Africa, and a small group of newly industrialized nations.[8] Most of these nations are increasing their primary energy consumption very rapidly—9.7 percent a year for newly industrialized nations and about 5.5 percent for China, the Middle East, Asia, and Africa. At these rates, their collective energy use will increase more than two and a half times just by the year 2025.[9] The developing nations (and former Communist nations) currently have energy efficiencies four times lower than the Western industrialized countries.[10] If these nations neglect efficiency and renewables for their growing energy needs, they will instead rely heavily on coal burning and other environmentally destructive central power plants.

The consequences will be nasty: more acid rain, air pollution, and global warming. Forests, world agriculture, species diversity, endangered habitats, and human health will suffer, as discussed earlier. Finally, the slower economic progress in developing countries will result in continued geopolitical instability and conflict related to poverty and the perpetuation of vast disparities between developed and developing nations. If this grim future transpires, we may need to get used to the added discomfort of hotter summers, more severe storms, coastal flooding, degraded ecosystems, and the energy-environmental costs of increased air-conditioning demands.

The developed nations have little moral ground to stand on and little credibility in trying to sound an alarm about all this now, due to our disproportionate national energy use and waste. We will simply be portrayed by developing nations as greedy, rich, self-serving people indifferent to their aspirations for a better life. In telling us to mind our own business, the developing nations would merely be asserting their prerogative to emulate us in squandering energy. For this, if for no other reason, the United States needs to "get religion" fast, by making maximum use of energy efficiency. If we become an energy-efficient society, we will have at least *some* credibility in advocating energy-efficient development and in encouraging others to leapfrog over the mistakes of our industrial history. Even with the best of examples, however, it will still be very difficult for the developing countries to build energy-efficient economies.

Bombay-born physicist Ashok Gadgil, who now works at Lawrence Berkeley Laboratory (LBL), knows this as well as anyone. Winner of a Pew Charitable Trust Award for advancing energy efficiency in developing countries, Dr. Gadgil has studied the potential benefits of energy efficiency there and how to overcome formidable obstacles to its utilization.

His research has shown conclusively that technologies such as compact fluorescent lights (CFLs) could have an enormous economic payoff in developing nations.[11] If India, for example, were to invest in a modern $7.5 million factory producing 6 million CFLs per year, he calculates that the country would save the energy equivalent of 3,700 megawatts of installed peak capacity. Coal-power stations of that capacity would cost India $5.6 billion, excluding fuel. Gas turbines would cost $2.8 billion. The CFL factory would be hundreds of times less costly.

The analysis can be extended to other energy efficiency technologies, as Dr. Gadgil has demonstrated.[12] A $10 million factory producing low-E windows in India can provide immense energy savings in the form of avoided air-conditioning loads over the windows' 30-year life. The 4 million megawatt-hours saved means that utilities can avoid building an 800-megawatt power plant costing $1.5 billion.

Unfortunately, many Asian utilities, industries, and governments are still indifferent or resistant to energy efficiency.[13] Indonesia is reportedly planning to build nine nuclear reactors—and in an earthquake-prone region.[14] Old ways of thinking and doing business die hard, and special interests can often skim fortunes from multibillion dollar power plant projects without public exposure or sanction. However, energy efficiency will ultimately flourish worldwide because of the powerful economic arguments in its favor.

GREEN LIGHTS:
A PROFITABLE PARTNERSHIP

In addition to the work of federal and nonprofit sector research scientists and private engineering consultants, government efforts that creatively explore the convergence of energy efficiency and private profit are also worthy of recognition.

Starting in the mid-1980s, John S. Hoffman, an urban economist, geographer, and systems modeler, directed efforts at the U.S. Environmental Protection Agency (EPA) to reduce destruction of stratospheric ozone by CFCs. While part of that program, he was impressed by the effectiveness of voluntary agreements between industry and government to find CFC substitutes. Thinking that the same approach might work to forestall climate change by promoting energy efficiency, Hoffman in 1991 established EPA's Global Change Division, now known as the Atmospheric Pollution Prevention Division.

Instead of merely aiming to control pollution, Hoffman's goal was to prevent it profitably, create jobs, and foster private initiative. The mechanism was to be a voluntary program with strong technical support that would alert participants to profitable energy efficiency investments. The initial budget was only $1 million, but Hoffman hoped that his demonstration program would reach beyond the immediate participants to catalyze a chain reaction in which successes would invite imitation and cascade through the whole economy.

The division's first initiative, the Green Lights Program, was designed to profitably increase the use of energy-efficient lighting in buildings by overcoming existing barriers to it. The program provided technical information, state-of-the-art computer software to aid decision making, a computerized database of available third-party financing options, a product information database, and valuable public recognition to participants. (Recognition can translate into competitive advantages.) Although participation was voluntary, participants were required to sign a memorandum of understanding committing them to survey their facilities and make cost-effective lighting changes without compromising lighting quality. Those who renege on their commitments can be expelled.

Green Lights began with industrial and commercial partners who now number more than 1,600 firms. Thus far, industrial customers have saved as much as 50–80 percent or more of their lighting electricity. Internal rates of return have been as high as 186 percent, with an average of 30–40 percent. Because of its success, the program is now expanding into the residential sector.

EPA believes that the Green Lights Program, if fully implemented, could save the nation $16 billion a year in electric bills; reduce utility emissions of carbon dioxide, sulfur dioxide, and nitrogen oxides by 12 percent; and create 220,000 new jobs by the year 2000. Resulting increases in private sector profits will also bring new tax revenues to the U.S. Treasury.

ENERGY STAR BUILDINGS
AND THE ENERGY STAR PROGRAM

Following the success of the Green Lights Program, the EPA has launched a more comprehensive Energy Star Buildings program that applies Green Light's voluntary, market-based approach to fostering efficiency in heating, cooling, and air handling in commercial buildings, using proven technologies. Participants enroll in a five-stage sequence of steps that results in the progressive adoption of profitable energy efficiency systems. The phased program maximizes savings and minimizes costs by first reducing loads to minimize later equipment-upgrade costs. The EPA provides technical support, including many of the tools needed for building energy audits and for evaluating efficiency measures. The EPA's Atmospheric Pollution Prevention Division also is helping to create a market for energy-efficient personal computers and other equipment by working with their manufacturers and consumers through the agency's Energy Star program.

Computing equipment now accounts for 5 percent of commercial electrical consumption and, as the fastest growing business load, could reach 10 percent by the year 2000. Energy-efficient computers and related equipment identified with the EPA Energy Star logo save energy by automatically switching to a low-power mode when inactive, and in other ways, yet they cost no more than conventional energy-wasting equipment. According to the EPA, Energy Star equipment could save the United States enough electricity by the year 2000 to power Maine, New Hampshire, and Vermont. The efficient equipment also lasts significantly longer than ordinary equipment, which gets hotter. Just by choosing more efficient PCs and other office machines, companies can significantly and profitably reduce their building cooling loads.

By President Bill Clinton's Executive Order, all U.S. Government agencies must purchase only equipment that meets Energy Star requirements. The EPA itself expects to save enough money through using the new energy-efficient computers, monitors, and printers to pay for the Energy Star computer program several times over. The EPA is now expanding the Energy Star product identification program to other office equipment, to residential appliances, and to new homes. It has already extended it to heating and cooling equipment.

Based on the success of the Golden Carrot super-efficient refrigerator program discussed earlier, the EPA has begun a Golden Carrot washer program and will operate similar programs for other large appliances and energy-saving equipment. A significant feature of the EPA's residential energy use program will be agreements with financial institutions to increase the availability of energy efficiency financing on favorable terms for homeowners. The EPA's intent is to lower the first cost of homeowners' investments in efficiency, so they become financially attractive.

THE EPA'S POLLUTION PREVENTION
BALANCE SHEET: +$37 BILLION

The EPA is also working to improve the efficiency of electric utility transformers, in which about 50 billion kilowatt-hours are lost each year, and to encourage the profitable recovery of methane from coal mines and landfills, while curtailing natural gas losses from natural gas transmission and distribution systems. Through its AgSTAR program, the EPA is simultaneously promoting the recovery of methane from livestock manure and the adoption of "best management practices" at thousands of the nation's farms. Recapturing methane not only provides a usable fuel, but is an important part of climate-stabilization programs, since methane is about 20 times as potent as carbon dioxide in producing global warming. When burned, however, methane is transformed to the more innocuous carbon dioxide.

The Atmospheric Pollution Prevention Division, begun on a shoestring budget in the Bush administration, by 1995 had grown to a $55 million program. The EPA estimates that its voluntary programs will produce a $37 billion profit for the U.S. economy by the year 2000 with a reduction of 215 million metric tons of carbon dioxide plus significant reductions of conventional air pollutants. This success demonstrates the enormous value that sound government programs can have, and is a bright spot on the environmental scene. By helping to break down barriers to energy efficiency through supplying information, technical expertise, motivation, and access to financing, and by creating opportunities for people to make money—but by not prescribing exactly how they ought to utilize those opportunities—the EPA has found a valuable formula.

John Hoffman has had the satisfaction of seeing his program's results spread beyond its participants themselves. Mobil Oil, for example, which participated in the Green Lights Program, brought many other companies into the program and went on to initiate a major energy plan for the whole company, using lessons learned through Green Lights. Hoffman's foray into promoting profitable energy efficiency has now convinced him that the energy efficiency industry is a sleeping giant. "I think it's going to be a gigantic money-making industry," he said.

In the two chapters that follow, we will glimpse another set of new trends and prodigious business opportunities unfolding in the transportation sector. There the applications of energy-efficient new technology are revolutionizing vehicles and creating dynamic new industries that will rule the road in the next millennium.

TECHNICAL FIXES:
NECESSARY BUT NOT SUFFICIENT

Although some of the energy efficiency gains described in Chapter 21 and in this chapter involve isolated, straightforward "technical fixes," changes in technology are only part of what's needed to create an energy-efficient economy. We must also deploy the advanced technology—computers, electronics, and new materials—in holistic integrated designs of buildings, communities, transportation systems, and other energy-using systems. While science and engineering have a central role in making efficiency possible, the efficient use of energy, broadly speaking, requires systems engineering. That, in turn, often involves public education, public policy initiatives, economic incentives, financing, and political and institutional reform. Cultural values and psychological idiosyncrasies can and do cause significant market failures, too. So major changes in energy use often also require shifts in cultural values and norms governing energy use.

23

ELECTRIC VEHICLES:
HAVE BATTERY, WILL TRAVEL

> In the year 2020, when a grandchild pushes your wheelchair
> around the auto museum, try to convince her those relics actually
> consumed nonreplenishable fuel, emitted pollutants, lived in traf-
> fic jams, and couldn't find parking spaces.
> —Paul MacCready, Chairman,
> AeroVironment, Inc.

Their skin is smooth and shiny. Their flanks are light and strong. Their breath is clean. They leap forward at your touch without a growl or roar. They do not belch, they leave no mess behind, and they are reliable travel companions. Meet the new high-tech electric vehicles—tomorrow's cars, yet they're here today. Although not yet fully mature, these lively adolescents are filled with exciting promise and can serve most travel needs, even in their slightly awkward youthful form. Moreover, they will have an increasingly important role in tomorrow's transportation system. Their technology is rapidly improving, their prices will fall with mass production, and customers are eagerly awaiting them. General Motors, Chrysler, Ford, Toyota, and Honda all are already selling or leasing small numbers of electric vehicles.

Susan Tierney, former assistant secretary of energy in the Clinton Administration, has gone so far as to predict that more than 20 million electric vehicles could be on the road by 2010, if their rapid technological and market-entry advances continue. The reasons are obvious. Electrics are clean, quiet, smooth, and efficient. They retain the best features of the personal auto—its convenience and privacy—while jettisoning most of the pollution.

ELECTRIC VEHICLES: NO PANACEA

Of course, electric vehicles are not a comprehensive solution to all our transportation problems. They do not eliminate traffic gridlock or traffic accidents, nor do they counter urban sprawl and a host of other problems created by auto dependency. That would be asking too much. Those important issues need to be addressed through comprehensive transportation reform efforts.[1] Fortunately, the development of electric vehicles need not undermine that pursuit. More effective public transportation systems can be built, and public policy initiatives can encourage less driving while we make the transition from internal combustion to electric vehicles. After all, we still live in an automobile-dependent society, so why shouldn't we use the best propulsion technology available to power our vehicles? Electric vehicle technologies—including battery cars, hybrid electrics (with fuel-engine "boosters"), fuel cell vehicles, and flywheel-assisted vehicles—are in many ways far superior to today's gasoline vehicles and can greatly alleviate some of the environmental problems they cause.

The substitution of an electric vehicle for a conventional gasoline vehicle in southern California's South Coast Air Basin would reduce all vehicular air pollutants by 99 percent or more, except for small particulates, which would fall by 80 percent, according to a study by the Union of Concerned Scientists (UCS).[2] Electric vehicles are therefore a powerful weapon in the battle against smog, global warming, and acid rain. They emit no tailpipe carbon or hydrocarbon compounds, no oxides of sulfur or nitrogen, and no particulates. They also reduce emissions from the entire petroleum fuel-cycle, including fuel-production, refueling, and evaporative emissions.

Naturally, when electric vehicles are connected to a power system, they are charged by electricity from whatever energy source is used to energize the grid. Since most grids include combustion power plants, some of the electric vehicles' tailpipe emissions are in effect displaced to plant sites. But controlling emissions in a stationary power plant (which operates at or near its thermodynamic optimum) is much easier than cleaning up the exhausts of millions of inefficient internal combustion vehicles. Emissions per mile are therefore far less for electrics charged with electricity from power plants than for gasoline and diesel vehicles. Moreover, even if the electric vehicle's displaced emissions at the power plant *were* equal to an equivalent gasoline vehicle's tailpipe emissions, the simple act of relocating them to remote and distant sites vastly improves urban air quality. Of course, emissions are avoided altogether when electrics are charged with fuel-free power from the sun, wind, and falling water.[3] In addition, electric vehicles can also be built to make their own electricity cleanly onboard in fuel cells, or by other technologies, as we will discuss later.

THE AVOIDANCE OF AIR POLLUTION

Each electric vehicle replacing a conventional vehicle in an urban area that is violating federal Clean Air Act standards can save society \$5,000–\$17,000 in avoided air-pollution abatement costs, according to the Union of Concerned Scientists. By 2003, the use of electric vehicles at levels initially prescribed by state mandates in 1990 would thereby save nearly \$2 billion.[4] (The first two deadlines for implementing the mandates were rescinded in 1996, however, in response to oil and auto industry pressure. The 2003 deadline remains in effect.)

Switching to electric cars will avoid the needless losses to our economy in damage and disease caused by vehicular air pollution. Mark DeLucchi and colleagues in *A Comparative Analysis of Future Transportation Fuels* blame internal combustion engines for more than half of all outdoor air pollution.[5] Despite all vehicular emission control devices, conventional vehicles still produce 62 percent of all carbon monoxide and 40 percent of nitrogen oxides emissions. The average car also spews tons of climate-destabilizing carbon dioxide from its tailpipe every year.[6] Serious air pollution takes years off the average life span and aggravates heart and respiratory problems. DeLucchi and colleagues put the human health costs of auto pollution in the billions of dollars. (Their estimates depend a great deal on how much economic value you impute to each human life.)

Use of petroleum-based products to fuel and lubricate car and truck engines does more than pollute the air. American drivers today dispose of 260 million gallons of used motor oil each year—24 times the oil spilled by the *Exxon Valdez* in Prince William Sound, Alaska. Much of the used motor oil finds its way into storm drains and landfills, polluting the water and soil. The problem is often even worse in developing countries with less environmental regulation.

The nation's first photovoltaic recharging station for electric vehicles, located at the University of Southern Florida. *Courtesy of Warren Gretz, National Renewable Energy Laboratory.*

THE EXPLODING WORLDWIDE
DEMAND FOR CARS

As living standards rise around the world, the Earth's automotive population is growing even faster than its human population. While there were about 50 million cars in the world in 1960, there are now more than 450 million, and about 600 million are expected by the year 2000, with no end to the growth in sight. Hundreds of millions of would-be drivers now aspire to own vehicles, and smog-control devices are not a solution to the pollution. Though the devices have improved air quality, the rapid proliferation of vehicles, and the increases in their average miles traveled, retard progress toward clean air and erode many hard-won gains. "End-of-the-pipe" technical fixes simply do not and will not eliminate air pollution. Flawed and complicated control devices merely reduce pollution without confronting its basic cause: combustion. Internal combustion engines and catalytic converters therefore lock us into a never-ending struggle with air pollution. Emission devices deteriorate with age, and after 100,000 miles of use (or less), catalytic converter performance is significantly impaired. (Cold starts and acceleration surges also often overwhelm converters.) With billions of people sharing the same limited atmosphere with half a billion cars and trucks and assorted polluting industries, alternatives to the internal combustion engine are clearly needed. Electrics when charged by renewables offer a practical way to "zero out" air pollution, and they provide dramatic reductions even when charged from conventional sources.

Power for electrics can come from conventional providers—utilities, independent power producers, power co-ops, or, more cleanly, from hydroelectric facilities, wind-energy plants, photovoltaic systems, and other renewable sources. Renewable energy and electric vehicles together can provide pollution-free transportation. Smog and grunge from today's "chariots of fire" can be entirely eliminated. That would help return one of our birthrights: clean, healthful air. But how practical are electric vehicles? How far do they go, and where do you charge them while public refueling stations are still scarce?

ELECTRIC VEHICLE PRACTICALITY AND PERFORMANCE

Electrics today can be plugged into an ordinary outlet, just like a washer or dryer, anywhere power is served. Much to many people's surprise, if the 20 million electric cars Susan Tierney envisions were already on the road, existing power plants could handle their charging requirements without having to add new capacity, provided only that the cars were charged overnight at off-peak hours.[7] Fortunately, plugging in an electric car in the garage at night usually is the most convenient way for homeowners to recharge them at a leisurely pace. Many people prefer it to refueling their gasoline vehicle at a service station. In the future, however, electrics could be fast-charged in minutes at commercial charging outlets.

As far as on-the-road performance goes, electrics can be built to accelerate rapidly—some are even faster than conventional vehicles. They cruise easily at freeway speeds and can travel far enough on each electrical charge to get most of us to and from work. The AC Propulsion Honda CRX, for example, gets 100 miles of range in city and highway driving.[8]

The performance of early commercial and prototype electrics indicates where the technology is heading. The GM Impact, for example, is an eye-catching, beautifully designed, two-seat vehicle first unveiled in 1990. The

The GM Impact, an all-electric record-setting two-door coupe that accelerates from 0 to 60 m.p.h. in 8.5 seconds, powered by a 137-horsepower AC motor. Although GM once announced it would mass produce the Impact in 1995, the company decided instead to conduct nationwide consumer-acceptance tests of the car. Subsequently, GM began leasing a commercial version of the Impact, the EV-1, in small numbers in late 1996. *Courtesy of General Motors Corporation.*

prototype went 120 miles on a fully charged battery set and reached 60 m.p.h. from a standstill in 8.5 seconds. Modified for racing competition, it hit 183 m.p.h. in a time trial. GM began leasing a commercial version of the Impact called the EV-1 in southern California and Arizona in December 1996. (Based on a sticker price of $33,995, leasing costs will be $480 to $640 per month, depending on federal and state incentive programs.)

While capable of speed and jackrabbit acceleration, electric vehicles are also efficient and energy-conserving. Whereas as much as 60–75 percent of the electricity delivered from an electric outlet to an electric car is ultimately available for moving its wheels, only about 15 percent of the energy in a tank of gasoline actually drives the wheels during a normal driving cycle. (And only 1 percent is actually used to move the driver.[9]) Measuring from the electric outlet or the nozzle of the gasoline pump, an electric vehicle is therefore four or five times as efficient as a gasoline vehicle.[10] That's because electric motors are inherently much more efficient than internal combustion engines, and when an electric car is stationary, the motor does not turn over, so energy is saved.[11] Finally, because of regenerative braking, when an electric car slows down, braking energy is recaptured and returned as electricity to the battery, instead of being wasted, as in a gasoline vehicle.[12]

ELECTRIC VEHICLES: A BRIEF HISTORY

Electric vehicles are neither a new nor an untested technology. They originated in Europe in the 1830s, half a century before gasoline-

The Solectria Corporation's all-electric, four-seat Sunrise, "purpose-built" with funding from the Pentagon and Boston Edison. Solectria says the all-composite Sunrise will have a range of 200 miles on advanced batteries. The car has achieved a range of 238 miles on a single battery charge under city driving conditions and has gone over 373 miles on a charge during the 1996 American Tour du Sol race. *Courtesy of Solectria Corporation.*

powered cars. Thomas Davenport, a Scot, built a nonrechargeable battery-powered electric in 1834. Englishman J. K. Starley produced the first light electric car in 1888, the same year Fred M. Kimball built an electric in the United States.

By 1897, the Electric Carriage and Wagon Company of Philadelphia was supplying New York City with electric taxis. The technology had improved so much in just over a decade that an aerodynamically designed racing electric exceeded 65 miles per hour in 1899.[13] Studebaker was making electrics in 1902,[14] and Ferdinand Porsche built an electric about the same time, followed by a "hybrid" vehicle in 1902 that used both an electric motor and an internal combustion engine.[15] Electric trucks were on sale by 1907, and by 1912, more than 34,000 electric vehicles were in use in the United States.

Between the introduction of electrics and their eclipse by gasoline-powered cars in the 1920s and early 1930s, more than 300 companies made electric cars in the United States. The invention of the electric starter, which replaced the hand-cranked starter, helped the internal combustion engine out-compete electrics. But it was the advent of the interstate highway system and long-distance vehicular travel opportunities that really favored the gasoline-powered car with its greater range. The electric vehicle survived in Europe, however, for various commercial applications. Paris and Bordeaux, for example, have been using electric garbage trucks since the mid-1930s.

General Motors began experimenting with electric vehicles in the 1960s and produced its Electrovair I and II, based on the Corvair, but these were never mass produced. Strong interest in electric vehicles did not revive until the 1970s, when two Arab oil embargoes and sharp oil price increases got the public's attention.

ELECTRIC VEHICLE PIONEERS

James Worden, an electric vehicle prodigy, was only a seventh-grader in Arlington, Massachusetts, when he built his first electric car in his parents' living room in 1979. Several more electrics soon followed, even before he graduated from high school, including a car that ran entirely on solar power. Such pursuits led Worden to the Massachusetts Institute of Technology, where he studied mechanical engineering. While still an undergraduate in 1986, he founded Solectria Corporation with a schoolmate and incorporated the company upon graduation in 1989, becoming CEO at only age 22.[16]

Solectria, a pioneering electric vehicle manufacturer based in Wilmington, Massachusetts, both converts gasoline vehicles to electric-drive and creates electrics "from scratch." The company's prototype Sunrise, a full-size, four-seat sedan, was designed from the wheels up as an electric. It will have a 120-mile range at 53 miles per hour with lead-acid batteries and a 200-mile range with nickel metal-hydride batteries. The car has a state-of-the art AC induction-drive system and will be sold with standard amenities and options, like power brakes, stereo, dual air bags, and cruise control, as well as an onboard battery-charging system. Government and utility company grants to Solectria totaling $1.13 million have enabled the company to produce a Sunrise prototype, which the company estimates would cost only $20,000 in mass production of just 20,000 vehicles a year, beginning in 1997.[17]

THE FIVE-MILLION-MILE MOTOR

The electric car's simplicity is one reason for optimism about its costs in mass production. In contrast to internal combustion engines with their hundreds of parts, electric cars have no pistons, cylinders, crankcases, starters, solenoids, distributors, spark plugs, carburetors, manifolds, fuel pumps, fuel injectors, radiators, fan belts, timing belts, camshafts, crankshafts, mufflers, catalytic converters, or exhaust pipes. In fact, electric motors themselves have only one moving part. Hydro-Quebec, the Canadian utility, is even developing a prototype vehicle that combines an electric motor in each wheel, eliminating conventional axles and transmission.[18]

Being so much simpler, electric propulsion systems are lighter and smaller than internal combustion engines. Naturally, they are also less costly to mass produce and, with far fewer parts, need far fewer repairs. If it's not there, you don't need to maintain or fix it. In addition, because electrics use no fuel, they need no tune-ups.

Electric motors are typically rated for 100,000 hours of operation. At three hours of operation a day, that's a working life of more than 90 years. Traveling at 50 miles per hour, a car driven by such a motor could in theory travel 5 million miles—if only a car body and driver could last that long. For comparison, American-made internal combustion vehicle engines usually don't go over 120,000 miles without a major overhaul. The typical gasoline-driven car goes 100,000 miles. Brakes in an electric vehicle last longer than in a gasoline vehicle, too, due to the regenerative breaking system that allows the battery to reabsorb energy instead of allowing braking friction to heat brake pads. All this is lucky, since certified repair facilities are not yet as ubiquitous as the corner gasoline station.

AEROVIRONMENT, EFFICIENCY, AND THE GENESIS OF THE IMPACT

GM appears to be leading the Big Three American automakers today in electric vehicle development, and its leadership position can be traced back to physicist-engineer Dr. Paul B. MacCready and the excitement generated by the Sunraycer, a prototype solar-electric vehicle that Mac-Cready's company, AeroVironment, Inc., developed with GM.

"Until you've designed and raced a solar car, you really don't know how to build an electric car," says J. Ward Phillips, who organizes and sponsors solar and electric vehicle races. In a solar car race, efficiency is at a premium, since cars have to operate at highway speeds on nothing more than the power of a handheld hair dryer, as Phillips puts it. The astoundingly efficient technology that makes this possible was specifically invented for solar electric car racing. But other electric vehicles, too, must extract every available kilowatt from their battery packs to minimize battery weight and cost.

The story of the GM Impact thus actually began when Howard Wilson, a now-retired Hughes Aircraft Company official, telephoned Dr.

MacCready to ask for his assistance in building Sunraycer, a job for which MacCready was exquisitely prepared. A half-century or so earlier, in the 1930s, MacCready as a child had designed and built model planes out of balsa wood. That interest in flight led him to a degree in physics from Yale University in 1947 and on to a Ph.D. in aeronautical engineering with honors from California Institute of Technology (Caltech) in 1952. Along the way, MacCready became interested in building ultralight, engineless sailplanes—paragons of efficiency that actually float on rising air currents. He eventually mastered the technology to become world soaring champion in 1956,[19] reaching a then-record altitude of 29,500 feet in one victorious ascent.[20]

In 1971, MacCready founded AeroVironment, Inc., an engineering-consulting company that focused on hazardous waste cleanup and problems pertaining to airborne propulsion. MacCready and his company made aviation history in 1977 by building the world's first successful human-powered aircraft, the Gossamer Condor. With a trained cyclist pedaling for all he was worth, the plane soared into the air and flew for a mile of controlled flight.[21] Built of balsa, aluminum tubes, corrugated cardboard, and mylar, all strengthened with piano wire, that 55-pound plane with a 96-foot wingspan conquered the problem of human-powered flight that had defeated Leonardo da Vinci. The bold and brilliant experiment also won MacCready the $95,000 Kremer Prize, a prestigious aviation award.[22]

The Gossamer Condor's success reaffirmed MacCready's ability to create energy-efficient machines through lightweight construction and attention to aeronautical design details, an approach Amory Lovins applied in designing Hypercars, described in Chapter 24. Using the same approach, MacCready in 1979 led the AeroVironment team that designed and built an improved human-powered plane, the Gossamer Albatross, which flew the English Channel to earn a second Kremer Prize worth $215,000. Two years later, MacCready crowned these successes by directing the team that built the Solar Challenger, a solar-electric airplane that flew from Paris to England on power produced by photovoltaic cells on its wings and tail.[23] Throughout all these projects, MacCready remained focused on minimizing weight to get the most of limited power supplies.

His aeronautical achievements led MacCready directly to the Sunraycer project and from it, straight to the GM Impact, although along the way, he led the team that designed a remarkable flying robotic model of a prehistoric pterodactyl for a Smithsonian IMAX movie.[24] Powered by 13 electric motors and equipped with airflow sensors and gyros to keep it headed into the wind, the amazing thing was not how well it flew, but that it flew at all. *Time* magazine described the beast as "an awesomely realistic, radio-controlled, computer-brained, wing-flapping replica of the largest creature ever to have flown."[25]

SUNRAYCER'S GLORIOUS TRIUMPH

MacCready's transition from airborne to terrestrial transportation problems came in 1986 when General Motors received an invitation to enter the World Solar Challenge, a 1,950-mile race across Australia for solar-powered electric vehicles. That was the invitation that then reached Howard Wilson, vice president of Hughes Aircraft, a $5 billion, technology-rich company that GM had acquired the previous year. Hughes's focus is the development and production of electronic propulsion systems and electric power conversion products.

Wilson knew Hughes could provide the electric power source required but had no experience building cars, and the World Solar Challenge was only ten months away, so he turned to AeroVironment. Fortunately, MacCready's staff included Alec Brooks, a young engineer who had unlimited enthusiasm for lightweight, human-powered vehicles and who had designed the world's fastest human-powered watercraft, the Flying Fish. As luck would have it, Brooks had already sketched plans for a World Solar Challenge car, but had put them away for lack of resources to enter the competition.

Wilson had to go to Detroit and buttonhole high-level GM officials to get the project approved. By the time GM agreed to participate, putting its corporate prestige and millions of dollars at stake, only seven months remained before the race.[26] Engineering teams led by Brooks, with MacCready's oversight, worked incessantly and, just in time, delivered a strange, sleek, 365-pound vehicle that looked like a flattened teardrop on bicycle wheels. This creation was so efficient and had so little aerodynamic drag, however, that it swept across the finish line two and a half days ahead of the nearest contender, averaging more than 41 miles per hour on less than two horsepower.[27] GM was ecstatic and showed the Sunraycer at 269 events before consigning it to the Smithsonian's National Museum of American History.[28]

BUILDING A PRACTICAL ELECTRIC CAR

Brooks now argued that the same principles that had brought the Sunraycer victory could be incorporated in a state-of-the-art electric vehicle.

He proposed that GM develop an electric car as a learning exercise and technology demonstration. The challenge would be to hold weight down and decrease drag and rolling resistance so that sufficient range and acceleration could be wrung out of a lead-acid battery pack.

Many automotive engineers in the past had failed to develop a practical electrical vehicle with satisfactory range and power. GM officials and engineers were skeptical of Brooks's new project at first, because of the company's past failures. But new technology was emerging. Electronics superstar Alan Cocconi had worked with AeroVironment on the flight controls for the pterodactyl and on the Sunraycer's electronics.[29] He had developed an electronic controller that managed battery power with unprecedented efficiency.[30]

The availability of the Cocconi Sunraycer controller helped persuade GM that the odds of success with an electric car were now in its favor. So, on the crest of the Sunraycer triumph—and with more of Howard Wilson's skillful in-house lobbying—GM management finally gave Wilson and Brooks a cautious commitment to a one-year development program in November 1988.[31]

As the R&D team progressed and the project evolved, GM Chairman Roger Smith, responding to pending clean air legislation, decided that the demonstration vehicle should be introduced to the world at the January 1990 Los Angeles Auto Show.[32] "Suddenly the team had barely six months to build a real car that would live up to its promises," wrote *U.S. News and World Report*.

Brooks managed the project for AeroVironment, which was the primary subcontractor. MacCready provided inspiration, guidance, and mastery of aeronautics. The GM-AeroVironment alliance combined the agility and technological prowess of the small company with the engineering and technology resources of 22 GM divisions and affiliates.[33]

Battles raged, however, between GM's designers, who wanted a beautiful car, and the AeroVironment team, which had to be preoccupied with keeping aerodynamic resistance low. A high-level GM official finally was obliged to mediate the dispute. The Impact wound up with a gently curved, aerodynamically slippery underbody and a slightly tapered tail, which kept both the stylists and the aerodynamicists happy.

Every detail of the car was examined to reduce weight. The engineers rethought the frame and provided structural integrity through better design, so they could shave unneeded material. The team went so far as to cut holes in interior panels. They also opted for light AC motors instead of using DC. That meant they would need an inverter, but conventional inverters built with switches and cooling devices were heavy. Alan Coc-coni, however, designed a 60-pound, digital-chip inverter for the car, a mere fifth the normal weight.[34] The all-electric Impact that resulted was striking in design, outstanding in performance, and dazzling on the race-track, where it put a gasoline-powered Nissan 300ZX sports car to shame.

The Impact is propelled by a computer-controlled, 137-horsepower AC motor powered by a maintenance-free, lead-acid battery pack containing 27 12-volt batteries.[35] The motor's snappy acceleration gives the Impact a sports car feel, yet the car can go 70 city miles or 100 miles of highway driving on a battery charge and takes only two to three hours to recharge from a 220-volt source.[36] According to Ovonic Battery Company projections, substituting Ovonic nickel metal-hydride (NiMH) batteries in the Impact would triple the car's range. Ovonic of Troy, Michigan, is involved in a joint venture with General Motors to develop NiMH batteries for electric vehicles.[37]

Typical electricity costs for the Impact are a scant penny a mile, as against 5 cents for a typical subcompact using gasoline at $1.50 per gallon.[38] The car has one of the most advanced, lightweight, low-aerodynamic drag designs of any vehicle on the road—its aluminum space-frame weighs only 295 pounds—yet, as we will see, cars can be made much lighter. The Impact also has regenerative braking and sufficient power for its hydraulic power steering, power brakes, cruise control, traction control, power windows, power door locks, and power side mirrors. An electric heat pump provides both heat and air-conditioning.

The car created such a sensation at the Los Angeles Auto Show that would-be customers started sending GM money as their down payments. GM then began seriously considering a mass-production version and announced three months later that the first 300–700 Impacts would be on the market in the 1994 model year. The company later retracted its pledge after deciding to make major design changes that would require extensive testing.[39] A nationwide consumer-acceptance test with the car produced an overwhelmingly positive response.

"The biggest problem we've faced so far," said GM Vice President Kenneth Baker, "has been getting the vehicles back from the test drivers. . . . They want to keep the car!"[40] Although the car was still not commercially available in November 1996, more than five years after its unveiling, the Impact has done much to advance electric vehicle technology and enhance its credibility. Paradoxically, the company's offical reason for not commercializing the car sooner is that battery technology is not adequate to satisfy consumers' needs. Since consumers obviously disagree, the true explanation may be different: GM may be protecting its profitable conventional business by delaying the commercialization of electrics while voicing support for them and conducting research and development. Once widespread introduction of electrics occurs, GM will be familiar with electric technology and will be poised to enter the market when the inevitable can no longer be postponed.

THE CRUISING RANGE

The main performance obstacle to rapid commercialization of electrics today is their limited range. While expensive electrics that go 100 miles or more on a charge are commercially available, some inexpensive electric conversions equipped with conventional lead-acid batteries can travel only 50–70 miles on each battery charge. Even this is fine for an "around town" vehicle, especially for commuting to and from mass-transit stations. PIVCO, the Norwegian manufacturer of a lightweight commuter car now being used in a California pilot program, says its car will sell for under $10,000. Electricité de France, the huge French utility, is planning to make electric commuter cars available on a self-service basis in a pilot program near Paris, with radio-tracking to prevent vandalism.

Range is being extended through the persistent efforts of engineers and battery designers around the world. At the American Tour de Sol race, the winning car in 1992 went 100 miles between charges; in 1993 the winning distance was 180 miles, and in 1996, the winner, a Solectria Sunrise, traveled 373 miles on a single battery charge. Further improvements are likely soon. Close observers of electric vehicle (EV) technology, such as the California Air Resources Board staff, project that new commercial electric vehicles in normal operation will have a range of 150 miles as early as 1998, and an energy efficiency equivalent to over 100 miles per gallon of gasoline, while small, high-efficiency electrics may go 200 miles by then.[41]

An optimistic assessment of recent scientific and technical data on battery development suggests that, within ten years, EVs in normal operation will have a range of 200–300 miles, similar to many gasoline-powered vehicles today.[42] Once EV range reaches 200 miles in normal use, further range improvements should be relatively unimportant; many drivers of gasoline vehicles are content with a similar range.[43] And range even today may not be the insurmountable problem it has been made out to be. A recent study of consumer acceptance of EVs by the University of California, Davis, concluded, "limited range is a relatively minor drawback in many, indeed most, multi-car households."[44]

BATTERIES REMAIN
THE MAJOR STUMBLING BLOCK

Present battery technology not only restricts range and power for today's EVs, but raises vehicle costs and causes inconvenience in the form of lengthy charging times. In the future, however, EVs using nonpolluting fuel cells that chemically combine fuels into electricity will not even require batteries. The fundamental cause of today's range limits and high battery costs is that conventional lead-acid batteries now hold only about 1 percent as much energy per pound as liquid fuel. To obtain desired performance, very large, heavy, and expensive battery sets therefore are needed. The huge mass itself then requires lots of energy for its transport and a relatively massive structure for its support. As lighter, more powerful batteries are developed and integrated with flywheels or ultracapacitors (or both), this problem will be greatly mitigated and ultimately solved. Until then, battery mass will penalize EV performance by preventing the development of ultralight vehicles capable of extraordinarily efficient operation. Apart from its batteries, a vehicle's range also depends on vehicle weight, aerodynamic resistance, tire rolling friction, and motor efficiency. (See Chapter 24.)

Contrary to the perception fostered by EV foes, dramatic breakthroughs in battery and other energy storage and conversion technologies appear imminent. The number of new battery technologies being developed, the heartening progress being made in battery engineering, and the scale of this international effort all suggest that battery-related performance problems will be mastered within five to ten years. Ten years may sound like a long time, but not on the scale of major technological developments with revolutionary global implications.

BASICS OF ELECTRIC VEHICLE BATTERY
TECHNOLOGY AND PERFORMANCE

The batteries of an electric vehicle are called traction, or deep-cycle, batteries, unlike the battery used in your gasoline-powered car. Rather than just providing a short burst of power for starting, traction batteries hold a great deal of energy, and supply it as needed for propulsion, until they are discharged. Many traction batteries are now sealed and maintenance-free: Virtually all are designed for recycling.

Batteries deliver their stored energy as a flow of electrical power over time. The amount of energy a battery contains per unit of mass is called its *specific energy* and is generally measured in watt-hours per kilogram. Similarly, the rate of instantaneous power delivered by the battery is given in watts per kilogram. (This is called *specific power.*) Battery performance is also described in terms of *energy density* (watt-hours per liter) and *power density* (watts per liter). Regardless of how it is specified, the battery energy content is a prime determinant of the vehicle's range, while the battery's ability to deliver power sets limits to vehicle acceleration. If energy is thought of as a volume of liquid in a flask, power can be regarded as the rate at which the liquid pours into or out of the flask.

Prototype batteries without toxic materials, like lead or cadmium, can already hold two to five times the energy of conventional lead-acid batteries. Some are expected to last ten years, operate through 1,000 charge-discharge cycles, absorb a full charge in minutes, require little maintenance, and use recyclable materials. Sony Corporation, for example, has announced development of a long-lasting lithium-ion battery for electric vehicles that has three times the specific energy of standard lead-acid batteries.

In May 1995, an inaccurate paper in the journal *Science* about the extent of increases in lead pollution from electric vehicles produced alarmist press reports about electric vehicles.[45] Like the *Science* piece, the misleading articles that followed it suggested that batteries are a toxic peril, due mainly to the mining and smelting of lead. "This could be the kiss of death for electric vehicles," said one "expert." Almost all lead for car batteries, however, is derived from *recycled* scrapped lead products. Moreover, the batteries for all new electric vehicles once mandated in the year 2000 would use less than 2 percent of all the lead already in lead-acid batteries in the nation's 188 million vehicles.[46] Most importantly, these articles left the false impression that lead-acid batteries are the only practical power sources for electric vehicles. In actuality, ordinary lead-acid batteries will soon be obsolete for all but the cheapest electric vehicles. Greatly superior new batteries under development today use a wide range of electrochemical "couples": nickel-cadmium, nickel metal-hydride, nickel-zinc, nickel-bromine, sodium-sulfur, lithium ion, lithium polymer, lithium aluminum iron disulfide, and metal-air. Other aspects of batteries design also vary.

BATTERY PERFORMANCE TRADEOFFS

Each of the new battery technologies has some shortcomings, and an unmistakable winner in the advanced battery derby has not yet emerged.

Some of the new batteries are too expensive or don't offer enough energy density. Others are unable to provide sufficient power, use materials in short supply, fail at low temperatures, must operate at hundreds of degrees, have short lifetimes, use highly reactive elements, or require replacement of electrodes at recharging stations. The challenge is to develop a technology that optimizes all key battery performance criteria simultaneously. Because of the complexity of these issues, many divergent views exist on what that battery will be made of. (See *Proceedings of the 12th International Electric Vehicle Symposium,* and publications of the International Energy Agency, and the Canadian Energy Research Institute cited in the Resources section.)

One view expressed by electric drive-train designer Alan Cocconi is that sealed lead-acid batteries are the best short-term solution. He believes they will soon improve to deliver 150 miles of range at 60 miles per hour with minimal maintenance. Such batteries would probably offer very stiff competition, especially at first, to the advanced batteries using zinc, nickel, lithium, and other elements. Already in mass production, conventional lead-acid batteries at perhaps $2,000 per battery pack (depending on vehicle weight) are still the least expensive energy storage commercially available for EVs. But today's lead-acid batteries have only about 30 watt-hours of energy per kilogram, and need to be replaced every two to three years, or after as few as 15,000–20,000 miles.

Though not yet the battery Cocconi seems to have in mind, the fully recyclable lead-acid Horizon battery, under development by Horizon Battery Technologies, Inc., holds the promise of being a lightweight, long-lived and affordable near-term battery. It can deliver 50–80 percent greater range than conventional lead-acid batteries, combined with a lifetime of 1,000 charge-discharge cycles—equivalent to an estimated 80,000 miles of driving—all for an eventual estimated cost of $150 per kilowatt-hour of energy storage. Moreover, the battery can be recharged in half an hour using high-power electronic charging equipment and can reach half-charge in only eight minutes. Horizon batteries are thus a significant battery improvement and are being commercialized more rapidly than advanced battery technologies that hold the prospect of far greater range and eventual lower cost. Testing of the Horizon battery is still in progress, and further improvements are necessary. Its costs are still $400 per kilowatt-hour, and its energy density is still 50 percent below commonly accepted target levels for so-called "midterm" battery technologies.

Horizon batteries are distinctly different from the traditional lead-acid battery that has scarcely changed in 100 years. The Horizon battery utilizes a lead alloy that is co-extruded as a wire on a fiberglass core. The wires are woven into a mesh on a continuous loom, coated with a proprietary paste, and stacked as horizontal cells. This horizontal bipolar design offers many electrochemical advantages over conventional cells, including less off-gassing and lower internal cell resistance. Nonetheless, with a maximum range of 120–130 miles, an electric vehicle powered by a Horizon battery would still be less attractive to some customers than a gasoline vehicle. To accelerate development of a satisfactory EV battery, the U.S. Department of Energy (DOE), the Big Three automakers (Chrysler, Ford, and General Motors), and the utility-sponsored Electric Power Research Institute are collaborating in a five-year, $262 million R&D effort called the U.S. Advanced Battery Consortium (USABC). Although USABC has been described as a 50-50 public-private cost-shared partnership, much of the automakers' contributions are in kind rather than cash. Yet the Big Three control the group's agenda, and with it, much public funding for battery research, along with research money from the electric utility industry. The consortium thus gives the Big Three substantial control over both U.S. battery research and when an affordable, high-performance EV battery will be commercially available. Critics have charged that the consortium has neglected to fund improvements in battery technology that might be available in the medium term in favor of developing advanced technology for the long term. Some see this as a gambit to slow electric vehicle commercialization by delaying the development of competitive battery technology until after the year 2000. To date, three-quarters of USABC research investments have been devoted to long-term battery technology.[47]

One of USABC's early contributions has been to formulate and standardize battery development goals. Specific medium-term USABC goals are a battery that holds 80–100 watt-hours per kilogram, provides a "specific power" of 150–200 watts per kilogram, lasts for 100,000 miles, and costs less than $150 per kilowatt-hour of stored energy. Such a battery would at best have modest appeal. For a battery pack large enough to power a sedan of the size and weight of a Ford Taurus, 45–50 kilowatt-hours of energy would be required at a cost of $6,750–$7,500. This is somewhat misleading, however, because vehicles designed especially for electric operation will have composite bodies that will be far lighter than those of today's gasoline-powered vehicles.

USABC's long-term goal is a battery that would result in a vehicle "competitive with today's internal combustion engine and capable of commercial production early in the next decade." USABC believes that requires a specific energy of 200 watt-hours per kilogram and specific power of 400 watts per kilogram. This combination would provide a range of about 200 miles in normal use and a 0–60 m.p.h. acceleration in nine seconds for a normal EV. The cost of a battery meeting these long-term goals would be under $4,000, but the battery would still be quite heavy.

THE PROMISE OF ZINC-AIR BATTERIES

One very promising battery technology is the rechargeable zinc-air battery developed by Zinc Air Power Corporation of Canton, Michigan. The firm is an early start-up venture whose commercialization efforts are based on technology owned by Dreisbach Electromotive, Inc. (DEMI), of Santa Barbara, California. This prototype technology offers very high specific energy, low-cost, low-toxicity materials (air, zinc, and potassium), combined with respectable power and durability. These batteries could hold six times the energy of conventional lead-acid batteries. Company test results on prototypes have significantly exceeded USABC midterm goals for most performance and cost criteria. Current battery lifetime is said to be 50,000 cycles, however, with 100,000 cycles needed to meet USABC standards. Also on the minus side, the zinc-air reaction is slow, and therefore the battery has not been as satisfactory at delivering large surges of power as in storing energy. Nonetheless, DEMI says an electric version of a GM Saturn auto-equipped with a zinc-air battery beat a gasoline Saturn in a one-mile race at the Phoenix International Raceway—the first electric ever to beat a gasoline car. A Honda CRX powered with the DEMI battery won the Solar & Electric 500 race at the same speedway in 1991.[48] Zinc Air Power Corporation projects that existing prototype technology scaled up for electric vehicles will provide a 250-mile range, which would be highly satisfactory. The company says that future versions of its battery may provide a range of 300–350 miles, which is greater than that of many gasoline vehicles.

The affordability of zinc, which in 1994 cost about 75 cents a pound for metal of sufficient purity to make a battery, is a tremendous asset for zinc-air technology. Materials costs are significant, since they may account for as much as 70 percent of battery costs. Competing advanced technologies require expensive nickel, which costs about $4 per pound, and lithium, which costs $40–$60 a pound; others use toxic cadmium.

ADVANCED BATTERIES OF
LITHIUM. SODIUM. AND NICKEL

Whereas both the sodium and sulfur of the competing advanced sodium-sulfur battery are not especially expensive as raw materials, the battery itself is expensive and has demonstrated only about 500 cycles of life. Its cost is still projected to be $200 per kilowatt-hour by 1998—$50 per kilowatt-hour above the USABC midterm battery criterion. And the sodium-sulfur battery also must be kept at an internal temperature of 500 degrees Fahrenheit. Meeting this requirement when the vehicle is parked consumes energy and adds cost and complexity to the system. Ford Motor Company, however, uses the sodium-sulfur battery in its prototype electric vehicle.

AEG Corporation is developing another sodium battery, this one using sodium-nickel-chloride reactions. Test batteries suggest the system will have a long cycle-life, but it is fairly expensive and is being redesigned to provide greater power. It too needs to be heated, at an energy cost of perhaps 20–30 cents per day.[49]

Westinghouse Corporation is pursuing advanced battery development and has a monopolar lithium metal-sulfide ($Li(Al)/FeS_2$) battery technology that it believes already meets the USABC's midterm goals. Lithium is a light, chemically active metal that promises high specific energy. The company says the battery could easily evolve into a more advanced bipolar version that could meet long-term USABC goals.

Many companies are enthusiastic about nickel metal-hydride batteries, despite the cost of nickel. This battery might provide as many as 2,000–3,000 cycles of 80 percent discharge,[50] will perform well in cold weather, and will absorb a fast charge (from 20 percent to 60 percent of capacity) in 15 minutes. The battery already meets the USABC midterm specific power goals, and the manufacturer believes that it will meet all of USABC's other goals. To do so, specific energy will have to be increased by about two-thirds, and cost will have to drop by about a third to a fourth.[51]

Nickel-cadmium batteries also have proponents. Battery maker SAFT claims the nickel-cadmium battery offers the highest energy and power density of any battery sold today. Other strengths of nickel-cadmium technology are its resistance to power loss at low temperatures and its long life—2,000 battery cycles, which implies a life expectancy of seven to ten years. In addition to its work with nickel-cadmium technology, SAFT is simultaneously investing in research on even more advanced batteries using a lithium-carbon couple and lithium aluminum-iron disulfide. Lithium-carbon could be recharged quickly and would have more than twice the specific energy of nickel-cadmium and three-quarters more than nickel metal-hydride. That would give it more than four times the specific energy of today's best conventional lead-acid batteries.[52]

In partnership with Hydro-Quebec and Argonne National Laboratory, the 3M Corporation is leading an effort to develop a lithium polymer battery. The battery uses ultrathin film laminates of lithium foil (the anode), a polymer electrolyte, and a cathodic film sandwiched between insulating and conducting films. The composite can then be rolled into cylinders, or formed into other shapes.

Duracell, Inc., has also been developing lithium-ion advanced batteries using a lithium-carbon compound at its Worldwide Technology Center in Needham, Massachusetts. The battery will have a negative electrode of lithium-carbon and a positive electrode of manganese dioxide. Duracell and VARTA Batterie AG of Hanover, Germany, recently received an $18 million contract from the USABC to work on the lithium-ion battery. Duracell is the largest maker of alkaline batteries for consumer applications, and VARTA Batterie is the largest European battery maker.

FLYWHEEL ENERGY STORAGE FOR LONG-RANGE, HIGH-POWER VEHICLES

Even in the improbable event that not one of the many companies now working on battery development succeeds in producing a high-power, affordable battery in the next ten years, flywheels of advanced design eventually are likely to supplement EV batteries. Flywheels can be set in motion by a charge of electrical energy from a wall outlet, a chemical battery, or electricity produced by an electric car's motor using energy recaptured by regenerative braking. While chemical batteries deliver and receive power relatively slowly and take a long time to absorb a charge, the mechanical flywheel battery charges and discharges rapidly.

Using advanced high-strength composite materials, tomorrow's flywheels will store large amounts of energy in high-speed spinning rotors, and are expected to fit easily in a vehicle. Spinning in a vacuum chamber on frictionless magnetic bearings in a protective containment vessel, flywheels could instantaneously deliver big surges of power. They are also far more efficient than chemical batteries in capturing the sudden surge of power produced during deceleration via regenerative braking. The technology could therefore provide vehicles with long range, by reducing the drain on batteries for acceleration, and with high power. Auto flywheels would operate cleanly and silently and are expected to have long lifetimes. Unlike most chemical batteries, they contain no corrosive or toxic materials.

Flywheels may thus overcome all the major objections to today's batteries. At first, the technology is likely to be used to boost the power and range of EVs or hybrid vehicles equipped with advanced chemical batteries, making battery energy last longer by meeting a car's peak power demands for acceleration and high speed. Later, flywheels could in theory serve as an independent energy storage device as well as a peak-power supply device. Edward W. Furia, chairman of American Flywheel Systems, Inc., of Seattle, Washington, says their proposed twelve-wheel system being developed with Honeywell, Inc., would accelerate a vehicle from zero to 60 m.p.h. in just seven seconds and would propel the car 400 miles.[53] The company hopes to build a prototype flywheel-powered vehicle called the AFS-20 and bring it to market by 1998.

SUPER FLYWHEELS, FUEL CELLS, AND ULTRACAPACITORS

While an ordinary flywheel revolves at 25,000–30,000 r.p.m., superflywheels, spinning at up to 100,000 r.p.m., "could triple the range of today's EVs," according to the Union of Concerned Scientists.[54] American Flywheel Systems, Inc., projects a range of 300–600 miles for flywheel vehicles, plus infinite rechargeability.[55] EV flywheels are still mainly confined to the laboratory workbench today, however, and have not even been conclusively proven there yet. Because both ultracapacitors and flywheels can deliver pulses of high power needed for rapid acceleration, they may not only improve performance in passing, hill climbing, and merging with high-speed traffic, but may extend battery chronological and cycle life. Other advantages include "super quick" emergency battery charging, more efficient recovery of braking energy, battery protection during fast charging, and decoupling vehicle performance from battery age or temperature.[56]

Lawrence Livermore National Laboratory (LLNL) is currently working on developing an electric vehicle that could be powered by a lightweight super-flywheel spinning at an astounding 200,000 r.p.m. Livermore scientists are also working on a high-energy, zinc-air battery that provides a 250-mile range and can be charged in just 15 minutes.[57] By coupling the zinc-air battery to the super-flywheel, lab engineers think they can exploit the battery's relatively long range while compensating for less satisfactory power delivery characteristics. Other companies involved in flywheel development are United Technologies Corporation at its East Hartford, Connecticut, research center; Satcon Technology Corporation of Cambridge, Massachusetts; and Trinity Flywheel of San Francisco, California, using technology from LLNL.[58] Chrysler is also testing flywheel technology in a hybrid gasoline-powered vehicle equipped with a SatCon flywheel using a high-strength, lightweight carbon-fiber rotor; BMW is likely to begin testing a flywheel vehicle, too, in the near future.

As an alternative to battery and flywheel cars, in the long term we may all be driving batteryless cars powered by hydrogen fuel cells that cleanly produce electricity. Steam would be the only by-product, and hydrogen can be produced from renewable as well as fossil fuels, or by using renewable sources of electricity to dissociate water into hydrogen and oxygen. Prototype fuel cell–powered buses are already on the road. By 1998 or so, they may be available as fleet vehicles. Fuel cells for passenger cars are expected to take somewhat longer to commercialize.

HOW MUCH RANGE IS REALLY ENOUGH?

From the preceding discussion of EV development, status, performance, and environmental benefits, it appears that new batteries and other advanced technologies will eventually resolve current concerns about EV range, now the most challenging performance issue. But even today, range limitations may not be the obstacle to EVs that some people think.

Many drivers who generally use their vehicles for errands, commuting, and short trips will accept reasonably priced but limited-range vehicles and will shift to other transport for their occasional long trips. As a bonus, they won't have to pay for and maintain ever more complicated internal combustion engines. For these drivers, albeit a minority, even today's electric may be quite suitable as a first car. But for drivers who must frequently drive long distances, or who must have that emergency capability, electrics still may make sense as a second or third vehicle. Some 40 million households in the United States today have two or more cars, and

90 percent of all second vehicles travel less than 30 miles a day, making them an ideal immediate target market for EVs.

According to the U.S. Department of Transportation, the average commute is only about 12 miles, and the average daily travel for all purposes is less than 30 miles—well within the range of even the feeblest road-worthy electric. Only 25 percent of all commutes exceed 50 miles. In fact, 90 percent of all vehicle trips are under 60 miles in length,[59] trips for which today's electric vehicle is already excellent. As the performance of electrics improves and their costs decline, the lines between the second or third car market and the first car market will blur. During this transition, aggregate EV demand is likely to swell into the tens of millions. A recent EV market acceptance study of 12,050 Phoenix, Arizona, households conducted by the Salt River Project utility found that a majority of respondents would accept an EV costing about $16,000, once EV range exceeds 120 miles.[60] As we have shown, that milepost is well within the capabilities of various advanced batteries already in early stages of commercialization.

In addition, for the minority of commuters who do have very long commutes, vehicles in many locales can be recharged during the day while the driver is at work. Recharging stations are beginning to pop up in company parking lots, public lots, or shopping centers. Boston Edison Company, for example, is already installing recharging facilities in its service area and later hopes to make credit card–activated recharging machines available at grocery stores, theaters, and restaurants. Orcas Power and Light Company is installing charging stations in the San Juan Islands of Washington State.

While limited range will not be a problem for many users, those who need cars for frequent long trips, or who want to pull trailers, or navigate lots of very steep terrain still will choose to stay with conventional engines until hybrid vehicles are available. Hybrids are electrics equipped with a small internal combustion engine (perhaps 25 horsepower) in addition to their battery-driven electric motor. They are therefore not limited in operation by the energy in conventional battery packs, since they carry fuel as well, which is high in specific energy. Hybrids offer all the advantages of gasoline vehicles—specifically, long range between refueling and recharging—with most of the advantages of electric drive.

24

HYPERCARS—
AND HYPE ABOUT CARS

Today's car . . . is the highest expression of the Iron Age.
—Physicist Amory B. Lovins,
Research Director and Vice President,
Rocky Mountain Institute

Today's hybrid, a car that uses both fuel and electricity, has an electric
motor plus a small auxiliary power unit—generally an internal combustion engine (ICE). These cars have none of the range or power limitations of the early-introduction electric vehicles (EVs), yet produce far
lower emissions than current ICE vehicles. Tomorrow's hybrid may substitute a low-emission gas turbine engine or fuel cell for that ICE. In either case, the engine or fuel cell is used to operate a generator that
produces electricity that can power the electric motor directly or charge
the vehicle's batteries. Therefore, unlike the pure electric, these vehicles
will not need to haul around a heavy battery pack and can be extremely
fuel efficient. The hybrid's engine can be built to run on any conventional
liquid fuel—diesel, gasoline, propane—or on alcohol, natural gas, biogas,
hydrogen, or hythane. Some hybrids can be driven exclusively on stored
electricity from the battery, which is desirable in urban areas where emissions must be minimized. If the hybrid has sufficient range in its all-
electric operating mode (say 60 miles), most drivers would have little
need for their ICE, and so the hybrid's emissions would approach those
of an all-electric vehicle.[1]

HYBRID TECHNOLOGIES MADE SIMPLE

Hybrids come in two basic designs, serial and parallel. In a parallel hybrid, the engine or fuel cell not only can operate the generator to charge the battery, but also can drive the wheels directly, much as in a conventional vehicle. Hybrid power systems use computer-controlled solid-state electronics to allocate power to motor, wheels, or batteries as efficiently as possible.

The serial hybrid uses the engine or fuel cell exclusively to spin the generator to produce electricity for either the motor or the battery. Because the engine therefore needs to run only at a single, constant speed and load, it can be optimally sized and adjusted to operate with much greater efficiency than a conventional ICE vehicle engine.

Hybrids can easily exceed today's battery-electric in range (and can even exceed typical ICE vehicles' range) because of the high energy density of their onboard liquid or gaseous fuel (in addition to the energy in their batteries). For the same reason, hybrids also have no significant power limitations and are not constrained by a lack of charging facilities: Those with gasoline- or diesel-fueled auxiliary engines can use existing gasoline stations for refueling. Thus they will neither be inconvenient for long trips, nor "97-pound weaklings" on hills and freeways. In addition, hybrids may use their engine or fuel cell to provide heating and defrosting, reducing demands on the battery for these energy-intensive activities.

Under terms of a 1992 solicitation for the development of prototype hybrid vehicles, the DOE's Office of Transportation Technologies is currently providing cost-sharing funds to two development teams led by General Motors and Ford, respectively. Under the five-year development-and-testing contracts, the DOE hopes to produce a hybrid vehicle ready for production by the year 2001 that will be twice as fuel efficient as today's average new car and will meet California's Ultra-Low Emission Vehicle requirements. (The average new U.S. car in 1995 got 28.2 miles per gallon.)

OFF THE DRAWING BOARD
AND ONTO THE HIGHWAY

Several international companies are already developing hybrid vehicles. BMW has built parallel and serial hybrid prototypes. The BMW 518i parallel hybrid is equipped with a four-cylinder internal combustion engine, an asynchronous electric motor, and nickel-cadmium batteries. The motor or engine can be used separately or together, since both supply power to the wheels, and the engine also powers the generator.

The parallel VW Golf Elektro-Hybrid uses a diesel engine for accelerating and cruising and an electric motor for traveling at speeds up to 35 miles per hour. By flipping a switch, the driver can operate the car in the all-electric mode.[2] The serial Volvo ECC hybrid uses a gasoline turbine instead of a diesel in conjunction with its electric drive system. The turbine is used only to run the generator to produce electricity for the motor or battery. The ECC is a stylish head-turner with an impressive range of 415 miles at 55 miles per hour.

THE MIND-BOGGLING REALM
OF ULTRALIGHT VEHICLES

Some energy experts are enamored with the hybrid because of the possibility for an elegantly engineered, lightweight, efficient vehicle with low aerodynamic drag. Amory Lovins has shown that by making hybrids sleek and ultralight, super-efficient four-seat cars are possible right now that will get 300–400 miles per gallon of liquid fuel—using readily available technologies.[3] These "Hypercars," as Lovins called them, will thus be ten times as efficient as today's cars and will be capable of going coast-to-coast on less than a tank of fuel. With technologies now under development, more than 600 miles per gallon would be possible.[4] It's hard to imagine, but that would mean a cross-country trip on only five gallons of fuel!

Similarly, extraordinary performance can be had simply by combining batteries and *un*-hybridized electric drive systems in ultralight, low air-resistance vehicles powered by nothing fancier than battery-powered electric motors. A 1,500-pound Swiss prototype along these lines reportedly gets the equivalent of 235 miles per gallon when cruising at city speeds.[5] Existing technology like this and the exciting possibilities inherent in advanced hybrids like Hypercars makes the Clinton administration's government-industry Partnership for a New Generation of Vehicles seem hopelessly antiquated from its inception. (The Partnership's goal is to triple the fuel economy of 1993 cars by the year 2004.)

As advanced batteries appear with four times the energy per pound of today's batteries, the new batteries could be used in tomorrow's ultralight vehicles—cars weighing half to perhaps a quarter of today's cars. Battery mass could therefore fall to as little as sixteenth the mass of today's battery packs, yet with equivalent performance. While critics of pure electric vehicles have derided the massive batteries that electrics must now use, surely no one will object to a battery pack weighing less than the average ten-year-old, especially since the electric vehicles carrying them will not need fuel-using engines at all. Moreover, flywheels and ultracapacitors could reduce battery demand and mass still further, as would solar cells on the car's skin. (Ultracapacitors are energy storage devices with high power capacity and long life.) While Hypercars would still be lighter than plain battery cars, the weight differential may not be crucial to customers. Cars today vary widely in weight and still enjoy customer loyalty. Moreover, for commuter or intracity electric vehicles, which don't require long-range capabilities, the battery mass could be reduced by another factor of four relative to the mass reductions just described.

THE MERITS OF HYBRIDS
VERSUS ALL-ELECTRIC VEHICLES

Hybrids and battery cars are likely to be in competition with each other in the years ahead, and the outcome rests on complex issues, including first cost, operation and maintenance costs, convenience, and performance.[6] The advanced hybrid will be somewhat lighter than the battery car, will have greater range, will fuel faster, and its performance will not suffer significantly in very cold weather, unlike a battery car. Initially, too, because of their high fuel efficiency, these hybrids may cause less air pollution than all-electrics charged with power from coal-fired plants, but more than those charged from hydroelectric sources or from modern natural gas plants.

On the other side of the ledger, the hybrid is more complex than the all-electric, will have to have fuel systems and emissions controls, and will require more maintenance. Making its small internal combustion engine thermodynamically efficient will also present challenges. In addition, even though a hybrid initially may have lower emissions than those attributed to a battery car charged by fossil fuel–fired plants, as the hybrid engine ages, or the driving cycle deviates from emissions test standards, the hybrid's emissions could rise significantly. Pure electrics will always appeal strongly to those who seek independence from oil companies and freedom from under-the-hood combustion of fossil fuels. How important all these issues will be to consumers is not yet known, just as the future costs of advanced battery cars and hybrids still involves guesswork. Although Hypercars conceivably may take a commanding market lead over all-electric vehicles, more likely the two technologies will share the new vehicle market. A pending decision by the California Air Resources Board (CARB) to accept hybrids meeting stringent emissions limits as equivalent to zero-emission vehicles would certainly boost hybrids' acceptance.[7]

Whatever objections one might have to hybrids as fossil fuel–burners are certainly going to vanish once advanced hybrids are fueled with renewably generated alcohol or hydrogen, which suddenly become much more affordable when only a tenth as much fuel energy is needed to propel a car as today. Just as EV electricity can come from the renewable sources we've discussed in this book, the fuel needs of a U.S. Hypercar fleet could easily be met from the ample biomass resources of the United States. Ultimately, both advanced hybrids and pure electric car technologies taken to their technological limits are likely to converge toward zero net emissions, a very encouraging prospect indeed. At that point, whether Americans eventually adopt ultralight pure electrics or Hypercars will matter little.

However, suggestions by some writers that ultralight hybrids are bound to be environmentally superior to purebred electric vehicles are misleading. "A properly engineered supercar," wrote one battery-car critic, "would be *cleaner* than a 'zero emission' battery car—whose pollution, after all is merely displaced from tailpipe to power plant."[8] The flaw here is to compare the perfectly operating ultralight hybrid of the future to a transitional EV of today that is likely to be charged in part by fossil fuel electricity. Let's not forget that EVs have the capability of charging entirely on renewably generated power with zero climate-destabilizing carbon emissions. Whether that renewable capacity is put in place or not is a social responsibility, not a reflection on all-electric vehicle technology.

Just as the hybrid might one day supplant the battery electric by virtue of its longer range, fuel flexibility, and greater independence from electric charging infrastructure, so it is conceivable that hybrids might prove to be a transition technology between today's conventional car and tomorrow's high-performance all-electric vehicle. Alternatively, both types of EVs may coexist in the market, filling different niches, as do gasoline and diesel engines. A key question is whether the marketing and promotion of fossil-fueled hybrids will accelerate or interfere with progress toward true zero-emission vehicles of whatever design. Thoughtful public policy is needed now to insure the former possibility and foreclose the latter.

OIL COMPANIES AND
UTILITIES WILL SQUARE OFF

Daniel Sperling, director of the Institute of Transportation Studies at the University of California at Davis thinks that 90 percent of all the cars in some states are likely to be EVs within 25 years. No huge technological change like the revolution in vehicle technology occurs without stirring up corporate rivalry—and usually a political battle or two. The coming demise of the gasoline vehicle as we know it will create tremendous economic opportunities for some companies and threaten others. For obvious reasons, electric utilities are enormously excited by the prospect that electric propulsion technology may enable them to wrest control of the multibillion-dollar private transportation market from the powerful oil companies. A monumental struggle between behemoths has in fact already begun, albeit rather quietly. The stakes in the battle are gigantic.

The electric utility industry, if successful, could eventually conquer a new market worth tens of billions of dollars nationally, ripping it away from the big oil companies. Southern California Edison Chairman John Bryson believes electrics can increase his company's sales by 5 billion kilowatt-hours annually by 2010. (That's about 7 percent of 1995 sales.) Moreover, the industry could benefit more from each new electric car sold than from the sale of appliances having equivalent energy requirements. That's because, unlike appliances, much of the electric vehicles' demand could be shifted to nighttime charging during off-peak hours. Each electric vehicle could be worth from a few hundred dollars to as much as $1,000 or more over its life, depending on usage and electricity rates.[9]

This would allow electric utilities to profitably increase their utilization of underused generating capacity by increasing night use—without having to make major capital outlays for new power plant construction. If that occurred, utilities would both make more money and would be able to offer lower rates to their daytime customers. Utilities will use time-of-day pricing to provide EV users with powerful incentives for nighttime charging, while providing charging stations for emergency use and for the minority of users who will wish to charge in the daytime at higher rates. As the utility industry continues its restructuring and is progressively deregulated in the late 1990s and beyond, this major load-balancing opportunity in the context of a vast new transportation market offers utilities an opportunity to become more competitive with aggressive independent power producers who now threaten their customer base.

THE EFFECT ON AUTOMAKERS

Electric propulsion technology is a threat to those automakers who fail to adopt and master it. In general, the more electrics sold, the smaller the remaining market will be for conventional vehicles. But many automakers would rather continue taking profits doing what they know how to do, than retool and risk vast amounts of money producing ultralight, composite-body electrics. Large carmakers would rather amortize their profitable multibillion-dollar investments in internal combustion technology before wholeheartedly supporting electric or hybrid propulsion. So they may persist in trying to slow the transition to electric vehicles, to ease their own financial adjustment, rather than to exert themselves in accelerating the arrival of the new technological order.

To conventional automakers, however, the risk of foot-dragging is that they may fail to adapt quickly enough to the inevitable competition from EV companies. Lovins calls this the "bet your company" strategy. Eventually, the Big Three *will* need to make the investments required to reorient their production lines from gasoline and diesel to electric vehicles, and from reliance on metal bodies to composites. If they do so belatedly, their rewards will be smaller, and they will unwittingly forfeit market share to more agile, astute competitors.

AUTOMAKERS' AMBIVALENCE—
AND RESISTANCE

John Wallace, Ford Motor Company's director of electric vehicle development, says, "Ford is still ambivalent about electric vehicles," and he points to the unavailability of high-performance, affordable batteries.[10] Equivocation can also be heard in the comments of GM Vice President Kenneth Baker, head of GM's research-and-development center. Baker said in late 1994, "We have a lot to learn . . . before we can determine if the electric vehicle has any chance to succeed in real-world markets."[11]

Detroit carmakers testifying before the California Air Resources Board in May 1994 expressed doubts that they could by 1998 make an electric car consumers would want, because of "inadequate batteries" and the scarcity of charging stations. A GM representative said the cars would have to be priced at $35,000 for GM to earn a return on its investment,[12] and indeed that is the price GM has set for its EV-1.

Though these prices are too high for most people, the public would be keenly interested in a competitively priced and mass-produced EV. More than 10,000 people in Los Angeles applied to participate in GM's EV test program. Public disappointment is likely, however, if the Big Three bring only poorly performing and expensive models to market, as if to prove their prophecy that EVs are not yet market-ready. While articulating support for EVs and conducting EV research and development, top auto industry officials have simultaneously opposed government zero-emission mandates, wherever proposed. Together with the oil industry, the auto-makers have challenged the mandates in court and in the California Legislature.[13] The mandates originated as part of the California Air Resources Board's 1990 Low Emission Vehicle and Clean Fuels regulations. The rules required that 2 percent of vehicles sold by large automakers in California by 1998 must be Zero Emission Vehicles (ZEVs). That quota was to have risen to 5 percent in 2001 and 10 percent by 2003.[14]

Chrysler Corporation, which was the first manufacturer to receive ZEV, Low Emission Vehicle (LEV), and Ultra-Low Emission Vehicle (ULEV) certification from California, mobilized its auto dealers and others to petition the California state government for repeal of the mandates. A Chrysler official told the dealers that "all vehicle prices will likely rise [by as much as $2,000] to cover the high costs associated with electric vehicles," and that dealers would be at a competitive disadvantage vis-à-vis smaller manufacturers who do not have to comply with the mandate.[15] Whether or not one believes this argument to be an exaggeration, in the long term, with mass production, EV costs will fall below those of gasoline vehicles.[16]

With strong support from California Governor Pete Wilson, the auto and oil industries in 1996 succeeded in getting the state's Air Resources Board to rescind the 1998 and 2001 deadlines. Their actions will slow EV research and development and reduce the "market pull" that is rapidly accelerating EV commercialization. Thus while automakers publicly voice support for EVs, they vociferously opposed the very rules that gave the electric vehicle industry its greatest impetus.[17] In California, the auto industry "fielded the largest lobbying forces in Sacramento that legislative regulars have ever seen," according to auto reporter Jim Motavalli of *E Magazine*.

Twelve northeastern states and the District of Columbia have petitioned the EPA to adopt California's air-quality legislation, and two of the twelve have already done so administratively. The Big Three lobbied frantically against this petition and appear to have thwarted it. If the northeastern states had adopted the California regulations, then 100,000 electric cars would have been sold nationwide in 1998, and by 2003, one in every ten new cars sold in the affected regions would have been electric. With the market pull of the Northeast and California stimulating electric vehicle technological development, electric vehicles would have been so prevalent, attractive, and advanced by 2004 that a tenth of all new U.S. vehicles might have been electric by then.[18]

Only the final mandate for 2003 and beyond survived the auto and oil industries' intensive opposition, but since 1.2 million passenger vehicles are sold in California every year, the mandate will require hundreds of thousands of EVs to be sold in California alone.[19] While resisting government mandates, and characterizing them as symptomatic of "Big Government," the large auto companies have no problem in the next breath asking government and utilities to make generous financial commitments to EV commercialization. Speaking for General Motors at the Twelfth Annual Electric Vehicle Symposium in Anaheim, California, GM Vice President Ken Baker said taxpayers should provide up to $10,000 per car in subsidies to retail EV customers and that government should buy half of all the EVs that California had said it would require automakers to provide in 1998. "Putting viable electric vehicles on the road here on planet Earth is a bigger challenge for us than it was to put the first electric vehicle on the moon," Baker claimed. The Apollo Project, of course, was 100 percent government financed. Baker also contended that government or utilities should rent EV batteries to relieve the EV purchasers from buying them, and that a public vehicle-charging network should be established. ". . . Government should be the early market," said Baker. "If government is really serious about electric vehicles, it should also offer incentives to early buyers, and possibly cover early development costs."[20] While government incentives to encourage EV commercialization would be helpful, the clamor for public money is incongruous coming from a former science adviser to President Ronald Reagan, whose administration was noted for its antitax, antiwelfare, and anti–Big Government rhetoric.

Automakers' protestations about the insuperable problems they face in EV technology should be put in historical perspective. The Big Three once staunchly opposed air bags, antilock brakes, fuel efficiency standards, seat belts, crash tests, and the emissions controls mandated by the federal Clean Air Act. What they historically *have* shown great enthusiasm for has been the marketing of big, flashy, over-powered cars, in preference to less profitable, fuel-efficient vehicles. It will be interesting to see their long-term response to advanced hybrid and other ultralight electric technology.

THE BIG THREE'S PRODUCTS AND PLANS

A cynical interpretation of the Big Three's behavior today is that, while voicing support for EVs, they are in the short run discouraging the public about EVs by publicizing overpriced, underperforming prototypes and lamenting their shortcomings. The fact that Ford and Chrysler chose to make their EV debuts with $100,000 minivans rather than small light vehicles doesn't allay the skepticism. (Chrysler actually sold its minivans to utilities at $120,000 each. Small wonder only 50 had been sold by mid-1994.[21]) Currently, the General Motors Corporation "G-Van," using a Vandura body, costs about $40,000 more than the typical gasoline-powered Vandura ($16,000–$20,000). These electrics, known as Conceptor G-Vans, were bought by the California Energy Commission and Pacific Gas and Electric Company, although they have a top speed of only 52 m.p.h. and a mere 60-mile range. By contrast, the $100,000 Chrysler Corporation TEVan with advanced batteries (either nickel-cadmium or sodium-sulfur) has a range of 185 miles and a top speed of 65 m.p.h. Also priced at $100,000 in 1995, the Ford Ecostar—a small, two seat utility van—can reach 70 m.p.h. and travel 100 miles on a charge. Unfortunately, the Ecostar takes 20 hours to recharge at 120 volts and 6 hours at 240 volts.

Chrysler's newest EV, the EPIC minivan based on the Dodge Caravan/Plymouth Voyager, is intended for fleets. It has a top speed of 80 miles per hour, a 60-mile range, and sealed lead-acid batteries that require eight hours to recharge (at 240 volts). The minivan comes with a regenerative braking system, power steering, power brakes, and dual air bags.

Regardless of the shortcomings of their early product introductions and despite their lackadaisical efforts to commercialize EVs, the major automakers have already spent hundreds of millions of dollars on EV research and are clearly making strategic alliances in an effort to position themselves for the long-term EV market. To date, General Motors has formed a joint venture with Ovonic Battery Company, developer of the Ovonic Battery, to produce nickel metal-hydride batteries that are expected to double the range of lead-acid batteries. GM has also established Delco Propulsion Systems as an internal joint venture to develop and market EV systems. GM Ovonic President John Adams foresees a $25 billion global market for electric propulsion systems by the year 2005, including the hybrid vehicle and golf cart markets. GM also owns Hughes Power Control Systems, which makes electric drive systems and a credit card–operated electric recharging station that resembles a gas pump in appearance.

Chrysler Corporation said it expected to have an EV available for the general public in 1996 but as of late 1996 had not done so. The company is also committed to building a minivan called the NS Electric and offering it in 1998. Chrysler is reportedly leaning toward an advanced nickel metal-hydride battery to provide the van with a 120-mile range, but says the battery may be nickel-iron.

Ford is now allowing other qualified companies to buy engineless gliders from it for conversion to electric drive. Until now, the Big Three had obliged independent electric car companies to buy the gasoline vehicles they were converting to electric drive, complete with gasoline engines. These smaller, often struggling electric carmakers then had to spend precious time and money to rip out the unneeded engine and equipment, raising their final product costs.

VAST POTENTIAL MARKETS
FOR ELECTRIC BUSES AND TRUCKS

Electric trucks and buses today are progressing rapidly toward mass market readiness. Modern and attractive electric buses offer passengers every convenience while replacing the roar and black fumes of polluting diesels with clean, almost-silent electric motors. Like electric autos, electric buses are durable, low-maintenance vehicles, and some have battery packs that can be rolled out of the bus and changed in minutes. Several companies are engaged in electric and hybrid bus manufacture. The El Dorado National Company of Chino, California, for example, offers heavy-duty hybrid electric transit buses that have a top speed of 50 m.p.h. and a 150-mile range when equipped with a 20 kilowatt-hour propane generator. An all-electric version is also available.

Specialty Vehicle Manufacturing Corporation of Downey, California, offers what it calls the world's most complete line of zero-emission buses and route vehicles. Products include passenger buses, school buses, trolleys, trams, and a step van being developed under contract to the Northeast Alternative Vehicle Consortium (funded jointly by Boston Edison and the U.S. Defense Department's Advanced Products Research Agency). The company also has a contract from the California Energy Commission and others to design and build an electric school bus "from scratch," rather than merely modifying the design of a conventional bus. The company's 28-passenger bus offers five-minute refueling through battery swapping and even kneels for elderly and handicapped riders.

Twenty-eight passenger, zero-emission electric bus made by Specialty Vehicle Manufacturing Corporation of Downey, California, and Advanced Vehicle Systems, Inc., of Chattanooga, Tennessee. *Courtesy of Specialty Vehicle Manufacturing Corporation.*

Preliminary studies of maintenance and fuel costs have shown that the energy cost for its 22-foot electric bus is only 5.7 cents a mile compared to 18 cents a mile for a diesel bus.[22]

The nation has 360,000 school buses in service today, and about half of their routes are 80 miles or less, so the school bus market is highly attractive to electric bus makers.[23] Thomas Built Buses, Inc., a leading school bus manufacturer, is in a long-term partnership to develop electric school buses with Power Control Systems, a business unit of GM Hughes Electronics. A prototype using sealed lead-acid batteries has already been produced. Westinghouse Electric Corporation, in partnership with Blue Bird Corporation, the world's largest bus manufacturer, will produce the electric-drive systems for electric and electric-hybrid school buses and transit buses.

Like Westinghouse, General Electric does not want to be left out of the EV market. General Electric's GE Drive Systems of Salem, Virginia, has been teaming with the Baker Equipment Engineering Company of Richmond, Virginia, a manufacturer of utility, firefighting, and construction vehicles. The partners are producing a full-size electric pickup truck conversion called the Baker EV100, powered by a GE 100-horsepower AC-drive system. While the EV100 can carry two passengers and 1,000 pounds of cargo, the truck has only a 60-mile urban driving range and costs a hefty $33,000, plus the cost of the original engineless GMC 2500 vehicle chassis.[24]

Financially troubled U.S. Electricar of Santa Rosa, California, builds a one-ton capacity electric delivery truck, the Electrolite 2000, for inner-city deliveries[25] and a 22-foot electric shuttle bus for community transportation. (The company now has a new CEO and is being financially restructured.) The 50–75-mile range and 35 m.p.h. maximum speed of the electric bus is fine for moving passengers locally to and from airports, mass-transit hubs, parks, and campuses.

A host of these special shuttle applications are virtually tailor-made for EVs. Airport vehicles typically have short predictable routes that bring them back to service centers for charging when needed. Los Angeles International Airport (LAX) is already operating a 32-passenger electric shuttle bus that will avoid the 5,000 pounds of smog-forming emissions that its diesel predecessor spewed every year. LAX plans to acquire more than 500 alternative-fueled vehicles. Chicago's O'Hare International Airport now has an all-electric baggage room, and San Diego International has an electric baggage carrier; Logan International employs a powerful electric "aircraft tug."[26]

Industrial electric vehicles are already in widespread use. One firm alone, 45-year-old Taylor-Dunn of Anaheim, California, has 120,000 of its EVs currently in use in airports, campuses, factories, military bases, and warehouses. The company makes a three-quarter ton ElecTruck delivery van that can go 50 miles and reach speeds of 32 miles per hour, which is suitable for smoggy, congested urban areas: Hundreds of the $15,000 ElecTrucks are currently in use in Mexico City.

ELECTRICS ABROAD: POWERFUL INCENTIVES FOR EARLY EUROPEAN ADOPTION

Electric vehicle development is occurring rapidly in Europe and the Far East. High gasoline prices abroad coupled with greater congestion, lower average speeds, and shorter average trip lengths make international markets very attractive to EV makers. In addition, carbon dioxide emissions have increased by a third in France just since 1980, and vehicular traffic there has swelled 25 percent, putting a premium on low- and zero-emission vehicles.[27] Just as European nations have shown more interest than the United States has in some renewable energy technologies, such as wind power, so they are also ahead in adopting EVs. Some 25,000 electric vehicles are now on Europe's roads, at least eight times more than in the United States.

Virtually all the major Japanese carmakers plus BMW, Fiat, Mercedes-Benz, PSA Peugeot-Citroën, Renault, Swatch, and Volvo are marketing or developing EVs. But if you're looking for stellar EV performance, don't look to France or Japan just yet, even though the French and Japanese are active in the field. PSA Peugeot-Citroën, the French carmaker, has been taking a leadership role in European EV development and has received a lot of publicity for its electric Peugeot 106 and Citroën AX "city cars." The company hopes to be selling 10,000 of these vehicles a year in Europe by the year 2000. Although relatively inexpensive, the 106 and the AX are small, uninspiring vehicles that offer modest performance at best: 45 city miles of range and a top speed of just 55 m.p.h. With subsidies of $4,000 per vehicle from the French government, the final cost to consumers will be $17,000, excluding taxes. The gasoline-powered model costs about $15,000. Or the purchaser can buy the car for $12,000 without batteries and then lease the costly nickel-cadmium battery pack for $120 per month.[28] However, with high French gasoline prices—about $4 a gallon with taxes—and low electricity prices, this monthly fee for batteries and electricity makes operating costs roughly equivalent to the cost of fuel for a comparable gasoline-powered vehicle.

Even if its vehicles are not yet technological marvels, PSA Peugeot-Citroën is commendable for the thoroughness with which it has conceived its step-by-step plan for commercializing EVs. The four-stage program began in 1989 with the introduction of electric versions of the Peugeot J5 and the Citroën C15 and C25 vans for corporate and government fleets. These customers, Peugeot reasoned, would be able to absorb the added cost of the vehicles and would be able to handle charging and maintenance in their own facilities. The electric Peugeot 106 and Citroën AX were scheduled for 1995 introduction. The AX was offered for sale in 1996 but the 106 was still being tested in 1996 in several mainly European markets. In stage three, at the end of the decade, the company will introduce the Citroën Cité, a subcompact concept car designed especially for electric drive. According to Peugeot, the Cité's motor will last a million kilometers (about 620,000 miles), and the car will have a nickel-cadmium battery that will last ten years. Peugeot intends to sell several tens of thousands of the Cité. Later, the company expects to produce a hybrid electric with an onboard turbine generator for long-distance highway driving. Peugeot's diesel-generator hybrid prototype unveiled in 1991 has already achieved a range of 465 miles at a speed of 62 miles per hour.

In an ongoing experiment being conducted by Peugeot and the city of La Rochelle, France, Peugeot 106 and Citroën AX sedans are leased to users for $182 and $164, respectively, a month, including batteries, maintenance, and insurance. Because of low electricity prices, the user then enjoys per-mile travel costs that are about 75 percent lower than for a comparable gasoline-powered car. The company in 1992 had signed a "framework agreement" with the French Ministry of Industry, the Ministry of the Environment, Electricité de France, and Renault to create a network of recharging and service facilities to encourage electric vehicle commercialization. The French government has authorized $91 million to equip 24 participating French cities with the recharging stations by 1995.[29] The utility partner, Electricité de France, itself owns a fleet of 450 electric cars and vans.[30]

Peugeot is not alone in the French electric vehicle market. The Société Européenne des Electromobiles Rochelaises (SEER) manufactures and sells a light commercial electric Volta microcar for $28,000. SEER/Volta also produces a minivan and pickup truck. These go 37–50 city miles carrying a 500-pound payload at up to 46 miles per hour. The sailboat maker Jeanneau produces a two-seat electric for $24,000, and Renault in 1995 started selling its four-door Electric Clio hatchback. But it costs about $30,000, less the $4,000 subsidy—for a vehicle that travels only 60 miles and can't exceed 55 miles per hour.

Switzerland, with about 2,000 electric vehicles, has one of the highest ratios of electric vehicles to people in the world. About two dozen types of mostly imported electric vehicles are sold on the Swiss market by about a dozen companies. Some recreation areas, such as Zermatt, allow only electric vehicles.

ELECTRIC VEHICLE
DEVELOPMENTS IN JAPAN

The United States still has an opportunity to consolidate electric vehicle technology leadership over rival Japanese auto manufacturers. U.S. electric vehicle technology in 1995 was far ahead, although the entire United States has less than 3,000 highway-capable electric vehicles. Japan had only about 2,000 electrics on the road, and the government has been reluctant to support commercialization.[31]

Toyota, however, now has an electric vehicle division, and outside the United States will be marketing the EV-50, an electric city commuter that currently uses a lead-acid battery and has solar panels on the roof to operate an interior fan. In Japan, Toyota was also marketing an unimpressive $65,000 minivan called the Lite Ace. By 1997, however, Toyota will begin selling about 300 RAV4-EV electric sport-utility vehicles for fleet use in the United States, with a nickel-metal-hydride battery (NiMH) and at least 120 miles of range, which the company is optimistic about increasing. The company's current marketing strategy is not to offer a purpose-built electric vehicle but to sell EVs based on existing gasoline-powered vehicle "platforms."

While Honda's electric Civic is even less exciting in range and acceleration than the EV-50, Honda plans to lease 300 new purpose-built electric vehicles with NiMH batteries to individuals and fleet buyers in California starting in 1997. These four-passenger sedans will be similar in range and acceleration to Toyota's RAV4-EV. Honda's lease program, however, will include battery replacement and other major items.

Of greater significance than these early vehicles is the entry of various Japanese companies into the race to develop a viable metal-hydride battery. Among the contenders, Toyota Automatic Loom Works and Matsushita Battery Industrial Company both have developed prototypes.[32] MITI, the Japanese Ministry of Trade and Industry, supports Japanese research into advanced battery technologies. While some of the Japanese companies, such as Toyota and its Daihatsu Motor Company affiliate, seem to be slow to relinquish lead-acid technology, Nissan has already developed a nickel-cadmium battery pack that can be fully recharged in just 15 minutes under laboratory test conditions. Because of concerns about Tokyo's air quality, the Tokyo Metropolitan government provides a 50 percent subsidy to purchasers of certified electric vehicles.

If a car is too big and a bicycle is too small for your needs, you may like the Japanese ES600 electric scooter, developed jointly by Kyushu Electric Power, Chubu Electric Power, and Tokyo R&D. Using a brushless DC motor, the scooter runs on sealed batteries underneath the seat. The 257-pound scooter has a 37-mile range at 19 m.p.h. and cost 596,000 yen in 1995 (about $6,800). Honda offers a three-wheeled GYRO EV scooter equipped to carry small cargo.

Unique Mobility, Inc., of Golden, Colorado, believes there is a market for millions of electric scooters and has reached an agreement with Kwang Yang Motor Company of Taiwan to develop a 180-pound scooter with a range of 35–50 miles at up to 35 miles per hour. Its permanent magnet, wheel-mounted AC motor will be roughly comparable to a three-horsepower gasoline engine and will come with a removable battery pack that can be hand-carried to the nearest wall socket for charging.[33]

To travel very inexpensively under electric power, you may want to consider an electric bike. Zap Power Systems of Forestville, California, sells an 8-pound motor and 24-pound battery that will take the rider 20 miles at up to 18 miles per hour (without pedaling). The Zap system can be installed on any two-wheeler and is far cleaner than the highly polluting two-stroke engines that power millions of conventional scooters and motorcycles around the world.

ACCELERATING EV COMMERCIALIZATION

Speeding the adoption of electrics will improve air quality, which benefits everyone. Yet virtually no one will voluntarily and single-handedly make financial sacrifices for it. Why should anyone, the economists argue, if others benefit without sharing the costs? Are government tax credits and exemptions for electrics justified to solve this problem? Government already gives subsidies to the coal, nuclear, farm, travel, and shipping industries, to name a few, and to the oil industry, which makes the gasoline with which electrics have to compete. EV subsidies are perfectly legitimate and will increase demand, jump-start mass production, increase competition, provide necessary public infrastructure, and then make themselves obsolete in short order. Government support should not be permanent.

Institutional support from both governmental and private entities is also helpful in overcoming obstacles to the early establishment of EV "infrastructure," specifically charging stations. That need has been addressed by a number of utilities and by CALSTART, a public-private consortium that develops EV incentives, including rebates on EV batteries and at-home electrical charging systems; access to carpool lanes; reserved parking with charging facilities at public and private lots; and special off-peak nighttime charging rates. CALSTART is also supporting the development of multivehicle charging kiosks for mass-transit-hub parking lots, and is leading development of a multiuser electric "station car" to take commuters between homes and transit hubs.

For most individuals, the best way to accelerate the commercialization of EVs is to buy one. If you are thinking of taking that step, a log of your daily trip lengths would be useful to assess your range needs. If buying a converted gasoline vehicle, check out the manufacturer thoroughly, make sure the car complies fully with all federal motor vehicle safety standards, and test the car uphill and under freeway driving conditions. While conversions can provide great satisfaction at a reasonable price, for the highest EV performance available, a vehicle designed from scratch for electric propulsion is de rigueur.

WHAT THE FUTURE WILL BRING

The torrent of technological activity now sweeping the EV field will radically transform trucks and buses, as well as cars. Just as the plethora of cable TV stations hurt the three major TV networks, so new aggressive electric car manufacturers and developers of flywheel, fuel cell, and hybrid technologies will reduce the Big Three automakers' market share, unless they excel at the new technology. The auto industry as we know it must change or die.

Far more important than the effect of EVs on the Big Three, however, is the enormous impact that electrics will have on the environment, the oil industry, the entire auto industry, and the utility industry—in which EVs will one day be a dominant load. Once high-performance electrics are mass produced at comparable costs to gasoline vehicles, consumers will stampede to the clean, quiet, low-maintenance, and economical technology. In a classic positive feedback cycle, falling EV manufacturing costs will increase EV market share, which will in turn permit economies of scale that will lower costs again, and so on. Electrics will in effect snatch the multibillion-dollar transportation market from the oil companies and deliver it to electric utilities and to those consumers who operate their own electric generating facilities, renewable or other.

While EVs naturally will not solve all our transportation problems— and the attractiveness and affordability of future electrics may even boost personal auto sales and traffic—the possible triumph of electric vehicles over internal combustion engines offers the United States a tremendous multifaceted opportunity to reduce oil imports, increase energy security and efficiency, improve our balance of payments, protect the environment, enhance global economic competitiveness, build new industries, and fill newly created jobs with American workers. When the electrics prevail at last, the din of internal combustion engines will be muted and the air will smell sweeter. We will be able to enjoy a quieter, more peaceful, healthier, and more pleasant world.

25

GETTING TO A
RENEWABLE FUTURE

Renewable energy sources already provide more energy than nuclear power and, with increased and sustained support, could become the dominant source of new energy production in the 21st
century.

—*America's Energy Choices: Investing in*
a Strong Economy and a Clean Environment

MAKING THE TRANSITION

This chapter outlines a set of energy policy measures to increase and accelerate the deployment of renewables and efficiency technology. Readers
looking for a quick, easy-to-implement formula may be disappointed here.
Bringing about a renewable energy economy will be a complicated and
contentious process. I have instead suggested basic principles, strategies,
and general policies that can facilitate the transition to renewables.

No single set of national policies will take best advantage of local renewable resource assets, political realities, and institutional conditions.
Ultimately, an intricate lattice of policies will be crafted by state, regional, and local stakeholders operating under broad federal guidelines.
The means to a renewable energy economy will emerge finely wrought
as from a crucible heated and pressured by the actions of Congress,
courts, legislatures, commissions, corporations, consumer advocates, researchers, and environmental organizations. Since this is a dynamic
process, no sooner will this chapter be completed than it will already be
somewhat outmoded, as technology changes and as political, economic,
legal, and regulatory landscapes evolve. Still, the underlying policy recommendations should remain valid.

THE CASE FOR RENEWABLES
AND EFFICIENCY REVISITED

The world's next energy transformation will be an inexorable shift to renewable energy. The impressive progress now occurring in renewable energy and energy efficiency is part of this process. What the Newcomen
steam engine was to the eighteenth and nineteenth centuries and electricity to the twentieth, the synergistic impact of renewable energy, energy efficiency, and electrified transport will be to the twenty-first century. The
emanations of the coming renewable energy era are already discernible,
just like the first shafts of light at daybreak. Why is the switch to renewable
energy inevitable?

The answer is not merely because the aggregate environmental impacts of a world fossil fuel economy on air, fresh water, and oceans will
become progressively more insufferable as world population expands and
as world economic activity continues to increase. The shift is also inevitable because renewable energy is abundant, domestically available,
generally fuel-free, very low to nonpolluting, climate-safe, and sustainable. In short, it is superior. Compare renewables with fossil fuels.

In 1994, the United States began importing more than half its oil—for
the first time ever. But imported oil supplies are liable to interruption—
either through market manipulation, political fiat, war, "acts of God," or
eventual resource depletion. By contrast to fossil fuels, renewables entail
no-to-low risk of fuel price escalation, and no risk of fuel embargo. Instead, they offer price stability and energy security. Slaking the nation's
growing thirst for oil already consumes 9 million barrels of foreign crude
a day at an annual cost of $56 billion. With our transportation sector al-
most totally reliant on petroleum, the U.S. economy is ever-more vulnerable to dislocations—including recessions—triggered by sudden fuel
price hikes. Even natural gas, so abundant today, historically, has been
volatile in price; supply bottlenecks and price spikes could again develop
in the future. A vigorous renewables industry could, in effect, provide the
nation with an insurance policy against sudden run-ups in natural gas and
oil prices.

Renewables have a whole suite of unique benefits: They have no adverse impact on our balance of trade, and they present no national security risk, far less risk to world biodiversity than petrochemical energy
systems, and little risk of acid rain. They can be installed quickly in convenient modular increments, so there is no need to make financially risky
guesses about the extent of future electrical demand. In contrast to large

nuclear or coal power plants, most renewables can be built in resilient, decentralized, efficient networks, close to users. They create few significant waste hazards and require no need for authoritarian political institutions to protect the public from "solar energy terrorism," or "silicon cell proliferation." Except for the risk of dam failures in the case of hydropower, renewables present no catastrophic accident risks.

The commercial prospects of renewable resources and technologies are exciting. Intensive development of domestic renewables could give the United States a valuable commercial edge in rapidly growing renewable energy industries, such as photovoltaics, that have huge worldwide market potential. Likewise, by making our economy as energy efficient as possible, we can improve our international competitiveness, both by lowering manufacturing costs and by selling the efficiency technology abroad that our domestic demand has nurtured. Simultaneously, the increase in energy efficiency would nourish our domestic economy. Greater energy efficiency will free capital now spent on wasted energy or on pollution control—or on trying to fix pollution's costly impacts.

Today we are a "throwaway society." We consume resources and discard the resulting waste at a furious rate. Conventional energy systems epitomize the throwaway approach: They burn fuel and produce copious waste. These practices presume both infinite resources and an infinite planetary capacity for assimilating waste. Creating a sustainable energy economy would be a major break with the throwaway approach. Living primarily off efficiently utilized renewable energy income rather than off fossil fuel capital would allow us eventually to achieve not only a clean energy economy but a clean environment and clean industries.

SHORTCOMINGS OF EXISTING NATIONAL ENERGY POLICIES

Renewables are popular in the United States today. Polls conducted by Republican pollster Vince Breglio in 1994 and 1995 showed that almost two-thirds of Americans believe that renewables should be the U.S. Department of Energy's highest or second-highest funding priority, and 73 percent said that funding for nuclear and fossil fuel development programs should be cut as a first step in reducing the DOE's budget.[1] Yet the DOE's budget is a mulligan stew of support for coal, oil, uranium, and natural gas technologies, with only a little dash of renewables and efficiency for garnish. The DOE spent over $3.3 billion in 1995 on applied energy R&D and energy-related basic research, but only $363 million on all renewable energy technologies combined.

The 1992 Energy Policy Act clearly illustrates the misplaced emphasis and diffuse focus of our national energy policy. While mildly supportive of efficiency—and including a valuable renewable energy production incentive in Section 1914—the Act fails to make renewables and efficiency its centerpiece, or to set our course unmistakably toward a renewable energy economy. Instead, the Act simultaneously tried to boost the development of a new generation of atomic power plants and provide tax relief to independent oil and gas drillers while sponsoring research on the extraction of oil from shale rock and promoting coal combustion technologies.

The decades-old practice of putting fossil fuels and nuclear power first in federal funding is clearly inappropriate to our national needs as we approach the next millennium, and is counter to the wishes of most Americans. For the United States and for developing nations, renewables and energy efficiency, rather than coal, oil, and gas, are the best hope of building secure, environmentally safe, and economically affordable energy systems. In particular, developing nations that lack an adequate national electricity transmission and distribution grid would find the construction of decentralized renewable energy systems to be economically competitive and generally preferable.

OUR INTEREST IN CLEAN ENERGY ABROAD

If renewables fail to take root in the developing nations, those countries will most assuredly resort to the cheapest combustion technologies they can find, including coal and wood (often cut in unsustainable ways from dwindling tropical forests). The United States today is neither modeling the rapid commercialization of renewables nor providing developing countries sufficient assistance in adopting them. The U.S. Agency for International Development's energy assistance budget for FY1996 was less than $20 million. Meanwhile, as developing countries get richer, their energy-use surges and their air pollution grows. China, India, and other developing countries are already choking on horrible air. Why should this concern us?

Apart from humanitarian concerns, if we do not help these countries to implement energy efficiency and acquire renewables cost-effectively, they will continue to expand their economies by vast increases in coal use. That would not only be at the expense of global air quality, but could accelerate global climate change, through increased carbon dioxide production.

China's carbon dioxide emissions have grown more than 200 percent between 1970 and 1990. At this rate, China will be the world's largest carbon dioxide producer in a mere 20 or 30 years. If we fail to bring the merits of efficiency and renewables vigorously to the attention of developing nations, they will instead make massive long-term commitments to fossil fuel plants that will likely spew pollution into the air for the next 30 to 40 years. Eventually, both developed and developing nations would suffer economically from the regrettable consequences.

Since an eventual global transition to renewables is inevitable, we have everything to gain by anticipating the trend and accelerating it. Instead of continuing to overemphasize fossil and nuclear energy, we need a coherent, sustainable energy strategy that enables us to concentrate precious research-and-development resources directly on renewables and efficiency, while minimizing the environmental impacts of energy use.

PUBLIC SPENDING TO
SPEED COMMERCIALIZATION

Public funding for renewables and efficiency is desirable for the following important reasons:

+ Pervasive market imperfections hinder some renewable and efficiency technologies. Public policy backed by public funding would help the technologies overcome the marketplace barriers discussed in Chapters 21 and 22.
+ Historical imbalances in federal funding patterns have favored fossil and nuclear power over renewable technologies. A modicum of public funding—properly leveraged with private development capital— now can help mitigate the effects of past favoritism shown to conventional energy technologies.
+ Energy research and development is often extremely costly and risky. Public support for renewables and efficiency can share in and thereby abate the risks to developers of these new and socially beneficial technologies.
+ Finally, early adopters of leading-edge renewable energy technology encounter relatively high initial costs, often because production volume is low and economies of scale are therefore not yet fully real-

ized. Temporary government support can reduce the costs of early adoption, increasing demand and shortening the time when increased production volume makes large economies of scale possible.

A NEW ENERGY POLICY

Renewable fuels and electric and hybrid vehicles could alleviate the energy supply risks and price risks of conventional fuels. Other advantages include climate protection, smog reduction, acid rain abatement, and the production of more jobs per dollar than fossil and nuclear industries provide. Moreover, building a renewable energy system sets us on the path toward an ecologically sound and sustainable world. Precisely because energy prices and interest rates are low today, this is an excellent time to build a new renewable energy infrastructure.

The shift to a renewable energy economy is not a "techno-fix," but a fundamental change in the way we generate and consume energy. It could lead not just to clean energy, but to the creation of a nonpolluting transportation system and a nonpolluting industrial sector. Accomplishing these long-term goals is worth significant public investment—especially since in the long run it will bring us cheap energy, too. But to guide that investment properly, we must reorient our energy policy to stop perpetuating the "throwaway society," in order to create an efficient, renewable energy economy.

A MAJOR SHIFT TO
RENEWABLES IS FEASIBLE

Earlier in this book, I showed that we have ample renewable energy resources and that various renewable technologies are already commercially available and competitive. We can have vast quantities of solar thermal electric, photovoltaic, wind, and biomass power as well as passive and active solar heat, if we choose to use them. Virtually all renewable technologies are dropping rapidly in cost and improving in efficiency and reliability. Most are already less costly than nuclear power, and several are within 2 cents a kilowatt-hour of new coal plants (even when the costs of coal pollution are neglected along with the multibillion-dollar subsidies nuclear and fossil fuels have received). Wind power, for example, is but a tenth the cost today that it was in 1981.

Photovoltaics, the most expensive of the commercial renewables, is already cost-competitive in 60 niche markets that include remote power and decentralized grid support. Wind at between 3 and 4 cents a kilowatt-hour in the best wind regimes, hydro at 4.5–7.5 cents, and geothermal power as low as 4.6 cents all are not far in price from new fossil fuel power. Solar thermal power is a few pennies more, and photovoltaics for Sunbelt rooftops and building-integrated photovoltaics both are not far behind. Biomass power is expected to be 5–7.5 cents by the year 2000, with whole-tree combustion technologies at 4.5 cents by then. All the remaining price barriers could be surmounted by well-formulated energy policies.

As for the ability of renewables to provide power on demand, although wind and solar are intermittent technologies, geothermal and biomass offer round-the-clock energy delivery; so does pumped storage, some hydro reservoirs, and solar thermal electricity backed by biomass. As the renewable technologies are deployed in the decades ahead, new utility-scale energy storage technologies will become available, greatly enhancing the value of solar and wind power for intermediate and baseload generation. Until then, utilities today can easily absorb at least 10 percent and perhaps 20 percent of their power from intermittent renewables with no storage and with no significant operational changes. A huge renewable energy capacity—75,000–150,000 megawatts—thus can be accommodated in the existing utility system before energy storage even becomes an issue.

Historical precedents suggest that major transformations in energy technology and in our energy supply mix have occurred swiftly and without much public management in the past. From about 1920 to 1970, U.S. coal use dropped from 70 percent of our energy supply to less than 20 percent as we shifted to oil. During the 1980s and 1990s, a shift from oil to natural gas occurred. For example, California—which got 25 percent of its electricity from petroleum in 1980—produces less than 1 percent of its power from oil now. Most Californians never noticed the difference. A more rapid and carefully planned shift from fossil fuels to reliance on energy efficiency and renewables could be just as acceptable today as the previous gradual shifts from wood to coal, coal to oil, and oil to gas, all of which occurred within the past 150 years. And with proper incentives and management, this switch could occur much faster. By not adopting renewables more quickly, we are paying an unnecessary price in environmental damage caused by fossil and nuclear fuels.

OUR SYSTEM OF REWARDS

How do we speed the shift to sustainable energy? First, let's be clear on what's slowing us down. The reason we do not get a larger proportion of our energy from renewables today is that the systems of compensation we currently use have not made renewables profitable enough, and the political, financial, and industrial framework we have created, based on utility monopolies, has actually impeded the development of renewables. Whereas the main overt obstacles to renewables are economic, when these economic obstacles are carefully analyzed, many are revealed to be inherently political barriers. Removing them is not easy, but it can be done. The examples in this book attest that when strong financial incentives are provided in the marketplace for renewables and efficiency, people willing to earn that money appear, and new generating capacity or efficiency resources become available. Market forces, skillfully guided and stimulated, can greatly accelerate the market penetration of energy efficiency and renewables. The decisions to influence those market forces, however, and the policies adopted to do so, are political actions and must be taken in the political realm.

We do know how to successfully develop and disseminate new energy technologies. Proven methods exist. More than 60 years ago, under the New Deal's Rural Electrification Administration, centrally generated electricity was brought to 90 percent of America's 30 million rural residents. Nowadays, we have far more sophisticated regulatory and fiscal tools available. They can steer the nation's energy production and consumption choices and provide us with superior energy technology.

Programs for the rapid commercialization of renewables and efficiency technologies should incorporate time-tested elements, such as the following:

- ✦ strategic goals for technology adoption, such as specific megawatt-scale targets of installed wind and solar generating capacity and "negawatts" of saved energy, through energy efficiency;
- ✦ preferential renewable energy production tariffs, investment cost-sharing, low-interest and guaranteed loans, investment tax credits, and accelerated depreciation;
- ✦ ample research, development, and demonstration support;
- ✦ expansion of renewable energy production through market demand aggregation (to lower costs via bulk orders);
- ✦ surcharges and rebates in order to direct energy investments and purchases to renewable technologies, as needed;

✦ long-term standard contracts issued by bulk power purchasers. This type of contract eliminates the need for fractious contract negotiations and provides developers of renewables with a fixed stream of predictable payments over time. Unfortunately, in today's increasingly competitive energy markets, such contracts would face stiff opposition.

Some of these elements are in place to varying degrees for certain technologies. Some need expansion, extension, or better funding. Fortunately, we currently have a textbook opportunity to deploy combinations of these measures. The United States currently has about 100,000 megawatts of electric generating capacity that is at least 40 years old and—when operating—produces a disproportionate share of utility air-pollution emissions. Utilities will probably keep at least half these plants operating for another 20 years, despite their inefficiency, because most of the plants will have been completely depreciated by 2000–2010. Instead, we could design strong incentives for utilities to voluntarily accelerate the retirement of this obsolete and polluting capacity and replace it with renewables. Unfortunately, the same pressures forcing the restructuring of utilities for greater market competitiveness may well provide perverse incentives for keeping the old, cheap, dirty power plants on-line. The rules of the new electricity market should be written to preclude that eventuality.

Over the next 15 years, the United States is expected to build the equivalent of 180 new large power plants (180 gigawatts) at a cost of $150 billion. Were this big new energy demand to be met by renewables, it would help bring down prices for these renewables, greatly enhancing their competitive positions. But unless we act decisively, renewables are expected to make little additional contributions to our energy supply in the next 15 years. The U.S. Department of Energy's Energy Information Administration estimates that under a "business-as-usual" scenario, without intensified support for renewables, the share of our electricity produced by renewables will grow by only 2 percent from 1990 to 2010 (rising from 11 percent to 13 percent, respectively).

HASTENING THE
ENERGY TRANSFORMATION

Some measures for speeding the adoption of renewables would be technically fairly simple to implement in a regulated utility environment. For instance, premium buyback rates for renewably generated electricity can be used to stimulate renewable energy production. Advantageous rates are justified because pollution-free power is a superior commodity. Today, however, even fair-market-value residential buyback rates are rarely offered.

In a regulated utility environment, the imposition of a "net metering" policy for renewably generated power is a straightforward regulatory decision. (With net metering, renewable power generated by a customer's photovoltaic array, for example, would be credited kilowatt-hour–for–kilowatt-hour against the customer's energy usage, giving each kilowatt-hour an imputed value equal to the relatively high residential price charged for power.) A national "net metering" policy, along with uniform national interconnection guidelines for renewable electricity, would provide an important additional stimulus to the premium buyback rates mentioned.

Most significantly, in a regulated environment, utilities can be required to do "integrated resource planning," an analytical process that mandates the selection of least-cost, long-term "packages" of energy supplies and energy efficiency, whose costs include imputed societal and environmental costs. Integrated resource planning has been a significant mechanism for getting renewables into utility portfolios.

The regulated and vertically integrated public utility we are all familiar with, however, is rapidly becoming an anachronism. Describing a path to a renewable energy economy is thus especially difficult because of the ongoing competitive restructuring of the electricity and gas utility sector. To become more competitive, some utilities have already scaled back their investments in efficiency, renewables, and R&D, because of anticipated competitive pressures. This represents a significant threat to the future health of the domestic renewables industry. Renewable energy advocates in a deregulated utility environment will need to be vigilant to assure that neither utilities nor the business entities that succeed them are allowed to dismantle demand-side management and renewable energy programs, or eliminate long-term energy R&D—a clear and present danger. In a deregulated utility market, the danger is also that power purchases will be made, and generating assets will be acquired, primarily according to short-term market incentives that fail to reflect long-term environmental costs and fuel price risks. The way to a long-term sustainable energy future might then be compromised. Short-term profit-maximizing stakeholders, and even power pool managers, may have little incentive to properly manage a diverse portfolio of energy sources with an eye to the future.

The paramount immediate challenge for renewable energy advocates is to see that deregulation is accompanied by the institutionalization of potent market incentives for renewables and efficiency. What should these measures be? A comprehensive national renewable energy policy should employ market development, capital development, technology development, and institutional development mechanisms.

MARKET DEVELOPMENT

Great emphasis should be given to market development actions to in- crease demand for renewables and to expand the scale of their produc- tion, since this quickly diminishes costs. Federal, state, and municipal government energy purchases also fall into the market development cate- gory, as do the net metering, premium buyback rates, and market aggre- gation measures already mentioned.

The federal government, as the nation's largest energy consumer and purchaser of energy-related products, could give renewables a big boost by gradually shifting the $9 billion or so it spends annually on energy to renewable energy sources. Similarly, the Department of Housing and Urban Development could require that federally assisted buildings make maximum feasible use of renewables and efficiency.

For electric vehicle market development, the federal government and electric utilities in concert could ensure that early electric vehicles cost their purchasers no more than internal combustion vehicles, by providing financial subsidies. The French government will be doing this for the next generation of French electric vehicles. As discussed in Chapter 23, electric vehicle sales mandates, so bitterly opposed by the makers of internal com- bustion engines, are a great tonic to the electric vehicle industry and should be widely adopted. In effect, mandates for zero-emission vehicles function quite like renewable energy set-asides for renewable energy technologies.

A renewable energy set-aside is that portion of a utility's generating capacity that, at the behest of some regulatory agency, must consist of re- newables. When a regulated utility or power pool acquires new generat- ing capacity, it can be required to meet a certain percentage of the new capacity with renewables. Although set-asides are an excellent mecha- nism to stimulate renewable energy industries, they have been protested by some prominent investor-owned utilities. In response, existing set- aside requirements were annulled by dint of a 1995 Federal Energy Reg- ulatory Commission (FERC) decision, as discussed in Chapter 15.

Under that FERC ruling, public utilities commissions will no longer be able to impose set-asides under provisions of the Public Utility Regu- latory Policy Act of 1978. Instead, renewables would have to compete on a presumed even footing with fossil fuels, even though the field on which they will compete is far from level. Thus new mechanisms for the implementation of set-asides will be needed. This need to relegit-imize set-asides has led to a concept known as the "portfolio standard," or "minimum purchase requirement"). Under a portfolio standard, any seller of retail electricity must buy a fixed percentage of its energy from renewable sources, or purchase tradable credits that will support the purchase by others of an equivalent amount of renewable energy. Utilities or other large-scale power providers thus rely on a "portfolio" of diverse energy resources that they combine hour-by-hour according to various economic and technical criteria in furnishing electricity to the grid. Renewable energy advocates are currently attempting to win utility, public utility commission, and legislative endorsement for the idea.

In a deregulated utility world, renewables should at a minimum remain at least as great a percentage of the generating portfolio and supply at least as many kilowatt-hours as before deregulation—and the percentage should increase systematically over time. For states with little-to-no renewable energy resources on-line, a baseline portfolio amount and implementation schedule should be set by legislation. A bill pending in the House called the Electricity Consumers' Power to Choose Act would institute a national minimum renewable energy supply requirement that could be met through a system of tradable credits.

Along those lines, a straightforward way to move toward a renewable energy economy would be to institute a 100 percent renewable energy set-aside/portfolio standard, requiring that, with a few exceptions, all new licensed capacity from now on be renewable, and all expanded power purchases be from renewable sources. In 20 to 40 years, once most existing capacity will have been retired, the cumulative effect of the renewable replacements and additions would be the creation of an electricity sector based almost entirely on renewables. Whereas this 100 percent standard for new capacity would be technically feasible, it is currently politically infeasible. A standard of 5–10 percent at the beginning, increasing to 25 percent by the year 2010, for example, is much more likely to be politically attainable. The Electricity Consumers' Power to Choose Act proposes a 2 percent standard that would peak at a mere 4 percent in 2010.

CAPITAL DEVELOPMENT:
FINANCING THE TRANSITION

As an alternative to the renewable portfolio standard, a modest tariff levied on all natural gas deliveries and on the transmission of all nonrenewably generated power could produce steady and substantial revenues for the sustained orderly development of renewables. Fees could be collected by a utility or through a power purchasing pool, or by a transmission and distribution company. These usage fees could be made available through state energy commissions or other public trust entities to support renewable energy and efficiency. The trustees could award the funds on an open, competitive basis as production incentives, cost-shared investments, and targeted purchases of innovative new technologies and bulk renewable power. All these expenditures would need to be made consistent with a commercialization plan for each renewable technology designed to help the technology overcome market-entry barriers. The agencies should also be authorized to purchase "packages" of verifiable projected energy savings from energy efficiency providers.

A variation on the tariff proposed would be to combine the renewable portfolio standard to expand the market for renewables with the surcharge on nonrenewable energy to support energy efficiency, research and development, and low-income customers. This formula was proposed in California by renewable energy advocates.

A usage charge levied exclusively on nonrenewables would certainly arouse fierce opposition from powerful fossil fuel and nuclear energy interests. Therefore, a broader "usage charge" on all electricity transmitted over distribution grids is much more likely to be instituted. Charges could be skewed to fall more heavily on fossil and nuclear power if political opposition could be overcome. Disincentives to fossil fuel use are justified in that fossil fuel prices don't reflect their full environmental or national security costs, nor their long-term price risks. Fossil fuels and nuclear energy have enjoyed decades of multibillion-dollar public subsidies, as discussed earlier. "Leveling the playing field" today by eliminating all current energy subsidies of any kind to any energy technology—even if politically possible—would hardly restore parity between the highly advantaged nonrenewable technologies and renewables, the neglected stepchildren of the energy field.

Though a carbon-emission charge would discourage pollution from fossil fuels and would help equalize the two sets of technologies, it is currently even more politically infeasible than other surcharges. Were even a very modest carbon fee instituted, it would generate enormous revenue and would make possible reductions in other taxes, including personal income taxes. Like surcharges, it would support investments in renewables and efficiency. The carbon fee would be levied at the point the carbonaceous fuel first entered the economy for conversion to energy; fuel used as a chemical ingredient in a final product would not be taxed.[2] Transitional impacts of fuel-price increases on low-income Americans could and should be offset by appropriate direct compensatory payments to affected people and regions.

By shifting taxation from labor to energy, a carbon charge or a BTU charge would also make labor less expensive relative to energy and would therefore tend to stimulate both employment and investment. This charge would therefore have a beneficial impact on the economy as a whole, although the fossil fuel and auto industries would oppose it vehemently. In conjunction with the tax, rebates and tariffs consistent with the General Agreement on Tariffs and Trade can be structured to compensate export industries whose costs would rise (and whose products might therefore become less competitive) as well as those industries for whom energy is a major production cost. The risks of instituting even a revenue-neutral pollution tax are that if the incidence of the tax were not properly offset, or if rates were set too high, or the program were too hastily implemented, it could harm the energy-intensive and export-oriented industries and, through them, damage the economy.

One solution to the political impasse over carbon taxes would be to impose progressive carbon-emission "caps" on fossil fuel burners in conjunction with annual tradable emission permits. The emission limits would be steadily lowered over time to bring emissions from current to target levels, and the permits for a set amount of carbon would expire if not used or traded within a year. This system—very similar in concept to the already proven SO_2 tradable allowances, with a declining cap—would make the air steadily cleaner while making renewables financially more attractive and accelerating the retirement of older, dirtier fossil power plants. A very attractive variation on this theme would be to impose the tradable emissions caps on regional transmission groups, a transmission network improvement supported by FERC.

Another way to foster "capital development" is for federal and state governments to provide expanded energy production tax credits considerably beyond the small 1.6 cents per kilowatt-hour credit for wind power and "closed-loop" biomass. In general, production credits are preferable; they reward actual performance. By contrast, investment credits are paid as a percentage of money invested.

Various forms of tax relief, however, could also be given to reduce the investment costs of certain capital-intensive renewable energy technology, such as photovoltaic and solar thermal electric power. States, for example, could grant renewable energy facilities selective sales and property tax exemptions or deferrals. This would serve as partial compensation for the huge tax benefits enjoyed by the fossil fuel industries on their fuel expenses, which can be almost 80 percent of the lifetime cost of building and operating a fossil fuel power plant.

An energy strategy that concentrates on efficiency alone in an otherwise "business as usual" world of rising population growth and ever-increasing material aspirations will merely buy time without solving the underlying problem of the throwaway society that consumes its energy capital instead of its income and in the process fouls its nest. We therefore must greatly expand our national investments apace in both efficiency and renewables.

INSTITUTIONAL DEVELOPMENT

To propel renewables and efficiency in tandem, a National Renewable Energy Trust Fund and a National Renewable Energy Bank to make low-interest revolving fund loans available should be established to help lower capital costs for renewable energy and efficiency investments. (The cost of all energy systems, evaluated on a lifetime cost basis, is extremely sensitive to the cost of borrowed money.) A parallel Renewable Energy Finance Program, possibly through an existing multinational lending entity, should also be established to increase financial support for the export of U.S. renewable energy products, especially to developing countries, augmenting the efforts of the Export-Import Bank and the World Bank's Global Environmental Facility. The increased foreign sales of renewables would benefit domestic renewable energy equipment manufacturers and enable them to bring down costs. With the renewable industry facing excess power capacity in the United States and stiff domestic cost competition here from conventional fuels, it is the export market that has almost entirely been responsible for the viability of the struggling but potentially very lucrative renewable energy sector.

On the diplomatic front, we should consider a propitious time to launch a cooperative international Renewable Energy Initiative to set specific timetables and targets for nations to attain in converting from fossil and nuclear to sustainable energy sources. This could eventually culminate in a renewable energy pact, much like the 1992 Earth Summit convention signed in Rio de Janeiro, Brazil.

TECHNOLOGY DEVELOPMENT

The states, through their energy commissions or offices, and the federal government, through the Department of Energy, could increase renewable energy R&D funding and employ "Golden Carrot" awards (like those discussed in the chapter on energy efficiency) to promote exceptional advances in renewable technologies. The development of needed infrastructure—such as charging networks for electric vehicles—could be given broad support, just as the federal government underwrote the interstate highway system for internal combustion vehicles. Similarly, state and federal cost-sharing is already provided for certain renewable energy demonstration projects under cooperative agreements between government and industry. These could be expanded to assist more participants.

MILKING THE AMERICAN MOTORIST'S
SACRED COW

Raising gasoline taxes is another measure that, like carbon taxes, would generate tremendous political opposition. Unpopular and politically difficult as it may be, a gradual increase in gasoline taxes to bring motor vehicle fuel more in line with fuel prices in western European industrialized nations would provide a valuable stimulus to vehicular fuel efficiency, liquid biomass fuels, the hydrogen economy, and electric vehicles. Even a 50 cent a gallon gas tax would generate a huge windfall of perhaps $60 billion a year (depending on price elasticities), while leaving U.S. gas prices less than half those in western Europe. Again, low-income groups would need to be buffered against price increases by compensatory payment programs. Fuel price increases should be coupled with strengthened Corporate Average Fuel Economy standards. These could be further augmented by a revenue-neutral system of fees for inefficient vehicles and rebates for efficient ones. In conjunction with higher gasoline taxes and higher national fuel efficiency standards, a zero-emission vehicle program resembling California's should be extended nationwide, but with an accelerated schedule so that 5 percent of all new cars sold in the United States would be electric by the year 2001 and at least 10 percent would be electric by 2003.

ACCOMPLISHING THE TRANSITION

The policy measures sketched in this chapter would help align economic incentives with society's interests in creating safe, efficient energy systems. Many studies have been done on this subject and scores of useful recommendations can be found in the September 1995 report *Renewing Our Energy Future,* by the U.S. congressional Office of Technology Assessment, and in the report *America's Energy Choices,* which is about to be re-released in an updated version.[3] The latter report explains how to increase the renewable share of our energy mix to one-half over a 40-year period merely by increasing the contribution of renewables to our energy supply by only 3.7 percent a year, while simultaneously adopting aggressive energy efficiency policies.

Although some of the policy elements presented in this chapter would arouse political ire from special interests, the nation and the states should not be deterred from prudent measures that will speed the transition to renewables. The overall process of converting our energy economy to renewables and making intensive use of efficiency opportunities will ultimately produce trillion-dollar benefits, as discussed earlier in the book. Gains on that order clearly are more than ample for compensating losers as well as for endowing winners—and for creating a "salable" policy package. Bold proactive programs are justifiable, both environmentally and economically.

An energy plan needs to reflect some consensus among important energy stakeholders so that key initiatives are not blocked. Therefore, renewable energy proponents should invite labor and business councils to join, for example, in planning for the retraining of coal, oil, auto, and atomic workers, and in facilitating their redeployment from fossil, nuclear, and conventional automotive industries to efficiency, renewable energy, and electric vehicle industries. This will be a gradual long-term change, not nearly as abrupt as the corporate downsizing of the past decade that has thrown many people out of work without retraining and replacement job opportunities. The attractive economic "multiplier" in the energy efficiency and renewable energy fields will create many more new jobs than are lost in the transition.

Presidential leadership is greatly needed. The President should initiate a national dialogue on energy policy, and once again elevate energy policy to the national prominence it had during much of the 1970s and early 1980s. We do not have the oil embargoes of earlier decades, but we do have more urgent social, economic, and environmental problems to solve. The President should project an inspiring vision of a sustainable energy economy and should challenge the nation to become the world's first renewably powered nation.

Once our national course is clearly set, administrative machinery exists to implement it. But implementing agencies that have been too closely allied with failed energy policies of the past will need to be redirected. The U.S. Department of Energy will need to focus more tightly on the commercialization of renewables and efficiency. The DOE's support for research and development on renewable energy should especially emphasize cost-effective energy storage and transmission technologies. Renewables could then provide a larger proportion of intermediate and baseload power, and over larger geographic areas.

Leadership of the Energy Department and Federal Energy Regulatory Commission should be refreshed with an infusion of talent from forward-thinking organizations, such as the American Solar Energy Society, the Union of Concerned Scientists, the Natural Resources Defense Council, the American Council for an Energy Efficient Economy, naming but a few here who, for the last generation, have been studying energy issues and campaigning effectively for safe energy. Inspiration for a National Renewable Energy Plan should also be drawn from the National Renewable Energy Laboratory, other national laboratories, NGOs, research institutes, universities, consumer groups, and renewable energy industries themselves, where a great deal of knowledge about renewables and efficiency resides.

Moving decisively as a nation to renewables and efficiency would have the beneficial effect of making our whole economy more competitive, creating new high-tech/high-wage jobs and industries, promoting strong competition among energy industries, and expanding U.S. energy technology exports. If we fail to lead in renewable energy and efficiency technology, we will by default lose the opportunity to compete effectively globally and to meet future domestic demand with domestic products. We will also cede the moral "high ground" on global environmental issues—leverage we need in international negotiations.

WHAT WE CAN DO:
THE PERSONAL TOUCH

Whereas many decisions about energy are made by large institutions, individuals in a democracy can vote, lobby, organize, and speak out. We can increase support for renewables and efficiency by working with government at all levels and with political parties. We can ensure that attainment of a clean, sustainable energy future occupies a prominent place on the public agenda at many levels. Similarly, we can work with the renewable energy and environmental organizations listed in the Resource Directory found in the appendix of this book to advance that goal, and we can create new local and regional organizations to promote effective local action. Likewise, within the workplace, school, and home, opportunities abound to practice what we preach. By becoming well informed, educating others, and working together in formal and informal alliances, we can expedite the transition to a sustainable society.

As this endeavor proceeds and results become more visible, a great deal of important but still-passive public and official support will coalesce and emerge to accelerate progress toward the affordable renewable and energy-efficient economy that most people really want. That economy is a prerequisite for a clean, healthy, sustainable, prosperous world. And that world is far more likely to be a secure and peaceful place for ourselves and posterity than a world wracked by pollution, energy waste, and dissension over insecure energy supplies.

APPENDIX:

RESOURCE DIRECTORY

Where to See Renewable Energy Systems at Work

To view demonstration solar homes and other examples of renewable energy technology at work in your vicinity, contact your local public utility, the American Solar Energy Society (which may have a local chapter in your area), or some of the other national organizations listed in the pages that follow, such as the Passive Solar Industries Council (page 343), and the renewable energy branch of your state energy office.

An excellent recent review of residential and commercial photovoltaic buildings can be found in *Solar Electric Buildings: An Overview of Today's Applications* (DOE/GO-10096-253, DE96000524), available through the National Technical Information Service, U.S. Department of Commerce, 5285 Port Royal Road, Springfield, VA 22161, (703) 486-4650.

Using Solar Energy at Home

For a simple introduction to easy and inexpensive solar energy applications for the home, see Scott Sklar and Kenneth Sheinkopf's *Consumer Guide to Solar Energy* (Chicago, Ill.: Bonus Books, 1991). A more recent sourcebook with practical ideas and construction details for building an energy-conserving passive solar home is Bruce Anderson and Malcolm Wells's *Passive Solar Energy: Homeowners Guide to Natural Heating and Cooling* (Amherst, N.H.: Brick House, 1993). Guidance for the homeowner on sizing and installing photovoltaics, with discussions of lighting and wiring, may be found in Paul Jeffrey Fowler's *Solar Electric Independent Home,* available through the American Solar Energy Society.

For an inexpensive, practical introductory guide to making your home more energy efficient, including a discussion of new energy-efficient appliances, see Alex Wilson and John Morrill's *Consumer Guide to Home Energy Savings,* published in 1995 by the American Council for an Energy-Efficient Economy (ACEEE) of Washington, D.C., and Berkeley, California. On the same subject, consult *Homemade Money: How to Save Energy and Dollars in Your Home,* available from the Rocky Mountain Institute (RMI). Contact ACEEE and RMI for their catalogs listing a large selection of popular books and technical research reports on energy efficiency.

For a practitioner's guide to creating energy-efficient buildings by taking advantage of local climate and site conditions, see Donald Watson and Kenneth Labs' *Climatic Building Design—Energy Efficient Building Principles and Practice* (New York: McGraw-Hill, 1993).

Let Your Views on Energy Be Heard

Congress

The Honorable _____
U.S. House of Representatives
Washington, DC 20515
(202) 225-3121

The Honorable _____
U.S. Senate
Washington, DC 20510
(202) 224-3121

The Administration

The White House
1600 Pennsylvania Avenue, N.W.
Washington, DC 20500
(202) 456-1111

The Vice President
17th Street & Pennsylvania Avenue, N.W.
Washington, DC 20501
(202) 456-2326

Secretary of Energy
Department of Energy
1000 Independence Avenue, S.W.
Washington, DC 20585
(202) 586-6210

Solar, Wind, and Energy-Efficient Design

The following professional organizations provide information on solar and energy-efficient architecture, construction, engineering, and contracting. Some of these listings are courtesy of the Energy Efficiency and Renewable Energy Clearinghouse of Merrifield, Virginia (see page 340).

THE ALLIANCE TO SAVE ENERGY

The Alliance is a nonprofit coalition of business, government, environmental, and consumer leaders dedicated to increasing the efficiency of energy use. The Alliance promotes greater investment in energy efficiency as a primary means of achieving the nation's environmental, economic, national security, and affordable housing goals. It strives to provide accurate information about the costs and benefits of energy efficiency and other resources, and works to make energy markets fairer and more competitive.

A publication catalog on various aspects of energy policy and energy efficiency is available.

The Alliance to Save Energy
1725 K Street, N.W., Suite 509
Washington, DC 20006
(202) 857-0666
FAX (202) 331-9588

AMERICAN INSTITUTE OF ARCHITECTS (AIA)

A good place to locate architects specializing in applications of solar energy or energy efficiency technology is AIA, the national professional society for architects, with 300 local and state organizations. The AIA fosters professionalism and accountability, promotes design excellence, sponsors educational and research programs, monitors legislation and operates a monthly news service, among many other activities.

American Institute of Architects
1735 New York Avenue, N.W.
Washington, DC 20006
(800) 365-ARCH & (202) 626-7300

AMERICAN INSTITUTE OF BUILDING DESIGN (AIBD)

Another good place to locate architects specializing in applications of solar energy or energy efficiency, the AIBD keeps its residential and light commercial building designers informed of techniques and principles of building design while promoting public interest in aesthetic and efficient building design. Activities include consulting services, legislative work, research programs, education, and compilation of statistics.

American Institute of Building Design
991 Post Road East
Westport, CT 06880
(203) 227-3640
FAX (203) 227-8624

AMERICAN COUNCIL FOR AN ENERGY-EFFICIENT ECONOMY (ACEEE)

The ACEEE is a nonprofit organization dedicated to advancing energy efficiency as a means of promoting economic prosperity and environmental protection. The organization conducts in-depth technical and policy assessments; advises government and utilities; works with business and other organizations; publishes books, conference proceedings, and reports; organizes conferences; and informs consumers. Program areas include national energy policy, energy efficiency and economic development, utilities, transportation, buildings, appliances, and equipment, industry, and international assistance with energy efficiency applications in developing nations and eastern Europe. A publications catalog is available.

American Council for an Energy-Efficient Economy
1001 Connecticut Avenue, N.W., Suite 801
Washington, DC 20036
(202) 429-8873
FAX (202) 429-2248

AMERICAN SOLAR ENERGY SOCIETY (ASES)

The ASES participates in educational and promotional activities on solar and other renewable energy sources. ASES fosters communication among various solar professionals and organizes an annual national solar energy conference. ASES has a broad selection of high-quality technical publications that cover the installation of solar energy systems and the construction of energy-conserving buildings. The society is an excellent source of hands-on advice. ASES also publishes *Solar Today,* a popular national magazine covering all solar technologies, and provides research "white papers" offering expert assessments of economic and technical issues in solar energy. A list of the local and regional chapters, publications, and membership information is available from its national headquarters:

American Solar Energy Society
2400 Central Avenue, G 1
Boulder, CO 80301
(303) 443-3130

AMERICAN WIND ENERGY ASSOCIATION (AWEA)

AWEA works to further the development of wind as a reliable, cost-effective source of clean energy. The association seeks to be the definitive source of wind energy–related information. AWEA publishes *Wind Energy Weekly* and *Windletter* and holds an annual wind energy conference. The association has a publications catalog with general resources, audiovisuals, conference proceedings, technical design and siting procedures, information for home and business users, energy and regulatory policy, plus educational and international resources. The *AWEA Membership Directory* describes the wind energy products and services of more than 200 companies. AWEA provides Internet access to information on wind technology and activities through the Wind Information Network, a conference bulletin board on EcoNet, an environmental information network available by subscription.

American Wind Energy Association
122 C Street, N.W., Fourth Floor
Washington, DC 20001
(202) 383-2500
FAX (202) 383-2505

ASSOCIATION OF ENERGY ENGINEERS

The Association offers referrals to certified energy professionals including engineers, energy managers, and energy auditors.

Association of Energy Engineers
4025 Pleasantdale Road, Suite 420
Atlanta, GA 30340
(404) 447-5083
FAX (407) 446-3969

BUSINESS COUNCIL FOR A SUSTAINABLE ENERGY FUTURE

This group was formed in 1992 by business leaders from the renewable energy, natural gas, energy efficiency, and utility industries. The organization is committed to furthering the development of a new national energy strategy for the twenty-first century based on the rapid deployment of more efficient and less polluting energy technologies. The Council therefore supports federal and state policies and programs that improve energy efficiency in all sectors of the economy, that accelerate the commercialization of renewable energy, and that promote greater use of natural gas.

The Council's four main areas of concentration are (1) promoting transportation alternatives that reduce vehicle emissions; (2) developing uniform policies for utility regulation; (3) assisting in achieving a reprioritization of the U.S. Department of Energy's research, development, and demonstration budget; (4) influencing the policy debate on the reduction of climate-destabilizing gases through advocacy of clean energy technologies.

Mr. John Hemphill, Executive Director
The Business Council for a Sustainable Energy Future
1725 K Street, N.W., Suite 509
Washington, DC 20006-1401
(202) 785-0507
FAX (202) 785-0514

CENTER FOR RENEWABLE ENERGY AND SUSTAINABLE TECHNOLOGY (CREST)

CREST uses advanced information and communication technologies to demonstrate renewable energy and energy efficiency technology interactively and to train people in their use. The Center works with other energy centers and consortia and provides "Solstice," a free on-line resource on renewable energy, energy efficiency, electric vehicles, the environment, climate change, policy, economics, legislation, education, social issues, computers, networking, and sustainable development. Solstice archives include case studies, databases, and a calendar.

CREST offers an interactive multimedia CD-ROM encyclopedia for the Macintosh on renewable energy and the environment called "The Sun's Joules," containing more than 60 video clips, interactive exercises, energy use by state, and more than 950 pages of illustrations, animations, and text.

Center for Renewable Energy and Sustainable Technology
777 North Capitol Street, N.E., Suite 805
Washington, DC 20002
(202) 289-5370
FAX (202) 289-5354

Internet: seref@digex.net
World Wide Web: http://solstice.crest.org
Gopher: gopher.crest.org
Anonymous FTP: solstice.crest.org
E-mail: info@crest.org

CRITICAL MASS ENERGY PROJECT (CMEP)

Critical Mass works with citizens' groups and individual activists to promote safe, economical, and environmentally sound energy alternatives. CMEP lobbies Congress, government agencies, and some state legislatures to promote safe and affordable energy alternatives. It also prepares studies and reports on topics such as innovative, energy-saving technologies, renewable energy, and small power systems. Information is provided to the public, and a list of publications is available.

Public Citizen's Critical Mass Energy Project
215 Pennsylvania Avenue, S.E.
Washington, DC 20003
(202) 546-4996

E SOURCE

E SOURCE is a membership-based, syndicated service providing independent information on energy efficiency technologies and their applications. The company has a worldwide range of utility, government, industry, and research clients, and its staff includes research professionals with diverse technical, analytical, and communication skills.

Products and services include a 6-volume, 2,200-page *Technology Atlas Series,* covering the major end-use areas: lighting, drivepower, space cooling, appliance and office equipment, water heating, and space heating. These regularly updated "encyclopedias" on efficient energy use provide an in-house reference library for application and training.

E SOURCE members receive a variety of concise bimonthly reports and "white papers" covering specific technology developments, strategic and regulatory issues, market overviews, energy management case studies, and individual product reviews. Individuals in E SOURCE member organizations receive a regular newsletter featuring brief news items on energy efficiency issues, as well as summaries of E SOURCE's latest research. The company sponsors the *Members' Forum,* an exclusive international conference on energy efficiency each fall, in addition to other unique events such as the Strategic Issues Retreat, the Corporate Energy

Managers Roundtable, and Advanced Technical Workshops. E SOURCE also offers on-site training seminars, conference calls, Fax Alert Memos, and direct consultation.

Annual membership fees range from $1,500 to $14,000 per year; call for specific pricing information.

E SOURCE, Inc.
1033 Walnut Street
Boulder, CO 80302-5114
(303) 440-8500

ENERGY DESIGN UPDATE (EDU)

EDU provides a link between qualified subscribers and people seeking the services of a builder, architect, designer, or consultant specializing in energy-efficient building design and construction. EDU offers a monthly newsletter, serves as a professional referral service, and maintains a list of subscribers, categorized by services provided and location.

Energy Design Update
P.O. Box 1709
Ansonia Station
New York, NY 10023
(212) 662-7248

ENERGY EFFICIENCY AND RENEWABLE ENERGY CLEARINGHOUSE (EREC)

EREC responds free to phone, mail, and electronic inquiries on energy efficiency and renewable energy technologies. Help is available to consumers, educators, experts, government, and businesses on topics from alternative fuels, photovoltaics, and wind energy to solar heating and cooling, buildings, and weatherization. Assistance includes engineering, scientific and technical assistance as well as business, financial, and marketing information. World Wide Web users on the Internet can also access EREC information through its sister service—the Energy Efficiency and Renewable Energy Network (EREN). EREN provides users a gateway to energy efficiency and renewable energy information from national laboratories and other organizations via the Uniform Resource Locator (URL) http://www.eren.doe.gov.

Energy Efficiency and Renewable Energy Clearinghouse
P.O. Box 3048
Merrifield, VA 22116

(800) DOE-EREC
TDD (800) 273-2975
Modem (800) 273-2955 (VT-100 [ANSI] terminal emulation, 8 databits, 1 stopbit, parity none)
Internet: energyinfo@delphi.com
FAX (703) 893-0400

ENVIRONMENTAL AND ENERGY STUDY INSTITUTE (EESI)

EESI is a nonpartisan, nonprofit research and educational institution. It publishes accurate, timely, and objective information about, and analysis of, national legislation with environmental or energy implications. Publications include the *Weekly Bulletin* and various special reports. Subscription to EESI publications provides readers with excellent coverage of actions by the House and Senate affecting energy and the environment.

Environmental and Energy Study Institute
122 C Street, N.W., Suite 700
Washington, DC 20001
(202) 628-6500
FAX (202) 628-1825

IRT ENVIRONMENT, INC.

IRT's Results Center tracks energy efficiency development in industry and highlights noteworthy programs. The company researches and then showcases domestic and international program successes in its 20-page, standardized published profiles, emphasizing how programs can be adapted for use elsewhere. IRT produces a monthly newsletter, *Energy Efficiency News & Views*, special reports on relevant research, utility developments, and program trends. A library of 120 profiles is maintained containing "a wealth of data and lessons from the cream-of-the crop of efficiency intiatives." Technical support is provided to dues-paying members.

IRT Environment, Inc.
The Results Center
P.O. Box 2239
Basalt, CO 81621
(970) 927-3155
(970) 927-9428
Internet: irt@irt.com

NATIONAL ASSOCIATION OF HOME BUILDERS (NAHB)
NAHB has information on the construction of solar homes and energy-efficient buildings.

National Association of Home Builders
15th & M Streets, N.W.
Washington, DC 20005
(800) 368-5242
(202) 822-0200

NATIONAL ENERGY INFORMATION CENTER (NEIC)
The Center provides statistical information on conventional fuel sources (petroleum, natural gas, and coal) and on nuclear energy. An inquiry unit answers questions over the telephone or refers callers to other information sources. NEIC also offers free Energy Information Sheets on a variety of topics. The U.S. Department of Energy publishes *The Energy Information Directory*, which is also helpful for locating government agencies involved in renewable energy, and for finding trade association contacts.

National Energy Information Center
1000 Independence Avenue, S.W.
MS EI 231 Forrestal Building
Washington, DC 20585
(202) 586-8800
TDD (202) 586-1181

NATIONAL RENEWABLE ENERGY LABORATORY (NREL)
Established by Congress as the Solar Energy Research Institute in 1977, NREL was designated a national laboratory in 1991 and is the only national laboratory dedicated primarily to renewable energy and energy efficiency. NREL performs research and development on technologies for every sector of energy use: electricity, transportation, industry, buildings, and utilities. The laboratory also appraises the extent of each major renewable resource and assesses the environmental and economic impacts of energy supply, conversion, and consumption. Renewable energy and energy efficiency publications and referrals are available through their Technical Inquiry Service.

National Renewable Energy Laboratory
1617 Cole Boulevard

Golden, CO 80401-3393
For technical inquiries: (303) 275-4099 and (303) 275-4065
FAX (303) 275-4091

THE NORTHEAST SUSTAINABLE ENERGY ASSOCIATION (NSEA)

NSEA, a vital 20-year-old regional chapter of the American Solar Energy Society, provides information on alternative energy and energy efficiency. NSEA offers one- and two-day conferences, seminars, and workshops for professionals and nonprofessionals in energy-efficient residential and industrial construction. NSEA also conducts the Solar and Electric Vehicle (S/EV) Symposium on sustainable transportation using solar and electric vehicles. Various publications are available, and visitors may use the Association's library free of charge.

NSEA annually sponsors the "American Tour de Sol," a five-day solar electric vehicle rally held in May. The rally brings together experimental solar vehicles from all over the country.

Northeast Sustainable Energy Association
23 Ames Street
Greenfield, MA 01301
(413) 774-6051

PASSIVE SOLAR INDUSTRIES COUNCIL (PSIC)

PSIC promotes the use of energy-efficient passive solar design and construction. They also address questions from builders, architects, engineers, and the public.

Passive Solar Industries Council
1511 K Street, N.W., Suite 600
Washington, DC 20005
(202) 628-7400

ROCKY MOUNTAIN INSTITUTE (RMI)

RMI is a nonprofit research and educational foundation dedicated to fostering the efficient and sustainable use of resources as a path to global security. The Institute conducts programs in the areas of energy, water, economic renewal, agriculture, and security. Its energy program works to speed the adoption of energy efficiency and renewables. Through RMI's COMPETITEK program, RMI helps utilities and large energy users expedite cost-effective energy efficiency improvements. Its energy outreach also provides critical energy information to policy makers, builders, designers, and the public. A catalog of superb and inexpensive publications is available, many authored by RMI cofounder and Research Director Amory B. Lovins.

Rocky Mountain Institute
1739 Snowmass Creek Road
Old Snowmass, CO 81654-9199
(303) 927-3851
FAX (303) 927-4178

SOLAR ENERGY INDUSTRIES ASSOCIATION (SEIA)

SEIA is the national trade organization of the photovoltaics and solar thermal manufacturers and component suppliers, comprised of over 550 companies. The Association provides information on solar manufacturing, contracting, and design. SEIA has an active government affairs program, organizes conferences and trade shows, gathers statistical information, conducts educational outreach, assists in solar industry standards development. It also publishes *Solar Industry Journal,* a quarterly, and has three directories of renewable energy system manufacturers, distributors, and products: (1) *Solar Electric Applications and Directory of the U.S. Photovoltaic Industry,* (2) *Directory of the U.S. Solar Thermal Industry,* and (3) *The Biofuels Directory—U.S. Manufacturers.* SEIA also has a directory with sources of technical assistance on renewable energy systems in each U.S. state and territory: *Networking: Renewable Energy in the States— Interstate Renewable Energy Council (IREC) State Renewable Energy Directory.*

Solar Energy Industries Association
122 C Street, N.W., 4th Floor
Washington, DC 20001-2109
(202) 383-2600
FAX: (202) 383-2670

SUN DAY

SUN DAY publishes the inexpensive bimonthly *Sustainable Energy Resources Newsletter,* which annotates published and organizational resources on a wide range of energy efficiency and renewable energy issues. It also includes model programs, book reviews, and a calendar of upcoming events.

SUN DAY: A Campaign for a Sustainable Energy Future
315 Circle Avenue, Suite 2
Takoma Park, MD 20912-4836
(301) 270-2258
FAX: (301) 891-2866

UNION OF CONCERNED SCIENTISTS (UCS)
Through an alliance forged of leading scientists and concerned citizens, UCS since 1969 has worked toward responsible stewardship of the global environment and life-sustaining resources, especially in areas where science and technology play a critical role. In addition to focusing on global resource issues, sustainable agriculture, transportation, weapons proliferation, and the demilitarization of international conflict, UCS has a multifaceted energy program.

Staff members work to influence public policy on renewable energy, energy efficiency, clean fuels and vehicles, and on ensuring the safety of the nation's nuclear power plants. UCS performs numerous technical studies, conducts public education, has an active publishing program, maintains a speakers' bureau, testifies often on legislation, and coordinates a Concerned Citizens Action Network and a Scientists' Action Network.

Union of Concerned Scientists
National Headquarters
Two Brattle Square
Cambridge, MA 02238-9105
(617) 547-5552

WORLDWATCH INSTITUTE
The Institute is a research organization that focuses on interdisciplinary approaches to solving global environmental problems. In the energy field, its interests include energy conservation, renewable resources, solar power, and energy use in developing countries. Worldwatch publishes the *State of the World Report, Worldwatch Papers,* diskette databases, and *World Watch Magazine.*

Worldwatch Institute
1776 Massachusetts Avenue, N.W.
Washington, DC 20036
(202) 452-1999
FAX (202) 296-7365

State Energy Offices

States have independent energy agencies or energy offices and a public utilities commission. Energy information may also be obtained from relevant energy committees of your state legislature or from your governor's office of research or science advice. The entry below is but one example of a state energy commission.

CALIFORNIA ENERGY COMMISSION
The California Energy Commission Communication Office answers questions on energy conservation and new technologies, including wind, photovoltaics, hydroelectricity, and geothermal energy, and makes referrals. The Commission prepares energy supply-and-demand forecasts, conducts energy research, and regularly announces public hearings, energy events, media issues, and general Commission information through its various telephone hot lines.

California Energy Commission
Local Assistance
1516 Ninth Street, MS 26
Sacramento, CA 95814
(916) 654-4008
FAX (303) 440-8502

Electric Utility Organizations

EDISON ELECTRIC INSTITUTE (EEI)
EEI serves the nation's investor-owned utilities and affiliates abroad. The Institute sponsors educational programs, compiles statistics, maintains a speakers' bureau and publishes *Electric Prospects, Electric Power Surveys, Electrical Reports, Statistical Reports, Rate Book,* and *Statistical Yearbook.* Among its many other activities, the Institute is heavily involved in advancing the commercialization of electric vehicles.

Edison Electric Institute
701 Pennsylvania Avenue
Washington, DC 20006
(202) 508-5000
FAX (202) 508-5786

ELECTRIC POWER RESEARCH INSTITUTE (EPRI)
EPRI conducts research and development on behalf of 700 electric power utilities. From their headquarters in Palo Alto, California, EPRI manages 350 scientists and 1,600 R&D projects. The Institute is often unreceptive to inquiries and requests for information from the public.

Electric Power Research Institute (EPRI)
207 Coggins Drive
Pleasant Hill, CA 94523
(415) 855-2411

Biomass Organizations

Biofuels Feedstock Development
  Program
Oak Ridge National Laboratory
P.O. Box 2008
Oak Ridge, TN 37831-5132
(615) 576-5132

Biomass Energy Alliance
1001 G Street, N.W., Suite 900 East
Washington, DC 20001
(202) 639-0384
(202) 393-5510
Internet: info@biomass.org

Biomass Processors Association
2251 Ralston Road
Sacramento, CA 95821
(916) 927-1770

Biofuels Systems Division
U.S. Department of Energy
1000 Independence Avenue, S.W.
Washington, DC 20585

California Biomass Energy Alliance
P.O. Box 1086
Foresthill, CA 95631
(916) 367-2332
FAX (916) 367-4422

Great Lakes Regional Biomass
  Energy Program
Council of Great Lakes Governors
35 East Wacker Drive, Suite 1850
Chicago, IL 60601
(312) 407-1077
FAX (312) 407-0038

National BioEnergy Industries
  Association
122 C Street, N.W., 4th Floor
Washington, DC 20001
(202) 383-2540
FAX (202) 383-2670

National Renewable Energy
  Laboratory
1617 Cole Boulevard
Golden, CO 80301-3393
(404) 231-1947

Northwestern Regional Biomass
  Energy Program
U.S. DOE Seattle Regional
  Support Office
800 5th Avenue, Suite 3950
Seattle, WA 988104
(206) 553-2079
FAX (206) 553-2200

Southeastern Regional Biomass
  Energy Program
Tennessee Valley Authority,
  CEB 3A
P.O. Box 1010
Muscle Shoals, AL 35662-1010
(Publishes an outstanding free
  newsletter, *SERBEP Update.*)

Western Regional Biomasss Energy
  Program

Western Area Power
  Administration, A7100
P.O. Box 3402
Golden, CO 80401-0098
(303) 275-1704
Internet: swanson@wapa.gov
(Publishes the very useful
  Biomass Bulletin and
  Biomass Digest, distributed
  free.)

Electric Vehicle Organizations

CALSTART
3601 Empire Avenue
Burbank, CA 91505
(818) 565-5600
FAX (818) 565-5110

Canadian Energy Research
  Institute (CERI)
3512 33rd Street, N.W.
Calgary, AL T2L 2A6
(403) 282-1231

Central EV Coaltion
2138 Welch Street
Houston, TX 77019
(713) 522-2606
FAX (713) 522-5702

Electric Vehicle Association of
  the Americas (EVAA)
601 California Street, Suite 502
San Francisco, CA 94108
(415) 249-2690

The Electric Vehicle Factfinder
501 14th Street, Suite 200
Oakland, CA 94612
(800) 438-3228

The Florida Alliance for Clean
  Technologies
1903 South Congress Avenue,
  Suite 340
Boynton Beach, FL 33426
(305) 552-4133
FAX (305) 552-3273

Hawaii Electric Vehicle
  Demonstration Project
  Consortium
531 Cooke Street
Honolulu, HA 96813
(808) 594-0100
FAX (808) 594-0102

Mid-Atlantic Regional Consortium
  for Advanced Vehicles
1450 Scalp Avenue
Johnstown, PA 15904
(814) 269-6807
FAX (814) 269-2799

Northeast Alternative Vehicle
  Consortia
205 Portland Street
Boston, MA 02114
(617) 371-1420
FAX (617) 371-1422

Sacramento Municipal Utility
District
Michael Wirsch, Manager
Electric Transportation
P.O. Box 15830
6201 S Street
Sacramento, CA 95852-1830
(916) 732-6754
FAX (916) 732-6839

Society of Automotive Engineers
(SAE), International
400 Commonwealth Drive
Warrendale, PA 15096-0001
(412) 776-4841
FAX (412) 776-5760
(The Society publishes a catalog
listing high-quality books on

topics in automotive
engineering, including electric
vehicle propulsion and design.)

South Central EV Consortium
Texas Engineering Experiment
Research Center
College Station, TX 77843
(409) 845-8281

Southern Coalition for Advanced
Transportation
395 Piedmont Avenue, N.E.
Atlanta, GA 30308
(404) 526-7228
FAX (404) 526-3452

Electric Vehicles Publications

Electric Vehicle Association of the Americas. 1995. *Proceedings of the 12th International Electric Vehicle Symposium, Dec. 5–7, 1994,* San Francisco, California. See also proceedings of EVS 1–11.

Electric Power Research Institute et al. May 1996. *Proceedings: 1995 North American EV & Infrastructure Conference.* TR-106554. Research Project 4858. Palo Alto, California.

Electric Vehicles: Technology, Performance and Potential. 1993. International Energy Agency (Organization for Economic Cooperation and Development/International Energy Agency). Paris, France.

Hart, S., and Spivak, A. 1993. *Automobile Dependence & Denial: The Elephant in the Bedroom: Impacts on the Economy and Environment.* Pasadena, Calif.: New Paradigm Books.

Hwang, R., et al. 1994. Rev. Ed. *Driving Out Pollution: The Benefits of Electric Vehicles.* Cambridge, Mass., and Berkeley, Calif.: Union of Concerned Scientists.

MacKenzie, J. 1994. *The Keys to the Car: Electric and Hydrogen Vehicles for the 21st Century.* World Resources Institute.

McCrea, S., and Minner, R. (eds.). 1992. *Why Wait for Detroit? Drive an Electric Car Today!* Ft. Lauderdale, Fla.: South Florida Electric Auto Association.

Nadis, S., and MacKenzie, J. 1993. *Car Trouble.* Boston, Mass.: Beacon Press (World Resources Institute).

Renne, Garth; Carr, Clyde; and Heath, Michelle. June 1994. *Electric Vehicles: Economic Costs, Environmental Benefits.* Canadian Energy Research Institute. Study No. 56. Calgary, Canada.

Sperling, D. 1995. *Future Drive: Electric Vehicles and Sustainable Transportation.* Washington, D.C., and Covelo, Calif.: Island Press.

Geothermal Energy Organizations

Earth Sciences and Resources
Institute
391 Chipeta Way, Suite C
Salt Lake City, UT 84108
(801) 584-4422
FAX (801) 584-4453

Geo-Heat Center, Oregon Institute
of Technology
3201 Campus Drive
Klamath Falls, OR 97601-8801
(541) 885-1750

Geothermal Education Office
664 Hilary Drive
Tiburon, CA 94920
(800) 866-4GEO

Geothermal Energy Association
2001 Second Street, Suite 5
Davis, CA 95617-1350
(916) 758-2360

Geothermal Energy Association
122 C Street N.W., 4th Floor
Washington, DC 20001
(202) 383-2550

Geothermal Resources Council
2001 Second Street, Suite 5
Davis, CA 95617-1350
(916) 758-2360

International Ground-Source Heat
Pump Association
409 Cordell South
Stillwater, OK 74078
(405) 744-5175

U.S. Department of Energy
Geothermal Division
1000 Independence Avenue, S.W.
Washington, DC 20585
(202) 586-5340

Environmental Organizations Involved in Energy Issues

ENVIRONMENTAL DEFENSE FUND (EDF)
EDF employs over 150 scientists, attorneys, economists, and other environmental professionals who specialize in proposing practical and economically feasible solutions to environmental problems. In the field of energy, EDF conducts work pertaining to energy efficiency, air quality, transportation, and climate change. The organization has five regional offices in addition to its headquarters in New York City.

Environmental Defense Fund
257 Park Avenue South
New York, NY 10010
(212) 505-2100

GREENPEACE
Among its other activities, Greenpeace has programs on nuclear energy
and climate change. The focus of the climate program is to decrease carbon dioxide and other global warming gas emissions and to increase energy
efficiency, while encouraging a switch from nuclear energy and fossil fuels
to renewables. Through its London office and the work of Dr. Jeremy
Leggett in particular, Greenpeace International has brought the risks of climate change to the attention of the insurance and banking industries.

Greenpeace
1436 U Street, N.W.
Washington, DC 20009
(202) 462-1177
FAX (202) 462-4507

THE NATURAL RESOURCES DEFENSE COUNCIL (NRDC)
NRDC is a national nonprofit organization dedicated to protecting the
world's natural resources and ensuring a safe and healthy environment.
Staffed by lawyers, scientists, and other environmental specialists, NRDC
uses law and science to influence energy policy. Offices are located in
Washington, D.C., San Francisco and Los Angeles, California, and New
York City.

The Natural Resources Defense Council
40 West 20th Street
New York, NY 10011
(212) 727-2700
FAX (212) 727-1773
World Wide Web: http://www.nrdc.org/nrdc

RENEW AMERICA
Renew America is a nonprofit membership organization that collects, ver-
ifies, and disseminates information to the public on solutions to the nation's environmental problems. It sponsors programs to encourage public
awareness of the uses of renewable energy technologies and energy
efficiency.

Renew America's Environmental Success Index (ESI) is a clearinghouse for information about successful environmental programs throughout the nation, including categories such as "Energy Pollution Control" and "Renewable Energy and Energy Efficiency." Renew America annually publishes the "State of the States" report, which evaluates and compares the environmental and energy policies and programs of each of the 50 states.

Renew America
1400 16th Street, N.W.
Suite 710
Washington, DC 20036
(202) 232-2252

SIERRA CLUB
Although created to protect wilderness and the natural environment, the Sierra Club's main concerns today include energy, pollution control, land use, growth, and resource management. It works on these issues through 4 national offices, 13 regional offices, and 21 major chapter offices.

Sierra Club Headquarters
85 Second Street, 2nd Floor
San Francisco, CA 94105
(415) 977-5500
FAX (415) 977-5799
World Wide Web: http://www.sierraclub.org

Renewable Energy Publications

Cole, Nancy, and Skerrett, P. J. 1995. *Renewables Are Ready: People Creating Renewable Energy Solutions.* Lebanon, N.H.: Chelsea Green.
The Global Cities Project. 1991. *Building Sustainable Communities: An Environmental Guide for Local Government. Energy: Efficiency and Production.* San Francisco: Center for the Study of Law and Politics.
Hollander, Jack M. (ed.). 1992. *The Energy-Environment Connection.* Washington, DC: Island Press.
Johansson, Thomas B.; Kelly, Henry; Reddy, Amulya K. N.; Williams, Robert H.; and Burnham, Laurie (eds.). 1993. *Renewable Energy: Sources for Fuels and Electricity.* Washington, DC, and Covelo, Calif.: Island Press.
Kozloff, K., and Dower, R. December 1993. *A New Power Base: Renewable Energy Policies for the Nineties and Beyond.* World Resources Institute.

Sterrett, Frances S. (ed.). 1995. *Alternative Fuels and the Environment*. Boca Raton, Fla.: CRS Press, Inc./Lewis Publishers.

U.S. Department of Energy/Energy Information Administration. September 1995. *Electricity Generation and Environmental Externalities: Case Studies*. DOE/EIA-0598. Government Printing Office: Washington, D.C.

U.S. Department of Energy/Energy Information Administration. September 1995. *Emissions of Greenhouse Gases in the United States, 1987–1994*. DOE/EIA-0573(87-94). Government Printing Office: Washington, D.C.

![sun icon]

NOTES

Introduction

1. U.S. Department of Energy. July 1995. *Sustainable Energy Strategy: Clean and Se-cure Energy for a Competitive Economy*. Washington, D.C.: U.S. Government Printing Office.

2. Hydropower was not included because it is a well-understood and long-established commercial technology that does not seem to offer the same technological challenges as other technologies described. Ocean technologies—currents, tidal and wave power, ocean thermal energy conversion, and ocean salinity gradients—were also excluded, as they do not appear to have large near-term potential for the United States for the follow-ing reasons.

Tidal power was not discussed because it requires construction of large, capital-intensive dams across bays or estuaries and is thus restricted to a relatively few sites with big enclosable areas and large tidal ranges. Because of these limitations, tidal dams are not going to be widely used.

Of all the oceanic resources studied over the years by the U.S. Department of Energy, the DOE has found that Ocean Thermal Energy Conversion (OTEC) could make the largest potential contribution to U.S. energy systems. OTEC power plants exploit the temperature difference between warm surface ocean water and the colder water found deep below the waves; the thermal gradient is used to operate a heat engine to produce electricity. But OTEC requires construction of a relatively high-cost central station power plant that is unlikely to be broadly commercialized anytime soon.

Wave power is a large resource, but is too diffuse to be cost-effective for major portions of the East and West Coasts of the United States, although it appears to have a large po-tential for parts of the United Kingdom, where it remains economically unproven for large-scale applications. British government studies of wave power facilities using large floating devices off the coast of Scotland were discontinued in 1982 after contested studies found costs would be uncompetitive. Development to date has been restricted to a few small prototypes, such as a 200-kilowatt plant in Norway, and early demonstration models.

Hydrogen, an energy carrier, is not discussed in depth, since my focus is mainly re-newable energy production and energy efficiency, rather than energy storage and trans-

356 ✦ NOTES

mission. Hydrogen releases energy when combined with oxygen to produce water. When that hydrogen is derived from a fossil fuel, hydrogen is considered nonrenewable. When the hydrogen is produced by using solar or wind power to electrolyze (separate) water, the energy liberated by recombining hydrogen with oxygen is exactly equal to the energy required for the separation of water into those two components. So the hydrogen, while renewable, is not itself a primary energy source. Yet it is a convenient way to store and transport energy produced from both renewable and fossil fuels. For similar reasons, fuel cells, which may use hydrogen-oxygen couples or other chemical reactions, are not discussed in detail except in passing in Chapter 23, as they relate to electric vehicles.

Solar heating and cooling are discussed briefly in Chapter 21, relative to energy efficiency. These valuable, widely accepted commercial technologies are not treated more systematically for reasons of space and because a vast literature is already available on them.

Chapter 1

1. Hydro and biomass together provided just under 7 percent of our primary energy consumption in 1994. But other renewables combined provided only two-tenths of a percent.

Chapter 2

Epigraph from Wald, Matthew L. October 17, 1993. After 20 Years, America's Foot Is Still on the Gas. *New York Times.*

1. Romm, Joseph J., and Lovins, Amory B. 1992. Fueling a Competitive Economy. *Foreign Affairs.* Vol. 72, no. 5.

2. Energy Information Administration. Office of Integrated Analysis and Forecasting. U.S. Department of Energy. January 1992. *Annual Energy Outlook 1992 with Projections to 2010,* p. 20.

3. Watson, Robert K. February 1991. Looking for Oil in All the Wrong Places: Facts About Oil, Natural Gas and Efficiency Resources. Natural Resources Defense Council.

4. Greene, David, and Leiby, Paul. March 1993. *Social Cost to the U.S. of Monopolization of the World Oil Market.* Oak Ridge National Laboratory.

5. Koplow, Douglas N. April 1993. *Federal Energy Subsidies: Energy, Environmental, and Fiscal Impacts.* Washington, D.C.: The Alliance to Save Energy.

6. Wald, Matthew L. October 17, 1993. After 20 Years, America's Foot Is Still on the Gas. *New York Times.*

7. Barbir, F.; Veziroglu, T. N.; and Plass, H. J. 1990. Environmental Damage Due to Fossil Fuels Use. *International Journal of Hydrogen Energy.* Vol. 15, no. 10.

8. and 9. Romm, Joseph J., and Lovins, Amory B. 1992. Fueling a Competitive Economy. *Foreign Affairs.* Vol. 72, no. 5.

10. Serpione, N.; Lawless, D.; and Terzian, R. 1992. Solar Fuels: Status and Perspectives. *Solar Energy.* Vol. 49, no. 4, pp. 221–234 [221].

11. Pace University Center for Environmental Legal Studies. Ottinger, Richard L., et al. *Environmental Costs of Electricity.* New York, London, Rome: Oceana Publications, Inc.

Chapter 3

Epigraph from Unsigned. Climate Change May Lead to Surge in Tropical Diseases. May 1995. *Windletter.*

NOTES ◆ 357

1. San Martin, Robert L. May–June 1989. Renewable Energy: Power for Tomorrow. *The Futurist,* pp. 37–40.

2. U.S. Department of Energy. Energy Information Administration. October 1995. Emissions of Greenhouse Gases in the United States, 1987–1993. Washington, D.C.: U.S. Government Printing Office.

3. Stevens, William K. September 10, 1995. Experts Confirm Human Role in Global Warming. *New York Times.*

4. Manabe, Syukuro, and Stouffer, Ronald. July 15, 1993. Century-Scale Effects of Increased Atmospheric CO_2 on the Ocean-Atmospheric System. *Nature.* Anderson, Christopher (ed.). July 30, 1993. Tales of the Coming Mega-Greenhouse. *Science.* Vol. 261, p. 553.

5. Stevens, William K. September 18, 1995. Scientists Say Earth's Warming Could Set Off Wide Disruptions. *New York Times.*

6. Stevens, William K. November 29, 1995. China's Inevitable Dilemma: Coal Equals Growth. *New York Times.*

7. Stevens, William K. September 10, 1995. Experts Confirm Human Role in Global Warming. *New York Times.* Harte, John, and Lashof, Daniel. January 10, 1996. Bad Weather? Just Wait. *New York Times.* Sullivan, Walter. May 2, 1995. New Theory on Ice Sheet Catastrophe Is the Direst One Yet. *New York Times.* Roemmich, Dean, and McGowan, John. March 3, 1995. Climatic Warming and the Decline of Zooplankton in the California Current. *Science.* Vol. 267. Cheer, Richard A. October 29, 1993. No Way to Cool the Ultimate Greenhouse. *Science.* Vol. 262, p. 648.

8. The last commercial nuclear power plant sale that was not subsequently canceled in the United States was made in 1973.

9. Pelline, Jeff. August 16, 1993. Nuclear Power Stalled in the U.S. High Costs, Safety Concerns Plague Industry. *San Francisco Chronicle.*

10. Pelline, Jeff. August 16, 1993. Nuclear Power Stalled in the U.S. High Costs, Safety Concerns Plague Industry. *San Francisco Chronicle.* Note concerning nuclear power: In 1994, the Tennessee Valley Authority, one of the last great bastions of enthusiasm for nuclear power, canceled another three large nuclear reactors that cost $6.3 billion. *Tennessee Valley News Release.* December 12, 1994. As their markets have shrunk in advanced industrial nations, nuclear reactor manufacturers have focused their sales efforts on developing countries. The truth about nuclear power's real costs is getting difficult to hide even there. As a 1995 editorial in *The Economist* observed, "For now, the case for nuclear power is full of holes. Asia should resist the temptation to throw money into them." *The Economist.* October 7, 1995.

11.–14. Komanoff Energy Associates. 1992. *Fiscal Fission: The Economic Failure of Nuclear Power. A Report on the Historical Costs of Nuclear Power in the United States for Greenpeace.* Washington, D.C.: Greenpeace USA.

15. and 16. Imbrecht, Charles. June 11, 1993. Chairman, California Energy Commission. Address at Responsive Energy Technology Symposium and International Exchange. San Diego, California.

17. Rashkin, Sam., Technology Assessments Manager, Energy Technology Development Division, California Energy Commission. "Is Sustained Orderly Development of Wind Technology a U.S. Success Story?" Address delivered July 14, 1993, at Windpower '93, the 23rd Annual Conference and Exposition of the American Wind Energy Association of Washington, D.C., in San Francisco, California.

18. Imbrecht, Charles. June 11, 1993. Chairman, California Energy Commission. Address at Responsive Energy Technology Symposium and International Exchange. San Diego, California.

Chapter 4

1. Yergin, Daniel. 1991. *The Prize: The Epic Quest for Oil, Money and Power.* New York: Simon & Schuster, pp. 685–687.

2. and 3. Mills, David, and Keepin, Bill. August 1993. Baseload Solar Power: Near-Term Prospects for Load Following Solar Thermal Electricity. *Energy Policy,* pp. 841–856.

4. and 5. Lotker, Michael. November 1991. *Barriers to Commercialization of Large-Scale Solar Electricity: Lessons Learned from the LUZ Experience.* Sandia National Laboratories. SAND91-7014.

6. The state solar energy credit shrank to only 6.6 percent in 1987 and vanished entirely for a year in 1988. The Federal Investment Tax Credit disappeared in 1988 and was not renewed during LUZ's corporate lifetime.

7. Lotker, Michael. November 1991. *Barriers to Commercialization of Large-Scale Solar Electricity: Lessons Learned from the LUZ Experience.* Sandia National Laboratories. SAND91-7014.

Chapter 5

1. Zweibel, Ken. 1990. *Harnessing Solar Power: The Photovoltaics Challenge.* New York and London: Plenum Press.

2. Maycock, Paul D. 1996. 1995 World PV Module Shipments 84 MW; 21 Percent Increase. *PV News.* Vol. 15, no. 1. Maycock, Paul D. 1996. 1995 World Module Shipments Changes. *PV News.* Vol. 15, no. 3.

3. Utility Photovoltaic Group. June 1994. *Photovoltaics: On the Verge of Commercialization.* Summary Report. Washington, D.C., p. 10.

4. For a thorough but very technical scientific and mathematical presentation on how light is transformed into electricity in a solar cell, see Sze, S. M. 1981. *Physics of Semiconductor Devices,* 2nd ed. New York: John Wiley & Sons.

5. Carlson, David E., and Wagner, Sigurd. 1993. Amorphous Silicon Photovoltaic Systems. Chapter 9 in Johansson, Thomas B.; Kelly, Henry; Reddy, Amulya K. N.; Williams, Robert H.; and Burnham, Laurie (eds.). 1993. *Renewable Energy: Sources for Fuels and Electricity.* Washington, D.C., and Covelo, Calif.: Island Press.

6. Kelly, Henry. 1993. Introduction to Photovoltaic Technology. Chapter 6 in Johansson, Thomas B.; Kelly, Henry; Reddy, Amulya K. N.; Williams, Robert H.; and Burnham, Laurie (eds.). 1993. *Renewable Energy: Sources for Fuels and Electricity.* Washington, D.C., and Covelo, Calif.: Island Press.

7. and 8. Boes, Eldon C., and Luque, Antonio. 1993. Photovoltaic Concentrator Technology. Chapter 8 in Johansson, Thomas B.; Kelly, Henry; Reddy, Amulya K. N.; Williams, Robert H.; and Burnham, Laurie (eds.). 1993. *Renewable Energy: Sources for Fuels and Electricity.* Washington, D.C., and Covelo, Calif.: Island Press.

9. Utility Photovoltaic Group. June 1994. *Photovoltaics: On the Verge of Commercialization.* Summary Report. Washington, D.C., p. 46.

10. Until recently, most economic comparisons of renewables with more traditional energy-generating technologies were based on conventional assumptions of utility economics. However, these assumptions and analytic procedures evolved for assessing the costs and benefits of fuel-based technologies operating in central power stations. The resulting analyses do not fully reflect the value of renewables to utilities, consumers, and so-

ciety. PV offers unique economic benefits due to transmission and distribution system savings resulting from its dispersed or "distributed" generation pattern. These benefits have significant financial value and must be explicitly accounted for in realistic economic analyses of renewables. The value depends on several other new variables not routinely factored into traditional utility economic models. Thus the old techniques used by utilities for calculating the value of a new power plant must be modified to reflect the unique characteristics and benefits of PV. For details, see works by Awerbuch, Shimon. October 1993. *Methodologies to Analyze the Economics of Competing Energy Technologies: A Framework for Evaluating Photovoltaics.* Report to Sandia National Laboratory. Contract 67-4226.

Conventional cash-flow utility accounting systems tend to gloss over hard-to-quantify impacts. Most comparisons between PV and old-style technologies generally do not fully recognize (1) the environmental benefits of PV (since markets are not valuing them); (2) the dollar value of distributed-energy generation; (3) the risks of future fossil-fuel price escalations, or future costly environmental regulations that may be imposed on fossil-fuel generation sources. Neither do they fully reflect the high external costs inflicted on society by fossil fuels. Hohmeyer (1988) has estimated that just the social costs of conventional electricity generation would add an average of roughly 10 cents per kilowatt-hour to the cost of conventional electricity generation. This would more than double the costs of power from conventional fuels. See Hohmeyer, O. 1988. *Social Costs of Energy Consumption.* New York: Springer-Verlag and Pace University Center for Environmental Legal Studies. 1990. *Environmental Costs of Electricity.* New York: Oceana Publications, Inc.

11. Despite Enron's sanguine cost projections, the plant evidently still will need federal assistance to be profitable: The company has asked the Department of Energy (DOE) to purchase the plant's power, and DOE is considering the request. Enron also anticipates leasing federal land in Nevada for the plant and using federal energy tax credits. The plant would be financed with the help of tax-free industrial development bonds. In addition, one analyst has suggested that Enron's business strategy is to escalate the PV power prices steeply as plants it builds expand toward full capacity.

12. Sklar, Scott. June 1994. Photovoltaics: Trends and Opportunities in the 1990s. *Real Goods News.*

13. and 14. Utility Photovoltaic Group. June 1994. *Photovoltaics: On the Verge of Commercialization.* Summary Report. Washington, D.C., p. 42 and p. 44.

15. Maycock, Paul D. November 1994. Egyptian PV, Wind Meeting Opens Critical Dialog. *Photovoltaic News.*

16. Sklar, Scott. June 1994. Photovoltaics: Trends and Opportunities in the 1990s. *Real Goods News.*

Chapter 6

1. Freeman subsequently left SMUD to become head of the New York State Power Authority, a post he left in 1995.

2. Osborn, Donald E. March 1994. *Utility Implementation of Grid-Connected Photovoltaics.* ASME International Solar Energy Conference. San Francisco, California.

3. Sacramento Municipal Utility District. 1993. *Advanced and Renewable Development Plan.* Sacramento, California.

Chapter 7

1. Siemens Solar, a German company, is the largest cell manufacturer in the United States with 47 percent of U.S. solar cell production. (The United States produced only about 30 megawatts of solar cells in 1994, two-thirds for export.) The next largest U.S. solar firm after Siemens Solar is Solarex Corporation of Frederick, Maryland, owned by Amoco, with 26 percent of U.S. production, followed by Solec. Source: Johnson, Bob. November 1994. Publisher. Strategies Unlimited, Inc. (industry consulting group). Personal communication. (U.S. 1995 solar cell production increased by another 5 megawatts.)

2. Unlike a conventional circular saw that has teeth on the outside of a revolving disk, the inside-diameter circular saw blade is shaped like a disk with a hole in it. The cutting action is performed by the sharpened inner diameter of the revolving disk.

Chapter 8

1. The article to which Bill Yerkes refers is by Spaeth, Anthony. November 12, 1979. Arco's Solar Chill. *Forbes.* Vol. 124, no. 10, p. 34.

2. and 3. A Carissa Solar Corporation spokesman declined to verify the plant's purchase price but said the company also acquired ARCO's 1-megawatt PV plant near Victorville in the deal as well, and that the two facilities together had cost well over a million dollars. One educated guess put the cost of the entire acquisition at $2 million.

Chapter 9

Epigraph from Boes, Eldon C., and Luque, Antonio. 1993. Photovoltaic Concentrator Technology. Chapter 8 in Johansson, Thomas B.; Kelly, Henry; Reddy, Amulya K. N.; Williams, Robert H.; and Burnham, Laurie (eds.). 1993. *Renewable Energy: Sources for Fuels and Electricity.* Washington, D.C., and Covelo, Calif.: Island Press.

1. Kelly, Henry. 1993 Introduction to Photovoltaic Technology. Chapter 6 in Johansson, Thomas B.; Kelly, Henry; Reddy, Amulya K. N.; Williams, Robert H.; and Burnham, Laurie (eds.). 1993. *Renewable Energy: Sources for Fuels and Electricity.* Washington, D.C., and Covelo, Calif.: Island Press.

2. Boes, Eldon C., and Luque, Antonio. 1993. Photovoltaic Concentrator Technology. Chapter 8 in Johansson, Thomas B.; Kelly, Henry; Reddy, Amulya K. N.; Williams, Robert H.; and Burnham, Laurie (eds.). 1993. *Renewable Energy: Sources for Fuels and Electricity.* Washington, D.C., and Covelo, Calif.: Island Press.

Chapter 10

1. Even the nominal market share enjoyed by thin films is a little misleading, since less than half of all the thin-film cell sales in 1992 were for outdoor terrestrial power production. Most of the cells were destined for use in consumer products, such as watches, calculators, and battery chargers, or for computer and copier applications. "These last two were driving the market," as recently as 1992, according to one analyst.

2. Stafford, Byron. June 15–18, 1992. The Emergence of Amorphous Silicon Photovoltaics. *Proceedings of Solar '92. The 1992 American Solar Energy Society Annual Conference.* Cocoa Beach, Florida. Boulder, Colo.: ASES.

3. and 4. Maycock, Paul D. 1993. *Photovoltaic Technology, Performance, Cost, and Market Forecast, 1990–2010.* Casanova, Va.: Photovoltaic Energy Systems, Inc.

5. Thus module power is initially higher than the nominal rating to allow for future degradation.

6. The measurement was made by PV USA, the government-supported photovoltaic test organization, which measured APS cell power output at the busbar (the electricity collection point).

7. Maycock, Paul D. 1995. Solarex Breaks World Record in Thin Film Efficiency. *PV News.* Vol. 14. No. 12.

8. Maycock, Paul D. 1996. Consumers Value Environmentally Benign Electricity. *PV News.* Vol. 15. No. 1.

9. Preliminary results of recent studies by Brookhaven National Laboratory and the National Institutes of Health have found that although CIS and CdTe are toxic compounds, they become insoluble and far less dangerous after processing into solar cells. Another way to put the cadmium toxicity in perspective is to recall that cadmium is a common pigment in yellow paint and a routine emission from coal power plants. But the amount of cadmium used per megawatt-year in sealed photovoltaic modules is less than the cadmium discharged to the atmosphere per megawatt-year of coal power.

10. Maycock, Paul D. 1995. NREL Awards 11 PV Manufacturing Contracts Totalling $22 Million. *PV News.* Vol. 14, no. 12.

11. Maycock, Paul D. March 1995. TI Terminates Spheral Solar™ Project, Offers It for Sale. *PV News.* Vol. 14, no. 3.

Chapter 13

1. U.S. Department of Energy. *Wind Energy Program Overview FY 1992.* Washington, D.C., p. 17.

2. The U.S. Energy Information Administration has reported that 240,000 square miles of land in the United States are within ten miles of a high-voltage transmission line and could accommodate 734,000 average megawatts of wind capacity, which would more than double current U.S. electrical power capacity. See *Wind Energy Weekly.* May 1, 1995. No. 644.

3. The Coriolis effect is a result of the Coriolis Force, which displaces any object moving over a rotating surface, such as the Earth, relative to the surface. The force is imaginary, because the deflection is an effect of inertia. The effect is most visible with objects, especially fluids (such as air or water), moving on the Earth's surface; these bodies are deflected at right angles to the direction of their movement, causing objects to curve counterclockwise in the Northern Hemisphere and clockwise in the Southern.

4. Woodcroft, Bennet (ed.; trans. from the ancient Greek by Joseph Gouge Greenwood). Reprint of first edition (1851). 1971. *The Pneumatics of Hero of Alexandria.* London: MacDonald & Co.

5. Golob, Richard, and Brus, Eric. 1993. *The Almanac of Renewable Energy.* New York: Henry Holt and Co., p. 129.

6. USW is a wholly owned subsidiary of Kenetech Corporation, an unregulated energy holding company. Kenetech was formed in 1986 and thereupon became USW's parent company. The name has been changed to Kenetech Windpower, Inc.

7. Starr, Chauncey. September 1971. Energy and Power. *Scientific American.* Vol. 225.

8. and 9. Heronemus, William E. January 12, 1972. The United States Energy Crisis: Some Proposed Gentle Solutions. Paper presented before a joint meeting of local sections of the American Society of Mechanical Engineers and the Institute of Electrical and Electronics Engineers.

10. The term *avoided cost* sounds deceptively simple, but has different technical meanings in the short versus the long term. In the short term, when a utility can meet its demand without capacity expansion and the capital costs of generating equipment are fixed or "sunk," the utility's avoided cost is the operating cost of the most expensive generating source in the system that can be "backed out," meaning either turned off or operated at reduced output, to "make room" for renewably generated electricity.

(Obviously the utility will shut down the plant with the highest operating costs, all else being equal.) In the long term, however, when the utility views the prospect of incurring both fixed and variable (operating) costs for any new generating technology, the avoided cost is now the total cost of the *least* expensive new capacity the utility could install or acquire. Renewable technology is cost-competitive when it can be provided at less than the overall (fixed and variable) costs of new generation or when it can be provided at less than the variable cost of the most costly generation currently on line and capable of "backing down." Source: personal communication with Dr. Jan Hamrin. See also Hamrin, Jan, and Rader, Nancy. 1994. *Investing in the Future: A Regulator's Guide to Renewables.* Washington, D.C.: National Association of Regulatory Utility Commissioners.

11. The power in a moving mass of air is proportional to the cross-sectional area of the air mass and the cube of the wind speed (the speed multiplied by itself twice). Double the radius of a turbine blade and the area is increased by the square of the radius, a factor of four. (This is a direct consequence of the basic formula for the area of a circle, $A = \pi r^2$. But double the wind speed and, because of the cubic relationship described, the power increases eightfold [$2 \times 2 \times 2$].) Wind speed, of course, is highly variable, moment to moment, seasonally, and on other cycles. As it varies, power naturally varies. Quadruple wind speed and power increase by a factor of 64 (the cube of 4). That means that a sudden high-velocity wind gust will transmit an enormous increase in power to a turbine, stressing its components. If controls do not act to relieve the stress by feathering or stalling the blades in time, turbine failure is likely. Turbines must also endure harsh weather extremes, such as ice and hail, or even driving sand in deserts. Because turbines are in use for thousands of hours a year and may perform billions of revolutions over their lifetimes, all their mechanisms must be very durable. Wind machines also must cope with wind shear, a rapid or abrupt wind speed change with height.

Chapter 14

1. Due diligence is the process of verifying all claims and salient characteristics of an investment prospect to identify elements of risk.

2. Asmus, Peter. Fall 1994. Hot Air, Hot Tempers, and Cold Cash. *The Amicus Journal.* Note: Where wind farms are proposed for siting in undeveloped areas, ecological consequences may extend beyond avian impacts. In conjunction with Kenetech's proposed 210-megawatt wind farm in Maine's Boundary Mountains, the company found itself facing four formidable environmental groups who became "intervenors" in the wind farm permitting process. USW mollified the opposition by agreeing to provide $300,000 for purchase of conservation easements in the high mountains of Maine, with funds to be

matched by a Boston foundation. Kenetech also pledged to fund bird migration studies and provide partial funding for a statewide wind farm siting survey.

Kenetech has been involved in at least one other notable controversy over the avian impact of its wind farms. The company's Cape Tarifa, Spain, wind farm was prominently linked in 1994 to the deaths of raptors and griffon vultures. European environmental groups, who support wind power technology, nonetheless called for a halt to expansion at the site pending further study. Photos of injured birds appeared in the widely distributed international wind energy journal *Windpower Monthly*. Kenetech and its partner, Abengoa, chose to expand their wind farm anyway. *Windpower Monthly* then editorialized, "When a wind company announces that it is not only going ahead with a project in the face of massive opposition from environmentalists, but that it would rather pay compensation for bird deaths than move to a site where they do not occur, then moral integrity starts wearing extremely thin. . . . The environmental argument for wind is one of its strongest. Take away wind's green credentials, lose the support of the environmentalists, then wind has a bleak future."

3. Asmus, Peter. Fall 1994. Hot Air, Hot Tempers, and Cold Cash. *The Amicus Journal.*

4. Parrish, Michael. February 1995. A Second Wind for Turbines. *Los Angeles Times.*

5. and 6. Asmus, Peter. Fall 1994. Hot Air, Hot Tempers, and Cold Cash. *The Amicus Journal.*

7. Bain, Donald A. Oregon Department of Energy, Salem, Oregon. Renewables Can Compete with Gas. Address delivered July 14, 1993, at Windpower '93, the 23rd Annual Conference and Exposition of the American Wind Energy Association of Washington, D.C., in San Francisco, California. Cost varies greatly with age of the system, technology, location, type of fuel, cost of fuel, and cost of financing or discount rate used.

8. Regan, Mary Beth. November 8, 1993. The Sun Shines Brighter on Alternative Energy. *Business Week.*

9. Parrish, Michael. February 5, 1995. A Second Wind for Turbines. *Los Angeles Times.*

Chapter 15

1. Marshall, Jonathan. May 23, 1995. Wind Blows Ukraine Some Good. *San Francisco Chronicle.*

2. Hermann, Henry, CFA. W. R. Lazard, Laidlaw & Mead. November 22, 1994. *The Wind Power Industry.*

3. Davidson, Ros. October 1993. Shares Rush as Wind Goes Public. *Windpower Monthly News Magazine.* Vol. 9, no. 10, pp. 24–26.

4. Although the California Public Utilities Commission (PUC) has in the past required utilities to buy renewable energy from independent power producers, such as Kenetech, California utilities have become increasingly restive over the requirement. Thus, in 1993, when the PUC held an auction of renewable energy generating contracts under provisions of a regulatory process called the Biennial Resource Plan Update (BRPU), Southern California Edison (SCE) and San Diego Gas & Electric Company (SDG&E) protested the whole process to the PUC. Their challenge rejected, they took their case to the Federal Energy Regulatory Commission, which decided in their favor in 1995, after an appeal. Wind power companies lost the contracts they thought they had won in the

auction and will face a much more competitive market as California and other states abandon renewable energy set-asides. SCE's situation illustrates why the company resents being forced to buy new renewable capacity. SCE in 1994 had 3,000 megawatts of excess capacity and actually was paying some wind power producers not to deliver power to their system.

5. $100,000,000 Kenetech Corporation 12.75% Senior Secured Notes due 2002. Prospectus. Merrill Lynch & Co. Morgan Stanley & Co., Incorporated. December 18, 1992.

6. The documents filed by FloWind contended that Kenetech grossly overstated its proposed wind project "capacity factors" to artificially lower its bid scores. Companies with the lowest bid scores won contracts. The bid scores are computed according to a PUC formula that accords lower scores to projects with higher capacity factors. (A wind power plant's capacity factor is the proportion of the plant's maximum theoretical electrical output that the project can actually deliver at different utility rate periods, such as the peak rate period or the off-peak period.)

Kenetech used capacity factors above 80 percent. In a sworn statement by consulting meteorologist and wind resource expert Ron Nierenberg, submitted to the PUC by FloWind, Nierenberg explains that the intermittent nature of the wind and the physical limitations of wind turbine technology make it *"impossible"* for wind producers to achieve capacity factors of more than 40 percent in California. FloWind attorneys charged that inflation of capacity factors by Kenetech would have provided it with $20 million in additional profits. Other FloWind exhibits include calculations showing that Kenetech would receive overpayments on its wind power contracts in the first operating year of more than $166 million, amounting to overpayments of 7.5 cents per kilowatt-hour—far from its promise of wind electricity for 5 cents a kilowatt-hour.

Sources for this note are as follows: Fraudulent bid charges against Kenetech, Sea-West and Zond by FloWind. Petition of FloWind Corporation for Rehearing of Decision No. 94-06-047. Before the Public Utilities Commission of the State of California. July 25, 1994; and Declaration of Michael McMullen before the Public Utilities Commission. Exhibit B. Kenetech's explanation of capacity factors. Miller, Eric. 1993. Wind Capacity Factor Bids in the 1993 FS04 Auction. Kenetech Windpower Internal Memorandum Submitted as Exhibit G to California Public Utilities Commission. December 20; PUC quote on wind bidders. Decision 94-12-051. December 21, 1994. Before the Public Utilities Commission of the State of California. Order Granting Limited Rehearing of Decision 94-06-047.

7. Public Utilities Commission's suspension of wind power contracts. Decision 94-12-051. Before the Public Utilities Commission of the State of California. Order Granting Limited Rehearing of Decision of 94-06-047. December 21, 1994.

8. Rashkin, Sam, Technology Assessments Manager, Energy Technology Development Division, California Energy Commission. Is Sustained Orderly Development of Wind Technology a U.S. Success Story? Address delivered July 14, 1993, at Windpower '93, the 23rd Annual Conference and Exposition of the American Wind Energy Association of Washington, D.C., in San Francisco, California.

9. Bain, Donald A. Oregon Department of Energy, Salem, Oregon. Renewables Can Compete with Gas. Address delivered July 14, 1993, at Windpower '93, the 23rd Annual Conference and Exposition of the American Wind Energy Association of Washington, D.C., in San Francisco, California.

10. Rakow, Sally. 1993. Commissioner, California Energy Commission. Wind Energy in California: A Decade of Progress. Address delivered July 14, 1993, at Windpower '93,

the 23rd Annual Conference and Exposition of the American Wind Energy Association of Washington, D.C., in San Francisco, California.

Chapter 16

1. Hermann, Henry, CFA. W. R. Lazard, Laidlaw & Mead. November 22, 1994. *The Wind Power Industry.*

2. FloWind is continuing to advance VAWT technology by designing, manufacturing, and installing a new utility-scale Extended Height-to-Diameter (EHD) class of VAWTs, which it will begin marketing in 1996 or 1997.

3. Marshall, Jonathan. June 14, 1995. Going with the Wind. *San Francisco Chronicle.*

4. December 1993. FloWind, AWT and Kaiser Sign Tubine Production Pact. *Windletter.* Vol. 20, no. 12. As of late 1993, Kenetech, too, had a preliminary agreement for joint wind plant development with the Siemens group, another very large industrial conglomerate. See Davidson, Ros. October 1993. Shares Rush as Wind Goes Public. *Windpower Monthly News Magazine.* Vol. 9, no. 10, pp. 24–26.

5. Ginsberg, Steve. March 17–23, 1995. Big Asian Wind-Energy Deals Turn Marin Company into Power Player. *San Francisco Business Times.*

Chapter 17

1. My treatment of European wind energy is based primarily on the excellent work by Jamie Chapman and Paul Gipe, to whom I am indebted. See next two notes.

2. Gipe, Paul. July/August 1993. The Race for Wind. *Independent Energy.*

3. Chapman, J. C. March 1993. European Wind Technology. TR-1-1301/ Research Project 1996-28. Final Report. Palo Alto, Calif.: Electric Power Research Institute.

4. The fact that WEG is a partnership of the major multinational Taylor Woodrow Construction Holdings Ltd., British Aerospace, and Britain's General Electric suggests a likely future trend to which I have previously alluded. As the wind power plant market expands and becomes more lucrative, very large, powerful aerospace and other industrial firms are likely to enter the field. If and when these megacompanies go shopping, companies like AWT or Carter Wind Turbines may be tempting acquisition targets.

Chapter 18

Epigraph from Morris, Gregory. February 1995. Legislative Initiatives Being Pursued by the California Biomass Energy Industry. *Bio Bulletin.* Vol. II.

1. To an ecologist, *biomass* is the amount of living matter produced per some metric, while *biomass* to a chemist would be but another word for organic matter.

2. National Renewable Energy Laboratory. U.S. Department of Energy. March 1994. *The American Farm: Harnessing the Sun to Fuel the World.* NREL/SP-420-5877. DE94000217.

3. The costs of biomass energy are currently being lowered by efforts to increase biomass crop yields. Dedicated biomass-to-energy crop systems are not likely to compete successfully with fossil fuel power until per acre yields of fast growing woody biomass are routinely and reliably doubled from current levels to 15–20 dry tons per acre. Conceiv-

ably it may take 15–20 years before wood-to-energy plants using these crops can produce power at a competitive 4 cents per kilowatt-hour.

4. and 5. Highlights of the Second Biomass Conference of the Americas. *SERBEP Update*. October 1995. A publication of the Southeastern Regional Biomass Energy Program, U.S. Department of Energy, Office of National Programs, TVA Environmental Research Center, Muscle Shoals, Alabama.

6. Morris, Gregory. December 1994. The Biomass Power Industry on the Precipice. *Bio Bulletin*. Vol. I. Issue 6.

7. Highlights of the Second Biomass Conference of the Americas. *SERBEP Update.* October 1995. A publication of the Southeastern Regional Biomass Energy Program, U.S. Department of Energy, Office of National Programs, TVA Environmental Research Center, Muscle Shoals, Alabama. Morris, Gregory. January 1995. Biomass Energy in the Brave New World of Restructuring. *Bio Bulletin*. Vol. II, no. 1.

8. Unsigned. April 17, 1995. Customer Interest Is Reported for Solar Zone Power Output. *Solar Energy Intelligence Report*. Vol. 21, no. 8.

9. A possibly more damaging FERC ruling was its decision to interpret "avoided cost" under the Public Utility Regulatory Policy Act of 1978 (PURPA) as meaning the least costly power one can buy. This means that one cannot just solicit renewable energy bids under PURPA exclusively for biomass power or for other distinct categories of renewables. Renewables of all kinds would be pitted against each other and the utility or power pool would have to take the lowest bid.

10. National Renewable Energy Laboratory. U.S. Department of Energy. January 1995. *Biofuels for Transportation: The Road from Research to the Marketplace.* NREL/SP-420-5439.

11. National Renewable Energy Laboratory. U.S. Department of Energy. January 1995. *Biofuels for Transportation: The Road from Research to the Marketplace.*

12. Rosillo-Calle, Frank, and Hall, David O. February 1992. Biomass Energy, Forests and Global Warming. *Energy Policy.*

13. Hall, D. O., and House, J. I. September–November 1993. Reducing Atmospheric CO_2 Using Biomass Energy and Photobiology. *Energy Conversion and Management*. Vol. 34, nos. 9–11, pp. 889–896.

14. Hall, David O.; Rosillo-Calle, Frank; Williams, Robert H.; and Woods, Jeremy. 1993. Biomass for Energy: Supply Prospects. Chapter 4 in Johansson, Thomas B.; Kelly, Henry; Reddy, Amulya K. N.; Williams, Robert H.; and Burnham, Laurie (eds.). *Renewable Energy: Sources for Fuels and Electricity*. Washington, D.C., and Covelo, Calif.: Island Press.

15. Reese, Randall A.; Aradhyula, Satheesh V.; Shogren, Jason F.; and Tyson, K. Shaine. July 1993. Herbaceous Biomass Feedstock Production: The Economic Potential and Impacts on U.S. Agriculture. *Energy Policy*. Vol. 7.

16. Pasztor, Janos, and Kristoferson, Lars. "Biomass Energy." Chapter 7 in Hollander, Jack M. (ed.). 1992. *The Energy Environment Connection*. Washington, D.C., and Covelo, Calif.: Island Press.

17. Rosillo-Calle, Frank, and Hall, David O. February 1992. Biomass Energy, Forests and Global Warming. *Energy Policy.*

18. For a discussion of tree planting as a means of greatly reducing the costs of stabilizing atmospheric carbon dioxide emissions, see Rosenthal, D. H.; Edmonds, J. A.; Richards, K. R.; and Wise, M. A. September–November 1993. Stabilizing United States Net Carbon Emissions by Planting Trees. *Energy Conversion and Management*. Vol. 34, nos. 9–11, pp. 881–887.

19. and 20. Union of Concerned Scientists. May 1994. *Biomass Power Commercialization: The Federal Role.* Cambridge, Mass.: Union of Concerned Scientists.

21. Southeastern Regional Biomass Energy Program. n.d. *Use of Wood Waste for Cogeneration of Electricity and Industrial Process Heat.* Muscle Shoals, Ala.: Tennessee Valley Authority.

22. Southeastern Regional Biomass Energy Program. n.d. *Cogeneration with Wood Residues as Fuel.* Muscle Shoals, Ala.: Tennessee Valley Authority.

23. Southeastern Regional Biomass Energy Program. n.d. *Production of Fireplace Logs from Wood Waste.* Muscle Shoals, Ala.: Tennessee Valley Authority.

24. Landfill Gas Back on Track. *SERBEP Update.* September 1995. A publication of the Southeastern Regional Biomass Energy Program, U.S. Department of Energy, Office of National Programs, TVA Environmental Research Center, Muscle Shoals, Alabama.

25. National Renewable Energy Laboratory. U.S. Department of Energy. January 1995. *Biofuels for Transportation: The Road from Research to the Marketplace.*

26. Hall, D. O., and House, J. I. September–November 1993. Reducing Atmospheric CO_2 Using Biomass Energy and Photobiology. *Energy Conversion and Management.* Vol. 34, nos. 9–11, pp. 889–896.

27. Pasztor, Janos, and Kristoferson, Lars. "Biomass Energy." Chapter 7 in Hollander, Jack M. (ed.). 1992. *The Energy Environment Connection.* Washington, D.C., and Covelo, Calif.: Island Press.

28. Lothner, David C. January–March 1991. Short-Rotation Energy Plantations in North Central United States: An Economic Analysis. *Energy Sources.* Vol. 13, no. 1.

29. Denzler, Eric. Fall 1993. Green Energy. *Nucleus.*

30. Biomass could be raised in biologically diverse polycultures interspersed with wildlife "corridors" that lead to natural areas. Portions of the energy forest or plantation also could be left unharvested cyclically to provide wildlife with continual habitat and sanctuary. For a discussion of possible detrimental environmental impacts of biomass plantations, see Cook, James H.; Beyea, Jan; and Keeler, Kathleen H. 1991. Potential Impacts of Biomass Production in the United States on Biological Diversity. *Annual Review of Energy.* Vol. 16, pp. 401–431.

31. Pasztor, Janos, and Kristoferson, Lars. Biomass Energy. Chapter 7 in Hollander, Jack M. (ed.). 1992. *The Energy Environment Connection.* Washington, D.C., and Covelo, Calif.: Island Press.

32. Reese, Randall A.; Aradhyula, Satheesh V.; Shogren, Jason F.; and Tyson, K. Shaine. July 1993. Herbaceous Biomass Feedstock Production: The Economic Potential and Impacts on U.S. Agriculture. *Energy Policy.* Vol. 7.

33. Phillips, Julie. 1994. Guidelines at Washington, D.C., Briefings. *Biologue.* 4th Quarter; and National Biofuels Roundtable. 1994. *Principles and Guidelines for the Development of Biomass Energy Systems.* Available from Overend, Ralph P., National Renewable Energy Laboratory. 1617 Cole Boulevard, Golden, CO 80401.

34. Southeastern Regional Biomass Energy Program. November 1993. Alkali Slagging Problems with Biomass Fuels: An Update. *SERBEP Update.*

35. Union of Concerned Scientists. May 1994. *Biomass Power Commercialization: The Federal Role.* Cambridge, Mass.: Union of Concerned Scientists.

36. Electricity from Agri-Forestry. *SERBEP Update.* September 1995. A publication of the Southeastern Regional Biomass Energy Program, U.S. Department of Energy, Office of National Programs, TVA Environmental Research Center, Muscle Shoals, Alabama.

37. Paper Pellet Fuels in the Southeast. 1994. *Biologue.* 4th Quarter.

38. Southeastern Regional Biomass Energy Program. n.d. *Use of Sewage Effluent to Cultivate Energy Feedstock.* Muscle Shoals, Ala.: Tennessee Valley Authority.

39. Southeastern Regional Biomass Energy Program. n.d. *Use of Wastewater to Cultivate Eucalyptus as a Biofuel.* Muscle Shoals, Ala.: Tennessee Valley Authority.

40. Methane's effect on climate. Each molecule of methane has an instantaneous radiative (warming) effect 25 times greater than each molecule of carbon dioxide; however, carbon dioxide remains in the atmosphere longer than methane, which is eventually removed from the atmosphere by oxidation to carbon dioxide. Gleick, Peter; Morris, Gregory P.; and Norman, Nicki A. July 22, 1989. *Greenhouse-Gas Emissions from the Operation of Energy Facilities.* Final Report. Prepared for the Independent Energy Producers Association, Sacramento, California. See also Morris, Gregory P. May 11, 1992. *Greenhouse-Gas Emissions From Wastewood Disposal.* Final Report. Prepared for the Independent Energy Producers Association, Sacramento, California. Berkeley, Calif.: Future Resources Associates, Inc.

41. Western Regional Biomass Energy Program. March 1994. Texas Tech Studies Use of Animal Wastes for Producing Methane, Microalgae, Fish. *Bio Tech Brief.* BTB-8-3/94 1660.

42. Pasztor, Janos, and Kristoferson, Lars. Biomass Energy. Chapter 7 in Hollander, Jack M. (ed.). 1992. *The Energy Environment Connection.* Washington, D.C., and Covelo, Calif.: Island Press.

43. National Renewable Energy Laboratory. U.S. Department of Energy. January 1995. *Biofuels for Transportation: The Road from Research to the Marketplace.* NREL/SP-420-5439. Note: In transesterification, oils are reacted with alcohols in the presence of catalysts, and products are then separated in stages.

44. Southeastern Regional Biomass Energy Program. *Economic Feasibility of Biodiesel Production in the Southeastern United States.* n.d. Muscle Shoals, Ala.: Tennessee Valley Authority.

45. Coast-to-Coast Round Trip on Idaho Biodiesel. 1994. *Biologue.* 4th Quarter.

46. Southeastern Regional Biomass Energy Program. n.d. Biodiesel from Beef Tallow. Muscle Shoals, Ala.: Tennessee Valley Authority.

47. Marin, Jonathan K. May 2, 1995. Why Diesel Particles Contribute to Cancer. Letter to the Editor. *New York Times.*

48. Southeastern Regional Biomass Energy Program. n.d. *Rapid Thermal Processing of Wood Residues to Produce "Bio-oil."* Muscle Shoals, Ala.: Tennessee Valley Authority.

49. Unsigned. 1994. Tecogen Reports Good Sludge-to-Ethanol Results; First Processor Sold. *Biologue.* 4th Quarter.

50. Biofuels Systems Program. 1994. U.S. Department of Energy. *Biofuels: At the Crossroads. Strategic Plan for the Biofuels Systems Program.* Washington, D.C. An added advantage of the SSF process is that hydrolysis and fermentation can be performed in the same vessel, reducing the number of process steps and increasing the process yield. Genetic engineering is also improving the yield by optimizing the biological makeup of the organisms responsible for the fermentation process. See National Renewable Energy Laboratory. U.S. Department of Energy. January 1995. *Biofuels for Transportation: The Road from Research to the Marketplace.* NREL/SP-420-5439.

51. For purposes of comparison, methanol produced commercially from natural gas in 1994 cost $1.30 to $1.80 a gallon, about two and a half to three and a half times as much.

52. Idaho National Engineering Laboratory, Los Alamos National Laboratory, Oak Ridge National Laboratory, Sandia National Laboratories and Solar Energy Research In-

stitute (now National Renewable Energy Laboratory). March 1990. *The Potential of Renewable Energy: An Interlaboratory White Paper. Appendix B: Biomass/Biofuels.* Office of Policy, Planning and Analysis. U.S. Department of Energy. SERI/TP-260-3674; DE90000322.

53. Williams, Robert H., and Larson, Eric D. 1993. Advanced Gasification-Based Biomass Power Generation. Chapter 17 in Johansson, Thomas B.; Kelly, Henry; Reddy, Amulya K. N.; Williams, Robert H.; and Burnham, Laurie (eds.). *Renewable Energy: Sources for Fuels and Electricity.* Washington, D.C., and Covelo, Calif.: Island Press.

54. Union of Concerned Scientists. May 1994. *Biomass Power Commercialization: The Federal Role.* Cambridge, Mass.: Union of Concerned Scientists.

55. Williams, Robert H., and Larson, Eric D. Advanced Gasification-Based Biomass Power Generation. Chapter 17 in Johansson, Thomas B.; Kelly, Henry; Reddy, Amulya K. N.; Williams, Robert H.; and Burnham, Laurie (eds.). *Renewable Energy: Sources for Fuels and Electricity.* Washington, D.C., and Covelo, Calif.: Island Press.

56. Western Regional Biomass Energy Program. February 1994. Integrated Tubine System Has Worldwide Potential. *Bio Tech Brief.* BTB-5-2/94 8318.

57. National Biofuels Roundtable. 1994. *Principles and Guidelines for the Development of Biomass Energy Systems.* Available from Overend, Ralph P., National Renewable Energy Laboratory, 1617 Cole Boulevard, Golden, CO 80401.

58. Morris, Gregory. February 1995. Legislative Initiatives Being Pursued by the California Biomass Energy Industry. *Bio Bulletin.* Vol. II. Morris, Gregory. March 1995. Regulatory Activities Being Pursued by the California Biomass Energy Industry. *Bio Bulletin.* Vol. II, Issue III.

Chapter 19

1. Lieberman, Lisa. April 1995. Fuel for Thought. *California Farmer.*

2. Those contracts, dating from the mid-1980s, had been tied to utilities' (then much-higher) avoided cost of alternative power. When those prices were recalculated in the mid-1990s based on utilities' avoided cost, which was the bargain basement cost of natural gas power, biomass plants could no longer cover their costs.

3. An estimated 17,000 new jobs would be created for every billion gallons of new ethanol production capacity; the use of that ethanol in place of gasoline would reduce the atmospheric carbon dioxide burden by 9.3 million metric tons, according to Hinman, Norman D. (Manager, Biofuels Program, U.S. Department of Energy). February 1994. "Ethanol from Biomass." Presentation at California Institute of Food and Agricultural Research, University of California, Davis. April 8, 1994.

Chapter 20

Second epigraph from Earth Sciences and Resources Institute (formerly University of Utah Research Institute). February 1995. *Geothermal Energy.* Brochure for Geothermal Energy Association.

1. For a thought-provoking, insightful discussion of sustainability in general in the context of geothermal energy resource sustainability in particular, see Wright, Phillip Michael. May 1995. The Sustainability of Production from Geothermal Resources. Submitted to the World Geothermal Congress in Florence, Italy.

2. U.S. Department of Energy. August 1994. U.S. Geothermal Resources. *Geothermal Fact Sheet.* Washington, D.C.

3. National Renewable Energy Laboratory. October 1994. *Geothermal: Clean Energy from the Earth.* DE94011884. DOE/GO10094-003.

4. U.S. Department of Energy. March 1995. Geothermal Direct Use. *Geothermal Fact Sheet.* Washington, D.C.

5. Palmerini, Civis G. 1993. "Geothermal Energy." Chapter 13 in Johansson, Thomas B.; Kelley, Henry; Reddy, Amulya, K. N.; Williams, Robert H.; and Burnham, Laurie (eds.). *Renewable Energy: Sources for Fuels and Electricity.* Washington, D.C. and Covelo, Calif.: Island Press.

6. U.S. Department of Energy. March 1995. Geothermal Direct Use. *Geothermal Fact Sheet.* Washington, D.C.

7. Earth Sciences and Resources Institute (formerly University of Utah Research Institute). February 1995. Geothermal Energy. Brochure for Geothermal Energy Association.

8. Office of Energy Efficiency and Renewable Energy. U.S. Department of Energy. Fall 1994. *Fiscal Year 1995 Program Plan for Climate Change Action No. 26: Renewable Energy Commercialization—Geothermal Heat Pumps.* Washington, D.C.

9. Palmerini, Civis G. 1993. "Geothermal Energy." Chapter 13 in Johansson, Thomas B.; Kelley, Henry; Reddy, Amulya K. N.; Williams, Robert H.; and Burnham, Laurie (eds.). *Renewable Energy: Sources for Fuels and Electricity.* Washington, D.C. and Covelo, Calif.: Island Press.

10. and 11. Palmerini, Civis G. 1993. "Geothermal Energy." Chapter 13 in Johansson, Thomas B.; Kelley, Henry; Reddy, Amulya K. N.; Williams, Robert H.; and Burnham, Laurie (eds.). *Renewable Energy: Sources for Fuels and Electricity.* Washington, D.C. and Covelo, Calif.: Island Press.

12. Pederson, Barbara L. 1990. *UNOCAL 1890–1990: A Century of Spirit.* Los Angeles: UNOCAL Corporation.

13. U.S. Department of Energy. April 1994. Geothermal Electric Power Systems. *Geothermal Fact Sheet.* Washington, D.C.

14. Anderson, David N. June 1986. B. C. McCabe and Magma Power Company. *Geothermal Resources Council Bulletin.* Vol. 15, no. 6.

15. Energy Information Administration. U.S. Department of Energy. September 1991. *Geothermal Energy in the Western United States and Hawaii: Resources and Projected Electricity Generation Supplies.* EIA/DOE-0544.

16. Hoch, Andrew. Board Chairman, Emeritus, Magma Power Company. A Brief History of the Magma Power Company. Unpublished manuscript.

17. Aidlin, Joseph W. March 29, 1995. Personal interview.

18. Anderson, David N. June 1986. B. C. McCabe and Magma Power Company. *Geothermal Resources Council Bulletin.* Vol. 15, no. 6.

19. Hoch, Andrew, Board Chairman, Emeritus, Magma Power Company. A Brief History of the Magma Power Company. Unpublished manuscript. Anderson, David. June 1986. B. C. McCabe and Magma Power Company. *Geothermal Resources Council Bulletin.* Vol. 15, no. 6.

20. Hoch, Andrew. Board Chairman, Emeritus, Magma Power Company. A Brief History of the Magma Power Company. Unpublished manuscript.

21. Aidlin, Joseph W., March 29, 1995. Personal interview.

22. Otte, Carel. March 29, 1995. Personal interview.

NOTES ✦ 371

23. and 24. Aidlin Joseph W., March 29, 1995. Personal interview.

25. and 26. Otte, Carel. March 29, 1995. Personal interview.

27. Pederson, Barbara L. 1990. *UNOCAL 1890–1990: A Century of Spirit.* Los Angeles: UNOCAL Corporation.

28. U.S. Department of Energy. April 1994. Opportunities in Developing Countries. *Geothermal Fact Sheet.* Washington, D.C.

29. Anderson, David N. March 14, 1995. Executive Director, Geothermal Resources Council. Personal interview. San Francisco, California.

30. Reed, Marshall J., and Renner, Joel L. 1994. Environmental Compatibility of Geothermal Energy. Chapter 2 in Sterrett, Frances S. (ed.). *Alternative Fuels and the Environment.* Boca Raton, Fla.: CRC Press, Inc./Lewis Publishers.

31. Wright, Phillip Michael. March 1995. Personal communication. Director, Earth Science Laboratory, University of Utah Research Institute.

32. Earth Sciences and Resources Institute (formerly University of Utah Research Institute). February 1995. Geothermal Energy. Brochure for Geothermal Energy Association.

33. Palmerini, Civis G. 1993. "Geothermal Energy." Chapter 13 in Johansson, Thomas B.; Kelley, Henry; Reddy, Amulya K. N.; Williams, Robert H.; and Burnham, Laurie (eds.). *Renewable Energy: Sources for Fuels and Electricity.* Washington, D.C. and Covelo, Calif.: Island Press.

34. Reed, Marshall J., and Renner, Joel L. 1994. Environmental Compatibility of Geothermal Energy. Chapter 2 in Sterrett, Frances S. (ed.). *Alternative Fuels and the Environment.* Boca Raton, Fla.: CRC Press, Inc./Lewis Publishers.

35. Like most renewable energy technologies, geothermal energy research and development has been receiving inadequate federal support—about $24 million (estimated) nationwide in fiscal year 1994. Fusion, by contrast, received about $344 million—14 times as much—though fusion is a far more problematic and longer-term energy option (see Chapter 2). See *Special Report. Administration Budget FY '96. Energy Department.* February 7, 1995. Washington, D.C.: Environmental and Energy Study Institute.

36. The regions where geothermal energy tends to be most concentrated are those in which tectonic forces disrupt the Earth's crust, through collision, separation, subduction, or other processes.

37. Wright, Phillip Michael. March 1995. Personal communication. Director, Earth Science Laboratory, University of Utah Research Institute.

38. U.S. Department of Energy. March 1995. Heat Mining the Earth's Largest Power Source. *Geothermal Fact Sheet.* Washington, D.C.

39. Diment, W. H.; Urban, T. C.; Sass, J. H.; Marshall, B. V.; Munroe, R. J.; and Lachenbruch, A. H. 1975. Temperatures and Heat Contents Based on Conductive Transport of Heat. In White, D. E., and Williams, D. L. (eds.). *Assessment of Geothermal Resources of the United States.* U.S. Geological Survey Circular 726, 84–103.

40. Wright, Phillip Michael. March 1995. Personal communication. Director, Earth Science Laboratory, University of Utah Research Institute.

41. Energy Information Administration. U.S. Department of Energy. September 1991. *Geothermal Energy in the Western United States and Hawaii: Resources and Projected Electricity Generation Supplies.* EIA/DOE-0544.

42. Mock, Ted. March 13, 1995. Telephone interview. The costs of geothermal power vary according to the characteristics of the reservoir and its fluids. Drilling and related infrastructure, such as roads, typically are the largest cost components at 30–60 percent of

total costs. Construction of the power plant itself is the second most important cost, ranging from 30–50 percent. See Palmerini, Civis G. 1993. "Geothermal Energy." Chapter 13 in Johansson, Thomas B.; Kelley, Henry; Reddy, Amulya K. N.; Williams, Robert H.; and Burnham, Laurie (eds.). *Renewable Energy: Sources for Fuels and Electricity.* Washington, D.C. and Covelo, Calif.: Island Press.

43. U.S. Department of Energy. April 1994. Geothermal: A Renewable Energy Success Story. *Geothermal Fact Sheet.* Washington, D.C. U.S. Department of Energy. March 1995. Geothermal Electric Power Systems. *Geothermal Fact Sheet.* Washington, D.C.

44. An exception to the geothermal industry's general domestic stagnation today is the partnership of Calpine Corporation of San Jose, California, with Trans-Pacific Geothermal Corporation of Oakland, California, to build a 30-megawatt geothermal power plant at Glass Mountain in northern California, for 1998 operation. (Glass Mountain is among the largest well explored but undeveloped geothermal resources in the United States.) California Energy Company, the successor to Magma Power, also holds a major interest in the Glass Mountain field, which Union Oil discovered in 1984. See Geothermal Field Near Oregon Nears Development. November 1994. *Geothermal Resources Council Bulletin.*

45. U.S. Department of Energy. March 1995. Heat Mining the Earth's Largest Power Source. *Geothermal Fact Sheet.* Washington, D.C.

46. Wright, Phillip Michael. May 1995. The Sustainability of Production from Geothermal Resources. Paper submitted to the World Geothermal Congress in Florence, Italy.

47. Walter, Ron. March 14, 1995. Vice President, Calpine Corporation. Address to Geothermal Program Review XIII, U.S. Department of Energy. San Francisco, California.

48.–50. Duchane, David. March 13, 1995. Telephone interview.

Chapter 21

1. National Renewable Energy Laboratory. March 1993. Energy Efficiency Strengthens Local Economies. *Tomorrow's Energy Today.* DOE/CH10093-172. DE9216427.

2. Bevington, Rick, and Rosenfeld, Arthur H. September 1990. Energy for Buildings and Homes. *Scientific American.* Vol. 263, no. 3.

3. Fickett, Arnold P.; Gellings, Clark W.; and Lovins, Amory B. September 1990. Efficient Use of Electricity. *Scientific American.* Vol. 263, no. 3. For exhaustive technical documentation with recent data, see the Technology Atlas Series and Strategic Issues Papers of E Source, Inc., Boulder, Colorado, and publications of the Rocky Mountain Institute, Snowmass, Colorado. E Source is a membership-based, syndicated information service serving utilities, industry, government, and other professionals with annual fees of $1,500 to $14,000. The Institute is a nonprofit organization with many excellent, inexpensive, and readable publications.

4. National Academy of Sciences. 1992. *Policy Implications of Greenhouse Warming: Mitigation, Adaptation, and the Science Base.* Washington, D.C.: National Academy Press.

5. Bevington, Rick, and Rosenfeld, Arthur H. September 1990. Energy for Buildings and Homes. *Scientific American.* Vol. 263, no. 3.

6. Lawrence Berkeley Laboratory. March 1995. From the Lab to the Marketplace: Making America's Buildings More Energy Efficient. University of California, Berkeley, California; Rosenfeld, Arthur; Atkinson, Celina; Koomey, Jonathan; Meier, Alan; Mowris, Robert J.; and Price, Lynn. January 1993. Conserved Energy Supply Curves for U.S.

Buildings. *Contemporary Policy Issues.* Vol. XI. Gadgil, Ashok; Rosenfeld, Arthur H.; and Price, Lynn. October 21–25, 1991. *Making the Market Right for Environmentally Sound Energy-Efficient Technologies: U.S. Buildings Sector Successes That Might Work in Developing Countries and Eastern Europe.* Lawrence Berkeley Laboratory Report LBL-31701. Berkeley, California. Presented at ESETT '91: International Symposium on Environmentally Sound Energy Technologies and Their Transfer to Developing Countries and European Economies in Transition, Milan, Italy.

7. Rosenfeld, Arthur H.; Atkinson, Celina; Koomey, Jonathan; Meier, Alan; Mowris, Robert J.; and Price, Lynn. January 1993. Conserved Energy Supply Curves for U.S. Buildings. *Contemporary Policy Issues.* Vol. XI.

8. Ervin, Christine. April 11, 1995. Assistant Secretary, Energy Efficiency & Renewables. U.S. Department of Energy. Do We Still Need Energy Efficiency? Speech to a meeting sponsored by Carrier Corporation in Los Angeles.

9. Romm, Joseph J., and Browning, William D. December 1994. *Greening the Building and the Bottom Line: Increasing Productivity Through Energy-Efficient Design.* Snowmass, Colo.: Rocky Mountain Institute.

10. Romm, Joseph J., and Browning, William D. December 1994. *Greening the Building and the Bottom Line: Increasing Productivity Through Energy-Efficient Design.* Snowmass, Colo.: Rocky Mountain Institute.

11. Browning, Bill. June 1992. NMB Bank Headquarters. *Urban Land.*

12. Bevington, Rick, and Rosenfeld, Arthur H. September 1990. Energy for Buildings and Homes. *Scientific American.* Vol. 263, no. 3.

13. Flanigan, Ted. August 1990. Energy Efficiency: *Economic and Environmental Profit.* IRT Quarterly Supplement. The Results Center.

14. Goldstein, David B., and Rosenfeld, Arthur H. December 24, 1975. *Projecting an Energy-Efficient California.* Lawrence Berkeley Laboratory Report LBL-3274. Berkeley, California. See also Goldstein, David B., and Rosenfeld, Arthur H. December 1975. *Conservation and Peak Power-Cost and Demand.* Lawrence Berkeley Laboratory Report LBL-4428.

15. One could argue that the avoided investments might have been a great deal larger. The Diablo Nuclear Power Plant alone cost $5 billion.

16. Geller, Howard S. 1995. *National Appliance Efficiency Standards: Cost-Effective Federal Regulations.* Washington, D.C.: American Council for an Energy-Efficient Economy.

17. Geller, Howard S. 1995. *National Appliance Efficiency Standards: Cost-Effective Federal Regulations.* Washington, D.C.: American Council for an Energy-Efficient Economy.

18. More information about SERP can be had from the office of the SERP Administrator, 2856 Arden Way, Suite 200, Sacramento, CA 95825, telephone (916) 974-3981.

19. Rosenfeld, Arthur H. April 17, 1991. *The Role of Federal Research and Development in Advancing Energy Efficiency.* Statement before Subcommittee on Environment, Committee on Science, Space, and Technology. Hearing on U.S. Department of Energy Conservation Budget Request. U.S. House of Representatives, Washington, D.C. See also Technologies Developed at the Center for Building Science, Lawrence Berkeley Laboratory Under the Direction of Arthur H. Rosenfeld. (Undated; unsigned.) Informal memorandum.

20. Lawrence Berkeley Laboratory. U.S. Department of Energy. March 1995. *From the Lab to the Marketplace: Making America's Buildings More Energy Efficient.* Berkeley, Calif.: Lawrence Berkeley Laboratory.

21. Lawrence Berkeley Laboratory. U.S. Department of Energy. March 1995. *From the Lab to the Marketplace: Making America's Buildings More Energy Efficient.* Berkeley, Calif.: Lawrence Berkeley Laboratory.

22. Lawrence Berkeley Laboratory. U.S. Department of Energy. March 1995. *From the Lab to the Marketplace: Making America's Buildings More Energy Efficient.* Berkeley, Calif.: Lawrence Berkeley Laboratory.

23. Rosenfeld, Arthur H.; Akbari, Hashem; Bretz, Sarah; Fishman, Beth L.; Kurn, Dan M.; Sailor, David; and Haider, Taha. April 1995. *Mitigation of Urban Heat Islands: Materials, Utility Programs, Updates.* Lawrence Berkeley Laboratory LBL-36587. Berkeley, Calif.: Lawrence Berkeley Laboratory. To appear in *Energy and Buildings,* 1995. See also Rosenfeld, Arthur H.; Akbari, Hashem; and Chen, Allan. April 27, 1995. *Climate Change Action Plan No. 9: Cool Communities.* Energy and Environment Division. Center for Building Science. Lawrence Berkeley Laboratory. For a nontechnical treatment, see Mestel, Rosie. March 25, 1995. White Paint on a Hot Tin Roof. *New Scientist.* Vol. 42, no. 3.

24. Similar results from experiments with reflective roof coatings were obtained by researchers from the Florida Solar Energy Center in Cape Canaveral.

25. Rosenfeld, Arthur; Gadgil, Ashok; and Chen, Allan. April 27, 1995. *Global Energy Efficiency and Technologies to Further Its Progress.* Unpublished draft.

26. Pacific Gas and Electric Company. April 1995. *Commitment to Environmental Quality: Pacific Gas and Electric Company 1994 Environmental Report.* San Francisco, California.

27. Information on the ACT2 program comes primarily from *Facts on ACT2,* issues 1–18, 1990–1995, published by Pacific Gas and Electric's ACT2 project and from interviews with Grant Brohard and George Hernandez, engineers in the Customer Systems Group, at the company's Research and Development Division, and from numerous company Fact Sheets about ACT2.

Chapter 22

1. With chillers costing about $1,500 a ton, that meant $1.2 million in capital savings. A ton of cooling capacity is the amount of cooling necessary to remove 12,000 BTUs.

2. Lovins, Amory B. August 1992. *Energy-Efficient Buildings: Institutional Barriers and Opportunities.* Boulder, Colo.: Competitek.

3. Lovins, Amory B. May 26, 1995. Telephone interview.

4. Lovins, Amory B. August 1992. *Energy-Efficient Buildings: Institutional Barriers and Opportunities.* Boulder, Colo.: Competitek.

5. Levine, M.; Gadgil, A.; Meyers, S.; Sathaye, J.; Stafurik, J.; Wilbanks, T. 1991. Energy Efficiency, Developing Nations, and Eastern Europe, A Report to the U.S. Working Group on Global Energy Efficiency. Washington, D.C.: International Institute for Energy Conservation.

6. Jechoutek, Carl. June 21, 1995. The World Bank. Telephone interview.

7. Clifford, Mark. August 1, 1991. Bright Lights, Dim Cities. *Far Eastern Economic Review.*

8. and 9. Gadgil, A. J. 1994. *Introducing Energy Efficient Technologies in Developing Countries.* In de Almeida, A. T., et al. (eds.). Integrated Electricity Resource Planning. Kluwer Academic Publishers. Printed in the Netherlands.

10. Rosenfeld, Arthur; Gadgil, Ashok; and Chen, Allan. April 27, 1995. *Global Energy Efficiency and Technologies to Further Its Progress.* Unpublished draft.

11. Gadgil, Ashok; Rosenfeld, Arthur; and Price, Lynn. *Making the Market Right for Environmentally Sound Energy-Efficient Technologies: U.S. Buildings Sector Successes That Might Work in Developing Countries and Eastern Europe.* Lawrence Berkeley Laboratory Report LBL-31701. Berkeley, Calif.: Lawrence Berkeley Laboratory. The energy savings are even greater than in the United States because of added inefficiencies in the Indian power sector. See Gadgil, Ashok; Rosenfeld, Arthur; Arasteh, Dariush; and Ward, Ellen. *Advanced Lighting and Window Technologies for Reducing Electricity Consumption and Peak Demand: Overseas Manufacturing and Marketing Opportunities.* Lawrence Berkeley Laboratory Report LBL-30389 (Rev.). Berkeley, Calif.: Lawrence Berkeley Laboratory.

12. Gadgil, Ashok; Rosenfeld, Arthur; and Price, Lynn. *Making the Market Right for Environmentally Sound Energy-Efficient Technologies: U.S. Buildings Sector Successes That Might Work in Developing Countries and Eastern Europe.* Lawrence Berkeley Laboratory Report LBL-31701. Berkeley, Calif.: Lawrence Berkeley Laboratory.

13. Gadgil, Ashok, and Anjali, Sastry M. February 1994. *Stalled on the Road to Market: Lessons from a Project Promoting Lighting Efficiency in India. Energy Policy.*

14. Suchontan, Cimi. March 8, 1993. Cities "Misuse" Water, Power. *Bangkok Post.*

Chapter 23

Epigraph from June 7, 1993. *Automotive News,* p. 10i.

1. These might include higher gasoline taxes, smog fees, pay-at-the-pump auto insurance, or other measures.

2. Hwang, Roland; Miller, Marshall; Thorpe, Ann B.; and Lew, Debbie. November 1994. Rev. Ed. *Driving Out Pollution: The Benefits of Electric Vehicles.* Cambridge, Mass., and Berkeley, Calif.: Union of Concerned Scientists.

3. All vehicles will cause some emissions produced by the energy consumed in their materials and manufacture, that is, unless the entire economy operates on fuel-free renewables.

4. Hwang, Roland; Miller, Marshall; Thorpe, Ann B.; and Lew, Debbie. November 1994. Rev. Ed. *Driving Out Pollution: The Benefits of Electric Vehicles.* Cambridge, Mass., and Berkeley, Calif.: Union of Concerned Scientists.

5. DeLucchi, Mark; Sperling, Daniel; and Johnston, Robert. 1987. *A Comparative Analysis of Future Transportation Fuels.* Research Report UCB-ITS-RR-87-13. Institute of Transportation Studies, University of California, Berkeley, California. (More recent versions of this study are now available.)

6. According to figures provided by the Union of Concerned Scientists, the average car, driven over an average vehicle lifetime of 13 years will produce nearly 20 tons of CO_2. Hwang, R., et al. November 1994. Rev. Ed. *Driving Out Pollution: The Benefits of Electric Vehicles.* Cambridge, Mass., and Berkeley, Calif.: Union of Concerned Scientists. The statistics used in the text come from the *Information Please Environmental Almanac.*

7. Speech by John E. Bryson, Chairman and Chief Executive Officer of Southern California Edison. Presented at the 12th Electric Vehicles Symposium (12-EVS) held December 5–7, 1994, in Anaheim, California, under the auspices of the Electric Vehicle Association of the Americas (EVAA).

8. State of California. California Environmental Protection Agency. Air Resources Board. Mobile Source Division. 1994. *Staff Report: 1994 Low-Emission Vehicle and Zero-Emission Vehicle Program Review.*

9. Lovins, Amory, and Lovins, Hunter L. January 1995. Reinventing the Wheel. *The Atlantic Monthly*. Vol. 273, no. 1

10. McCrea, S., and Minner, R. (eds.). 1992. Why Wait for Detroit? Drive an Electric Car Today! *EV Myths*. Ft. Lauderdale, Fla.: South Florida Electric Auto Association.

11. While the efficiency of an AC electric motor may be roughly 95 percent, the range of efficiency losses in the electronic motor controller, transmission (or gearing), charger, and battery of the car reduce overall vehicle efficiency to about two thirds, measured from charging plug to wheels, a value that rises to about three quarters when credit is given for energy recovered from braking. Efficiency can also be measured from the oil wellhead or power plant to the vehicle's wheels for an efficiency comparison over the entire gasoline and electric fuel cycles. Most notably, if the energy for producing the electricity that charges an electric vehicle is produced by the combustion of fossil fuels, as much as two-thirds of the energy in the fuel is wasted in converting the fuel to electricity in older thermal power plants. However, even under these conditions, the electric vehicle is still one and a half to about two times as efficient as a gasoline car over their whole fuel cycles. By contrast, when the electricity for an electric vehicle is renewably produced by wind, there are no Carnot losses due to combustion, and the electric vehicle is then once again four to five times as efficient as the gasoline vehicle compared over their respective fuel cycles.

12. The system temporarily converts the vehicle's electric motor to an electric generator each time the car's brakes are applied. The car's kinetic energy (energy of motion) is imparted to the armature of the motor to produce electricity, which is sent to the battery, where it is stored, extending the vehicle's range. Regenerative braking captures roughly 15 percent of the battery energy used in urban driving, although the amount can be quite variable, depending on the details of the design of the braking and battery systems and on the requirements of the driving cycle. (The driving cycle is characterized by the acceleration, speed, and braking requirements of the course.) Regenerative braking systems in the future may recapture as much as 70 percent of a car's kinetic energy.

13. and 14. Adler, Dennis. Winter 1994. A Brief Look . . . at Automotive Man's Investigations into Electricity. *Electric Car*. Vol. 1, no. 1.

15. Frank, Len. Winter 1994. Driving the Future. *Electric Car*. Vol. 1, no. 1.

16. Wald, Matthew L. March 6, 1994. In Quest for Electric Cars, He Adds the Power of Faith. *New York Times*.

17. This would be a great step beyond the firm's 1995 offerings, a converted Geo Metro known as the Solectria Force, and a converted Chevy S-10 pickup. The least expensive version of the Metro equipped with sealed lead-acid batteries costs $31,495 and can go only 60 miles at 45 m.p.h., while a Force with nickel-cadmium batteries can travel 100 miles at 45 m.p.h. but costs an imposing $48,494. The young company has declined to release sales information but reportedly now does about $4 million worth of business a year.

18. Wald, Matthew L. December 2, 1994. Advances in Electric Cars for Masses. *New York Times*.

19. Pearlman, Chee. November–December 1990. Inventing Efficiency. *International Design*.

20. and 21. Jaroff, Leon. June 11, 1990. He Gives Wings to Dreams. *Time*. June 11.

22. Pearlman, Chee. November–December 1990. Inventing Efficiency. *International Design*.

23. Unsigned. April 23, 1990. Paul MacCready: Driving Ahead. *U.S. News and World Report*.

24. Freedman, David H. March 1992. Batteries Included. *Discover.* Vol. 13, no. 3.

25. Jaroff, Leon. June 11, 1990. He Gives Wings to Dreams. *Time.*

26. Cook, William J. August 26–September 2, 1991. The Soul of a New Machine. *U.S. News and World Report.*

27. Freedman, David H. March 1992. Batteries Included. *Discover.* Vol. 13, no. 3.

28. Cook, William J. August 26–September 2, 1991. The Soul of a New Machine. *U.S. News and World Report.*

29. Considine, Tim. Winter 1994. The Cocconi Factor. *Electric Car.* Vol. 1, no. 1.

30.–32. Cook, William J. August 26–September 2, 1991. The Soul of a New Machine. *U.S. News and World Report.*

33. McCarty, Lyle H. August 26, 1991. Lessons from an Aeronautics Master. *Design News.* Vol. 47, no. 16.

34. Freedman, David H. March 1992. Batteries Included. *Discover.* Vol. 13, no. 3.

35. Hampson, Bruce. Winter 1994. Making an Impact on the Electric Car Market. *Electric Car.* Vol. 1, no. 1.

36. The car is charged by an inductive coupling system that imparts a charge without direct contact between a plug and socket. Hampson, Bruce. Winter 1994. Making an Impact on the Electric Car Market. *Electric Car.* Vol. 1, no. 1.

37. Unsigned. Winter 1994. Edison's Electrics. *Electric Car.* Vol. 1, no. 1.

38. Hampson, Bruce. Winter 1994. Making an Impact on the Electric Car Market. *Electric Car.* Vol. 1, no. 1.

39. Considine, Tim. Winter 1994. The Cocconi Factor. *Electric Car.* Vol. 1, no. 1.

40. Baker, K. R. December 5, 1994. The EV Commercialization Challenge: Meeting Customer, Company, and Infrastructure Requirements. Address to the 12th International Electric Vehicle Symposium held in Anaheim, California.

41. and 42. State of California. California Environmental Protection Agency. Air Resources Board. Mobile Source Division. 1994. *Staff Report: 1994 Low-Emission Vehicle and Zero-Emission Vehicle Program Review.* The California Zero Emission Vehicle Program. El Monte, California.

43. Small, fuel-efficient gasoline vehicles today, of course, can go as much as 350 miles on a tank.

44. Kurani, K.; Sperling, D.; and Turrentine, T. December 12–14, 1995. Consumer Attitudes Towards EVs. Institute of Transportation Studies, University of California, Davis. North American EV and Infrastructure Conference. Electric Power Research Institute (and other sponsors).

45. Lave, L. B.; Hendrickson, C. T.; and McMichael, F. C. May 19, 1995. Environmental Implications of Electric Cars. *Science.* Vol. 269. Passell, Peter. May 19, 1995. Lead-based Battery Used in Electric Car May Pose Hazards. *New York Times.*

46. Electric Vehicle Association of America. 1995. Environmental Impacts of Lead-Acid Batteries in Electric Vehicles. Technical Brief TB1995-1. San Francisco, California.

47. Mader, Jerry. September 1994. Jerry Mader & Associates. *Electric Vehicle Battery Development Status.* District Contract Number 95020.

48. Wald, Matthew L. July 11, 1993. Coaxing More Miles from the Electric-Car Battery. *New York Times.*

49. State of California. California Environmental Protection Agency. Air Resources Board. Mobile Source Division. 1994. *Staff Report: 1994 Low-Emission Vehicle and Zero-Emission Vehicle Program Review.* The California Zero Emission Vehicle Program. El Monte, California.

50. Ovshinsky, S. R.; Fetcenko, M. A.; and Ross, J. April 9, 1993. A Nickel Metal Hydride Battery for Electrical Cars. *Science.* Vol. 260.

51. State of California. California Environmental Protection Agency. Air Resources Board. Mobile Source Division. 1994. *Staff Report: 1994 Low-Emission Vehicle and Zero-Emission Vehicle Program Review.* The California Zero Emission Vehicle Program. El Monte, California.

52. SAFT. 1993. *The Electric Vehicle Comes of Age.* SAFT Communications Department. Romainville, France.

53. Wald, Matthew L. June 22, 1994. Flywheels to Power Vehicles. *New York Times.*

54. Ris, Howard. Executive Director, Union of Concerned Scientists. April 1994. Letter to Sponsor.

55. Unsigned. June 7, 1994. Flywheel Funds. *Automotive News,* p. 2i.

56. Brusaglino, G. December 6, 1994. Flywheel Benefits. *EVS 12 Automotive News.* Vol. 1, no. 2. Anaheim, California. Unsigned. December 6, 1994. Toybox: Flywheels & Ultracapacitors. *EVS 12 Automotive News.* Vol. 1, no. 2. Anaheim, California.

57. Perlman, David. September 30, 1994. Weapons Labs Start Working on Cars. *San Francisco Chronicle.*

58. Wald, Matthew L. June 29, 1994. A Battery Built on Wheel Power. *New York Times.*

59. Unsigned. Spring 1994. Electric Vehicles: Batteries, Cars & Supercars. *Rocky Mountain Institute Newsletter.* Vol. 10, no. 1.

60. Rechtin, Mark. December 6, 1994. How Many People Will Buy? *EVS 12 Automotive News.* Vol. 1, no. 2. Anaheim, California.

Chapter 24

1. Cackette, T., and Evashenk, T. December 12–14, 1995. A New Look at Hybrid-Electric Vehicles in Meeting California's Clean Air Goals. North American EV and Intrastructure Conference. Electric Power Research Institute (and other sponsors).

2. Sly, James. Winter 1994. The Best of Both Worlds. *Electric Car.* Vol. 1, no. 1.

3. Lovins, Amory, and Lovins, Hunter L. January 1995. Reinventing the Wheel. *The Atlantic Monthly.* Vol. 273, no. 1. Lovins, Amory. Spring 1994. Supercars: Advanced Ultralight Hybrid Vehicles. Peer-review reprint of article in the 1995 *Encyclopedia of Energy Technology and the Environment.* New York: John Wiley & Sons.

4. Lovins, Amory, and Lovins, Hunter L. January 1995. Reinventing the Wheel. *The Atlantic Monthly.* Vol. 273. Lovins, Amory; Barnett, John W.; and Lovins, Hunter L. February 1993, Rev. 1995. *Hypercars: The Next Industrial Revolution.* The Hypercar Center, Rocky Mountain Institute, Snowmass, Colorado.

5. Lovins, Amory, and Lovins, Hunter L. January 1995. Reinventing the Wheel. *The Atlantic Monthly.* Vol. 273, no. 1. Lovins, Amory et al. February 1995. *Hypercars: Materials and Policy Implications.* The Hypercar Center, Rocky Mountain Institute, Colorado.

6. Lovins, Amory. Spring 1994. Supercars: Advanced Ultralight Hybrid Vehicles. Peer-review reprint of article in the 1995 *Encyclopedia of Energy Technology and the Environment.* New York: John Wiley & Sons, see pages 17–18.

7. The decision would need to be based on a finding that Hypercars' net contribution to air pollution was less than a "ZEV" battery car charged from California's grid.

8. Electric Vehicles: Battery Cars & Supercars. Spring 1994. *Rocky Mountain Institute Newsletter.* Rocky Mountain Institute, Snowmass, Colorado. Vol. 10, no. 1.

9. Clark W. Gellings, Vice President, Electric Power Research Institute stated in more optimistic remarks to the North American Electric Vehicle and Infrastructure Conference, December 12, Atlanta, Georgia, that electric vehicles offer utilities a profit potential of $2,000 per car over five years.

10. Unsigned. December 5, 1994. *EVS 12 Automotive News.* Vol. 1, no. 1.

11. Meyer, Harvey. Fall 1994. Charged Up over Electric Cars. *GEICO Direct.*

12. Unsigned. May 13, 1994. Doubts on Electric Car Rise. *New York Times.*

13. Commenting on the automakers' position in a front-page feature article, a *Wall Street Journal* reporter wrote, "To try to prevent the electric-car movement from establishing itself in the Northeast, the Big Three have been promoting their own clean-air plan, one they maintain would be more palatable to consumers and easier for the car companies to meet. And in California, a bill has been introduced [with auto industry support] that attacks the state's zero-emission mandate." Suris, Oscar. January 24, 1994. Californians Collide with Folks in Detroit over the Electric Car. *Wall Street Journal.*

14. State of California. California Environmental Protection Agency. Air Resources Board. Mobile Source Division. 1994. *Staff Report: 1994 Low-Emission Vehicle and Zero-Emission Vehicle Program Review.* The California Zero Emission Vehicle Program. El Monte, California.

15. Auto Dealers Meet over Economic Impact of Electric Vehicles Mandate. March 22, 1995. *E-Wire.*

16. and 17. Sperling, Daniel. Winter 1994–1995. Gearing Up for Electric Cars. *Issues in Science and Technology.*

18. Hwang, Roland; Miller, Marshall; Thorpe, Ann B.; and Lew, Debbie. November 1994. Rev. Ed. *Driving Out Pollution: The Benefits of Electric Vehicles.* Cambridge, Mass., and Berkeley, Calif.: Union of Concerned Scientists.

19. State of California. 1996. Department of Motor Vehicles. Vehicle registration data via telephone interview.

20. The federal government already offers a 10 percent tax credit for the purchase of an electric vehicle up to a maximum of $4,000. The credit applies to vehicles bought between June 30, 1993, and January 1, 2005; however, the amount of the credit diminishes by one-quarter a year, beginning in 2002.

In California, the state offers a credit worth up to $1,000 on the *incremental* cost of a low-emission automobile or motorcycle (or installation of a retrofit device) *over* the cost of a gasoline-powered model. Similarly, vehicles larger than 5,750 pounds are eligible for up to $3,500 in state tax credits on incremental purchase and retrofit costs. These state credits are reduced by the amount of the federal low-emission vehicle tax credit if both are in effect simultaneously. The state also offers a sales tax exemption for low-emission vehicles, based again on their incremental costs. Oregon offers a 30 percent tax credit to businesses that purchase electric vehicles.

21. Motavali, Jim. July–August 1994. California Dreamin'. *E Magazine.*

22. Unsigned. May 16, 1994. In Chattanooga, Electric Buses Are Here Now. *Passenger Transport.* Vol. 52, no. 20.

23. Pearson, Curtis, and Young, Mack. n.d. "Electric and Electric Hybrid Drives for Bus Applications." Unpublished paper produced by Westinghouse Electric Corporation, Baltimore, Maryland.

24. The Baker EV100 can carry a 1,350 pound payload and can travel 70 m.p.h.; however, it takes 12 seconds to get from 0 to 50 m.p.h. Unsigned. December 5, 1994. *EVS 12 Automotive News.* Vol. 1, no. 1.

25. Unsigned. December 5, 1994. *EVS 12 Automotive News.* Vol. 1, no. 1.

26. Lamarre, Leslie. October–November 1994. airport rEVolution. *EPRI Journal.*

27. French Technology Press Office, Inc. Release 94-02-022 citing a report by the French Environmental Institute.

28. Gates, Max. December 5, 1994. Acid Test. *Automotive News,* p. 1i.

29. French Technology Press Office. July 5, 1994. Release 94-02-022.

30. French Technology Press Office. September 21, 1994. Release 94-08-107.

31. and 32. Schreffler, Roger. July–September 1993. Japan Gets the Lead Out, Slowly. *Tomorrow.* Vol. 3, no. 3.

33. Wald, Matthew L. January 2, 1994. A Cleaner Motor Scooter. *New York Times.*

Chapter 25

Epigraph from Alliance to Save Energy, American Council for an Energy-Efficient Economy, Natural Resources Defense Council, Union of Concerned Scientists. 1991. *America's Energy Choices: Investing in a Strong Economy and a Clean Environment.* Cambridge, Mass.: Union of Concerned Scientists.

1. The poll "Energy: Post-Election Views" was conducted by Dr. Vincent J. Breglio, of Research/Strategy/Management, a leading Republican consulting firm. Breglio directed polling for the Reagan/Bush campaigns of 1980 and 1984. Copies of the poll may also be obtained from the SUN DAY Campaign of Takoma, Maryland.

2. Lashof, Daniel A. May 31, 1993. The BTU Tax: A Revenue Source That Fights Pollution. *Tax Notes.* Given immense political opposition from the fossil fuel industry, neither a carbon tax nor a BTU tax is likely in the foreseeable future. If political conditions change, it would be less problematic initially to implement the proposal as an across-the-board BTU tax that would simply exempt renewable noncombustion technologies. The tax could be coupled with a package of benefits aimed at low-income households so as to offset the income effect of the tax. Alliance to Save Energy, American Council for an Energy-Efficient Economy, Natural Resources Defense Council, Union of Concerned Scientists. 1991. *America's Energy Choices: Investing in a Strong Economy and a Clean Environment.* Cambridge, Mass.: Union of Concerned Scientists.

3. Alliance to Save Energy, American Council for an Energy-Efficient Economy, Natural Resources Defense Council, Union of Concerned Scientists. 1991. *America's Energy Choices: Investing in a Strong Economy and a Clean Environment.* Cambridge, Mass.: Union of Concerned Scientists. U.S. Congress. Office of Technology Assessment. September 1995. *Renewing Our Energy Future.* Washington, D.C.

SOURCES

Chapter 5

Annan, Robert H. April 1989. Photovoltaic Solar Approaches Role as Peaking Power Producer. *Power Engineering*.

Ashley, Steven. January 1992. Solar Photovoltaics: Out of the Lab and onto the Production Line. *Mechanical Engineering*. Vol. 114, no. 1, pp. 48–55.

Awerbuch, Shimon. October 1993. *Methodologies to Analyze the Economics of Competing Energy Technologies: A Framework for Evaluating Photovoltaics.* Report to Sandia National Laboratory. Contract 67-4226.

Braun, Gerald W.; Suchard, Alexandra; and Martin, Jennifer. 1991. Hydrogen and Electricity as Carriers of Solar and Wind Energy for the 1990s and Beyond. *Solar Energy Materials.* 24. North Holland, pp. 62–75.

Carlson, D. E. 1990. Photovoltaic Technologies for Commercial Power Generation. *Annual Review of Energy 1990.* 15, pp. 85–98.

Green, Martin A. 1989. Highlights of the Fourth International Photovoltaic Science and Engineering Conference (PVSEC-4). *Solar Cells.* 27, pp. 3–10.

Henderson, Breck W. October 23, 1989. Boeing Achieves Major Advance in Space Solar Cell Efficiency. *Aviation Week & Space Technology.*

Hubbard, H. M. April 21, 1989. Photovoltaics Today and Tomorrow. *Science.* Vol. 244, pp. 297–304.

Moore, Taylor. December 1992. High Hopes for High Power Solar. *EPRI Journal.* Vol. 17, no. 8, pp. 16–25.

Pool, Robert. August 19, 1988. A Bright Spot on the Solar Scene. *Science.* Vol. 241, pp. 900–901.

Pool, Robert. August 19, 1988. Solar Cells Turn 30. *Science.* Vol. 241, pp. 900–901.

Swan, Christopher C. March 1988. Light-Powered Architecture. *Architectural Record.* Vol. 176, no. 3, pp. 126–131.

Unsigned. June 24, 1994. Photovoltaic Bidders Please SMUD with Lower Prices Than Last Year. *The Solar Letter.* Vol. 4, no. 14.

Williams, Neville; Jacobson, Ken; and Burris, Harold. April 22, 1993. Sunshine for Light in the Night. *Nature.* Vol. 362, no. 6422.

Chapter 9

Tilford, C.; Sinton, R. A.; Swanson, R. M.; Crane, R. A.; and Verlinden, P. J. May 10–14, 1993. Development of a 10 kW Reflector Dish PV System. In the *Conference Record of the Twenty Third IEEE Photovoltaic Specialists Conference—1993*. Louisville, Kentucky.

Verlinden, P. J.; Swanson, R. M.; and Crane, R. A. April 1994. 7000 High Efficiency Cells for a Dream. *Progress in Photovoltaics: Research and Applications*. Vol. 2, no. 2, pp. 143–152.

Verlinden, P. J.; Swanson, R. M.; Sinton, R. A.; Crane, R. A.; Tilford, C.; Perkins, J.; and Garrison, K. May 10–14, 1993. High-Efficiency, Point-Contact Silicon Solar Cells for Fresnel Lens Concentrator Modules. In the *Conference Record of the Twenty Third IEEE Photovoltaic Specialists Conference—1993*. Louisville, Kentucky. *SunPower Corporation Company Profile*. n.d. SunPower Corporation, 430 Indio Way, Sunnyvale, CA 94086.

Verlinden, P. J.; Swanson, R. M.; and Crane, R. A. October 7–11, 1991. Single-Wafer Integrated 140 W Silicon Concentrator Module. In the *Conference Record of the Twenty Second IEEE Photovoltaic Specialists Conference—1991*. Las Vegas, Nevada.

Wald, Matthew L. December 4, 1994. Thirsty New Solar Cells Drink in the Sun's Energy. *New York Times*.

Chapter 11

Lotker, Michael. n.d. *A Structure for Commercialization of Utility Scale Photovoltaic Systems*. Ormat Incorporated. Westlake Village.

McEvoy, A. J. 1993. Outlook for Photovoltaic Electricity. *Endeavor*. New Series. Vol. 17, no. 1. pp. 17–20.

Strong, Steven J. March–April 1983. Photovoltaics Down to Earth. *Public Power*, pp. 18–21.

Chapter 20

Geothermal Division, U.S. Department of Energy. March 3, 1995. *Geothermal Energy Draft Multi-Year Program Plan, FY 1996–2000*. Washington, D.C.

McLarty, Lynn, and Reed, Marshall J. 1992. The U.S. Geothermal Industry: Three Decades of Growth. *Energy Sources*. Vol. 14, pp. 443–455.

Roane, Kit R. September 21, 1994. Magma Shares Surge 26 Percent on Takeover Offer by Rival. *New York Times*.

Wright, Phillip Michael. May 1995. "The Sustainability of Production from Geothermal Resources." Paper submitted to the World Geothermal Congress, Florence, Italy.

Chapter 22

Note: The references below pertain to the Green Lights and Energy Star Programs, the Golden Carrot, and the AgSTAR program.

U.S. Environmental Protection Agency. Office of Air and Radiation. August 1994. *The Climate Is Right for Action: Voluntary Programs to Prevent Atmospheric Pollution*.

EPA 430-K-94-004. More than a dozen other brief EPA publications were also consulted, several of which are described in this reference. Further information can be obtained from Green Lights and Energy Star Programs, U.S. EPA, 401 M Street, S.W., Washington, DC 20460.

U.S. Environmental Protection Agency. Office of Air and Radiation. July 1993. *Green Lights: An Enlightened Approach to Energy Efficiency and Pollution Prevention.* EPA 430-K-93-001.

Chapter 25

U.S. Congress. Office of Technology Assessment. September 1995. *Renewing Our Energy Future.* Washington, D.C. A superb report by a valuable government agency that was abolished by Congress in 1995.

ACKNOWLEDGMENTS

I have been exceedingly fortunate in having had the help of many friends, colleagues, and family while writing *Charging Ahead*. My most obvious debt is to the renewable energy pioneers who graciously shared their personal and professional histories. This book could not have been written without their accomplishments. The following busy individuals were exceptionally generous with their time and information: Dale W. Osborn, former President, Kenetech/U.S. Windpower, Inc.; Steven J. Strong, President, Solar Design Associates, Inc.; Ishaq M. Shahryar, Chairman and CEO, Solar Utility Company, and former President and CEO, Solec International, Inc.; and J. W. (Bill) Yerkes of Teledisk Corporation, former Manager, Electronics/Prototype Laboratory, Boeing Corporation and former President, ARCO Solar, Inc.

Carl Weinberg, former director of research for Pacific Gas and Electric Company, was gracious and kind enough to act as my principal technical adviser, reading and critiquing the entire manuscript. Whereas his broad interdisciplinary knowledge provided a bulwark against error, any remaining mistakes are my sole responsibility.

I am especially grateful for research support received from the Compton Foundation and from the Max and Anna Levinson Foundation, as well as for the sponsorship of the Pacific Environment and Resources Center.

I appreciate the cheerful optimism of my agent, Virginia Barber, who encouraged me in writing this book, provided insightful editorial guidance, and placed the book with Henry Holt and Company, where editors

Jack Macrae and Alison Juram shaped the book through astute editorial recommendations and patient developmental editing.

My warm thanks to James Levine, P.E., President, Levine-Fricke, Inc., at whose offices most of this book was written. Those congenial surroundings and friendly colleagues made for an excellent writing environment.

A number of high-quality energy and transportation conferences to which I was invited provided valuable information and contacts. These meetings were convened by the American Solar Energy Society, the American Wind Energy Society, the Electric Vehicle Association of the Americas, the Northeast Solar Energy Association, the Department of Energy's Geothermal Program Review, and the Responsive Energy Technology Symposium and International Exchange (RETSIE).

The following knowledgeable individuals were kind enough to provide me with helpful advice, information, or interviews: Joseph Aidlin, Esq.; Ross Ain, Esq., Van Ness Feldman Company; David C. Allen, California Biomass Energy Alliance; David Anderson, Geothermal Resources Council; Earl M. Anderson, Jackrabbit Corporation; Peter C. Aschenbrenner, AstroPower, Inc.; Dr. Shimon Awerbuch, College of Management, University of Massachusetts–Lowell; Jared Barlage, American Wind Energy Association; Newton D. Becker, Becker CPA Review; Michael L. S. Bergey, Bergey Windpower, Inc.; John Bigger, Electric Power Research Institute; Gerald Braun, Pacific Gas and Electric Company; Grant J. Brohard, P.E., Pacific Gas and Electric Company; Andrew F. Burke, Institute of Transportation Studies, University of California, Davis; Roy Castillo, Jackrabbit Corporation; Tony Catalano, National Renewable Energy Laboratory; Stuart Chaitkin, California Public Utilities Commission; Jamie Chapman, OEM Development Corporation; Anthony Chargin, Lawrence Livermore National Laboratory; Allan Chen, Lawrence Berkeley National Laboratory; Alan Chertok, formerly of U.S. Windpower, Inc.; Dr. G. Alan Comnes, Morse, Richard, Weisenmiller & Associates; Joel Davidson, Solec International, Inc.; Michael DeAngelis, California Energy Commission; Mark DeLucchi, Institute of Transportation Studies, University of California, Davis; Edward DeMeo, Electric Power Research Institute; Christine T. Donovan, C. T. Donovan Associates, Inc.; Dr. David Duchane, Los Alamos National Laboratory; Alvin Duskin, The Bering Company; Lance Elberling, Pacific Gas and Electric Company; Hap Ellis, Kenetech, Inc.; Michael Elliston, Carisso Plains Solar Corporation; Richard Farrell, Altamont Group; Richard G. Ferreira, Sacramento Municipal Utility District; Ted Flanigan, IRT Environment, Inc.; Dr. Harvey Forrest, Amoco-Enron/Solarex Corporation; Dr. Ashok Gadgil, Lawrence Berkeley National Laboratory; Dr. Charlie Gay, National Renewable Energy Laboratory; Paul Gipe, Paul Gipe and Associates; Finn Møeller Godtfredsen, Risø National Laboratory (Denmark); Arnold Goldman, LUZ International; Dave Goldstein, Program Development Associates; Dr. David B. Goldstein, Natural Resources Defense Council; Clarence (Bud) Grebe, Kenetech, Inc.; Derrick Grimmer, Iowa Thin Films; Christopher Gronbeck, Center for Renewable Energy and Sustainable Technology; Robert Guertin, Kenetech, Inc.; Garth C. Hall, Pacific Gas and Electric Company; Denis Hayes, Esq., Bullitt Foundation; George R. Hernandez, P.E., Pacific Gas and Electric Company; Dr. William Heronemus; Thomas C. Hinrichs, Pacific Energy Consultants, Inc.; John S. Hoffman, U.S. Environmental Protection Agency; Dr. William E. Holley, Kenetech, Inc.; Darcel L. Hulse, UNOCAL Corporation; Glenn Ikemoto, Kenetech, Inc.; Mary A. Ilyin, Pacific Gas and Electric Company; Michael B. Jacobs, Massachusetts Department of Public Utilities; Dr. Alan Jelacic, U.S. Department of Energy; Peter Hjuler Jensen, Risø National Laboratory (Denmark); John Jordan, Photon Energy; Tim Kasmerski, National Renewable Energy Laboratory; Kent Kaulfuss, Wood Industries Company; Dr. David W. Kearney, P.E., Kearney and Associates; Tami Kelly, U.S. Windpower/Kenetech, Inc.; Angelo Koniares, Advanced Photovoltaic Systems, Inc.; Daniel A. Lashoff, Natural Resources Defense Council; John Lewelling, Kenetech, Inc.; the late Dr. Joseph Lindmeyer, Quantex Corporation; Roger Little, Spire Corporation; Lee Eng Lock, Supersymmetry Services Pte. Ltd.; Michael Lotker, Ormat, Inc.; Amory B. Lovins, Rocky Mountain Institute; Robert Lynette, Advanced Wind Turbines, Inc.; Jerry Mader, Mader & Associates; Louis Manfredi, Kenetech, Inc.; Alvin Marks, Advanced Research Development, Inc.; Paul D. Maycock, Photovoltaic Energy Systems, Inc.; Kathy McGee, Solar Energy Industries Association; Dr. Charles R. McGowin, Electric Power Research Institute; Maureen McIntyre, *Solar Today;* Tandy McManess, Manager, Kramer Junction Operating Company; Steve R. Meyers, Lawrence Berkeley National Laboratory; William G. Miller, Biomass Processors Association; Ted Mock, U.S. Department of Energy, Geothermal Division; Dave Modisette, California Electric Transportation Coalition; Norman H. Moore, Kelmoore Investment Co.; Rob Morgan, Pacific Environment and Resource Center; Guy Nelson, U.S. Department of Energy, Area Power Administration; Donald E. Osborn, Sacramento Municipal Utility District; Dr. Carel Otte, former President, UNOCAL Geothermal Division; Ray Paz, Southern California Energy Co.; Ron Perkins, Supersymmetry Services USA; J. Ward Phillips; Nancy Rader, Independent Consultant (utility regulation); Mary Raftery, California Public Interest Group; Michelle Robinson, Union of Concerned Scientists; David Roodman, WorldWatch Institute; Dr. Art Rosenfeld, U.S. Department of Energy; Armin Rosencranz, Esq., Pacific Environment and Resource Center; Steve Rubin, Independent Consultant (energy information); Jeffrey Serfass, Utility Photovoltaic Group; Daniel S. Shugar, P.E., New World Power Co.; Scott Sklar, Solar Energy Industries Association; Ed Smeloff, Sacramento Municipal Utility District; Paul Smith, Kenetech, Inc.; Dr. Daniel Sperling, Institute of Transportation Studies, University of California, Davis; Dr. Woody Stoddard, Second Wind, Inc.; Dr. Raymond A. Sutula, U.S. Department of Energy; Dr. Richard M. Swanson, SunPower Corporation; Randy Tinkerman, Windpower Management/ Bluestar Products; Michael Totten, Center for Renewable Energy and Sustainable Technology; Tim Townsend, Endecon; Dr. Edward S. Van Dusen, Composite Engineering, Inc.; Dr. Peter F. Varadi, Consultant, Amoco/Solarex Corporation; Howard J. Wenger, Pacific Energy Group; Eric Wills, LUZ International; Dr. Robert Wills, P.E., Solar Design Associates; Dr. C. Edwin Witt, National Renewable Energy Laboratory; Dr. John Wohlgemuth, Solarex Corporation; Dr. Phillip Michael Wright, Earth Science Laboratory, University of Utah Research Institute; Dr. Kenneth Zweibel, National Renewable Energy Laboratory.

I would especially like to thank my research assistants, James Catlin, Louis Suarez-Potts, and Grant Weeden, for electronic database searches, gathering research material, manuscript preparation; and my administrative assistant, Angela Bruckel-Lichtenoecker, who secured photo permissions, corrected manuscripts, and prepared photos for submission. Thanks also to my two excellent professional transcribers: Janis Goodall and Avril Tolley, and to Doris Madden, who generously did some *pro bono* transcribing.

More than a dozen people read portions of the manuscript; the early readings by master manuscript analyst Lester Gorn were extraordinarily helpful and instrumental in securing timely publication. Critical readings of the book's concluding policy recommendations by Dr. Allan R. Hoffman, Associate Deputy U.S. Assistant Secretary of Energy, and by Dr. Donald W. Aitken, Senior Scientist, Union of Concerned Scientists, were also extremely valuable. Thoughtful and supportive readings were also given to other portions of the book by Dr. Jerry Bass, Dr. Agnes P. Berger, Caryn Diamond Bosson, Ernest Callenbach, Dr. Eugene P. Coyle, Dr. Bruce Gordon, Dr. Nancy P. Gordon, Robert Lee Hall, Charles Komanoff, Paul D. Maycock, Virginia Morgan, Laura Nelson, J. Ward Phillips, Robert Roat, Dr. Alexandra von Meier, and special thanks to Robert Masterson, former head of Technical Editing at Lawrence Berkeley National Laboratory, for both reading and meticulously editing several chapters.

Last but not least, my wife, Nancy, was especially helpful in providing me with the utmost support for which a husband and writer could hope. *Charging Ahead* would unquestionably not be what it is without her love and support. Nancy and my sons, Daniel and Michael, gracefully accepted my many late arrivals home for dinner and my all-too-frequent working weekends.

Contacting the Author

Communications regarding environmental science and policy, energy policy and natural resource restoration, corporate and professional communications and publications, and corporate marketing of environmental services may be directed to

John J. Berger, Ph.D.
Consultant
c/o LSA Associates, Inc.
157 Park Place
Point Richmond, CA 94801
(510) 231-7714
HFWS90A@prodigy.com

INDEX

Los Alamos National Laboratory (LANL), 237–38
Lotker, Mike, 39, 40, 42
Lovins, Amory, 255, 257, 262, 263, 297, 301
Low-E (low-emissivity) windows, 251–52
LUZ, named after a utopian city, 25
LUZ International Ltd., 23
bankruptcy, 45
contract with Southern California Edison, 30–31
plant construction, 1984–1991, 24
plant operation, 33–34
projected electric output of, 30
projected revenues of, 30
and solar electricity for utilities, 30
solar plant design, 33
solar trough technology, 46
technology to track the sun, 33
Lynette, Robert, 152, 173–77

MacCready, Paul B., 279–82
Manfredi, Louis, 148, 149
Manure
as fertilizer, 192
as methane gas, 192
Markets
electric power, 7
Marks, Alvin M., 107–9
Marks Polarized Corporation, 107
Massachusetts Institute of Technology
Energy Laboratory, 235
Maycock, Paul, 97, 98, 99, 101
McCabe, Barkman, 224, 229
Merrill Lynch, Pierce, Fenner and Smith, 152
Mills, David, 37
Mitsubishi engineers and turbine, 33
Mock, Ted, 235, 239
Mojave Desert, 23
Montreal Protocol
for reduction in chlorofluorocarbons (CFC), 249
Moore, Norman, 146–48, 150–51, 164
Motavalli, Jim, 302
Multiple-junction systems, 53

Nanotechnology, 107
National Appliance Energy Conservation Act of 1987, 249
National Biofuels Roundtable, 201

National Energy Policy Act of 1992
with tax credit, 169
National "net metering" policy, 322
National Renewable Energy Laboratory (NREL)
aid to AstroPower Corporation, 104
and methods for making ethanol, 206
National Resources Defense Council (NRDC), 246
Natural gas-fired heater, 37
New World Power Corporation, 171–72
Noncrystalline ceramic film cell
by Pilkington/Flachglas, Inc., 106
Nonrecourse debt as financial strategy
for investors of Kenetech, 167
for LUZ International, 39
Northeast Alternative Vehicle Consortium, 305
Nuclear energy
problems of, 3, 17–19
Nuclear fusion
problems of, 17–19
Nuclear reactors
risks of, 18
Nuclear waste
disposal of, 17

Oil
and pollution, 11
Oil, domestic
reserves of, 10
Oil, foreign
costs of importing, 9
and U.S. balance of payments, 244
Oil prices
reduction worldwide, 10
Oil versus utilities
for the electric vehicle (EV) market, 30
O'Leary, Hazel
and nuclear reactors, 18
Osborn, Dale, 163–64
Osborn, Donald E., 63
Otte, Carel, 228–30
Ovonic Display Systems, Inc., 103
Ovonic Synthetic Materials Company, 103
Ovshinsky, Stanford R., 84, 103
Ozone layer, 17

Pacific Gas & Electric Company (PG&E)
ACT II program, 255
CEE program, 255

ABOUT THE AUTHOR

JOHN J. BERGER is an independent consultant on environmental science and policy who has worked for the National Research Council of the National Academy of Sciences, private environmental engineering firms, nonprofit groups, and governmental organizations.